Introduction to Communication Electronic Warfare Systems

For a listing of recent titles in the *Artech House Information Warfare Library*,
turn to the back of this book.

Introduction to Communication Electronic Warfare Systems

Richard Poisel

Artech House
Boston • London
www.artechhouse.com

Library of Congress Cataloging-in-Publication Data
Poisel, Richard.
  Introduction to communication electronic warfare systems / Richard Poisel.
    p. cm. — (Artech House information warfare library)
  Includes bibliographical references and index.
  ISBN 1-58053-344-2 (alk. paper)
  1. Military telecommunications. 2. Information warfare. 3. Computer
networks—Security measures. 4. Electronics in military engineering. I. Title. II. Series.

UG590 .P65   2002
355.3'43—dc21   2001056572

British Library Cataloguing in Publication Data
Poisel, Richard
  Introduction to communication electronic warfare systems. —
  (Artech House information warfare library)
  1. Electronics in military engineering 2. Communications,
  Military
  I. Title
  623.7'3

ISBN 1-58053-344-2

Cover design by Igor Valdman

© 2002 ARTECH HOUSE, INC.
685 Canton Street
Norwood, MA 02062

All rights reserved. Printed and bound in the United States of America. No part of this book may be reproduced or utilized in any form or by any means, electronic or mechanical, including photocopying, recording, or by any information storage and retrieval system, without permission in writing from the publisher.
  All terms mentioned in this book that are known to be trademarks or service marks have been appropriately capitalized. Artech House cannot attest to the accuracy of this information. Use of a term in this book should not be regarded as affecting the validity of any trademark or service mark.

International Standard Book Number: 1-58053-344-2
Library of Congress Catalog Card Number: 2001056572

10 9 8 7 6 5 4 3 2 1

Contents

Preface	xiii
Acknowledgments	xv
Chapter 1 Communication Electronic Warfare Systems	1
1.1 Introduction	1
1.2 Information Warfare	1
1.3 Electronic Warfare	3
1.3.1 Electronic Support	3
1.3.2 Electronic Attack	4
1.3.3 Electronic Protect	4
1.4 Electronic Support	5
1.4.1 Low Probability of Detection/Interception/Exploitation	7
1.4.2 Future Communication Environments	8
1.4.3 Wired Communications	9
1.4.4 ES Summary	9
1.5 Electronic Attack	10
1.5.1 EA Summary	11
1.6 Typical EW System Configuration	11
1.6.1 System Control	11
1.6.2 Antennas	12
1.6.3 Signal Distribution	13
1.6.4 Search Receiver	13
1.6.5 Set-On Receiver	13
1.6.6 Signal Processing	14
1.6.7 Direction-Finding Signal Processing	14
1.6.8 Exciter	14
1.6.9 Power Amplifier	15
1.6.10 Filters	15
1.6.11 Communications	15
1.7 Concluding Remarks	16
References	16
Chapter 2 Electromagnetic Signal Propagation	19
2.1 Introduction	19
2.2 Signal Propagation	19
2.3 RF Band Designations	20

2.4 Polarization	22
2.5 Power Density	22
2.6 Free-Space Propagation	24
2.7 Direct Wave	27
2.8 Wave Diffraction	33
2.9 Reflected Waves	35
2.10 Surface Wave	40
2.11 Ducting	41
2.12 Meteor Burst	41
2.13 Scattering	42
2.14 Characteristics of the Mobile VHF Channel	43
2.15 Propagation Via the Ionosphere	46
2.15.1 Ionospheric Layers	46
2.15.2 Refraction	48
2.15.3 Near-Vertical Incidence Sky Wave	50
2.15.4 HF Fading	50
2.15.5 Maximum Usable Frequency and Lowest Usable Frequency	51
2.15.6 Automatic Link Establishment	52
2.16 Concluding Remarks	52
References	53
Chapter 3 Noise and Interference	**55**
3.1 Introduction	55
3.2 Thermal Noise	56
3.3 Internal Noise Sources	56
3.4 External Noise Sources	57
3.5 Cochannel and Multipath Interference	59
3.6 Concluding Remarks	60
References	60
Chapter 4 Radio Communication Technologies	**61**
4.1 Introduction	61
4.2 Modulation	63
4.2.1 Amplitude Modulation	65
4.2.2 Angle Modulation	67
4.2.3 Orthogonal Signaling	71
4.2.4 Access Methods	73
4.2.5 Duplexing	77
4.2.6 Digital Signaling	78
4.2.7 Spread Spectrum	97

4.3 Coding of Communication Signals	110
4.3.1 Source Coding for Data Compression	110
4.3.2 Channel Coding for Error Control	115
4.4 Modems	134
4.5 Facsimile	140
4.6 Communication Security	140
4.6.1 Data Encryption	141
4.6.2 Public Key Encryption	141
4.6.3 Digital Signatures	142
4.6.4 Data Encryption Standard	144
4.6.5 Pretty Good Privacy	145
4.6.6 Fortezza Encryption System	146
4.6.7 Escrow Encryption System	147
4.6.8 Over-the-Air Rekeying	148
4.7 Concluding Remarks	148
References	149
Chapter 5 System Engineering	**151**
5.1 Introduction	151
5.2 System Engineering	151
5.2.1 Performance Characteristics	155
5.2.2 Environmental Characteristics	155
5.2.3 Reliability and Availability	160
5.2.4 Human-Factors Engineering	171
5.2.5 System Cost	174
5.3 Concluding Remarks	175
References	176
Chapter 6 Electronic Support	**177**
6.1 Introduction	177
6.2 Intercept	177
6.2.1 Internals Versus Externals	178
6.2.2 Propagation Loss	178
6.3 Geolocation	179
6.4 Triangulation with Multiple Bearings	180
6.5 Deployment Considerations	184
6.6 Electronic Mapping	185
6.7 Common Operational Picture	185
6.8 Operational Integration with Other Disciplines	186
6.9 Support to Targeting	187
6.10 Concluding Remarks	187
References	187

Chapter 7 Electronic Attack	189
7.1 Introduction	189
7.2 Communication Jamming	189
7.3 Jammer Deployment	193
7.4 Look-Through	194
7.5 Analog Communications	195
7.6 Digital Communications	196
7.7 Narrowband/Partial-Band Jamming	197
7.8 Barrage Jamming	198
7.9 Jamming LPI Targets	201
7.10 Follower Jammer	202
7.11 Concluding Remarks	203
References	203
Chapter 8 Antennas	207
8.1 Introduction	207
8.2 Isotropic Antenna	209
8.3 Antenna Gain	210
8.4 Wire Antennas	211
8.4.1 Dipole	212
8.4.2 Monopole	214
8.4.3 Loop	215
8.4.4 Biconical/Discone	215
8.4.5 Yagi	216
8.4.6 Log Periodic	216
8.4.7 Helix Antennas	218
8.4.8 Spiral Antennas	220
8.5 Active Antennas	220
8.6 Aperture Antennas	222
8.6.1 Parabolic Dish	222
8.6.2 Antenna Arrays	223
8.7 Genetically Designed Antennas	225
8.8 More on Antenna Gain	226
8.9 Concluding Remarks	228
References	228
Chapter 9 Receivers	231
9.1 Introduction	231
9.2 Receivers	232
9.2.1 RF Amplification and Filtering	234
9.2.2 Sensitivity	236

9.2.3 Dynamic Range	237
9.2.4 Mixing/Frequency Conversion	240
9.2.5 Intermediate Frequency Filtering and Amplification	241
9.2.6 Detection/Demodulation	242
9.3 Types of Receivers	246
9.3.1 Narrowband Receivers	246
9.3.2 Wideband Receivers	250
9.4 Concluding Remarks	263
References	264
Chapter 10 Signal Processing	265
10.1 Introduction	265
10.2 Orthogonal Functions	266
10.3 Transforms	267
10.3.1 Trigonometric Transforms	268
10.3.2 Haar Transform	274
10.3.3 Wavelet Transforms	276
10.3.4 Fast Transforms	285
10.4 Cyclostationary Signal Processing	286
10.5 Higher-Order Statistics	289
10.6 Applications	291
10.6.1 Signal Detection	291
10.6.2 Signal Classification	302
10.6.3 Recognition/Identification	317
10.7 Concluding Remarks	328
References	329
Chapter 11 Direction-Finding Position-Fixing Techniques	331
11.1 Introduction	331
11.2 Bearing Estimation	331
11.2.1 Circular Antenna Array	332
11.2.2 Interferometry	333
11.2.3 Monopulse Direction Finder	347
11.2.4 Amplitude Direction Finding	352
11.2.5 Doppler Direction Finder	359
11.2.6 Array-Processing Bearing Estimation	362
11.2.7 Line-of-Bearing Optimization	375
11.3 Position-Fixing Algorithms	379
11.3.1 Eliminating Wild Bearings	380
11.3.2 Stansfield Fix Algorithm	381
11.3.3 Mean-Squared Distance Algorithm	384

11.3.4 Combining Error Contours	388
11.4 Single-Site Location Techniques	391
11.5 Fix Accuracy	392
11.6 Fix Coverage	396
11.7 Concluding Remarks	399
References	401
Chapter 12 Hyperbolic Position-Fixing Techniques	**403**
12.1 Introduction	403
12.2 Time Difference of Arrival	403
12.3 Differential Doppler	410
12.4 Cross Ambiguity Function Processing	414
12.4.1 Position Fix Accuracy	416
12.5 Time of Arrival	418
12.6 Concluding Remarks	422
References	422
Chapter 13 Exciters and Power Amplifiers	**425**
13.1 Introduction	425
13.2 Exciters	425
13.2.1 Oscillators	426
13.2.2 Synthesizer	428
13.2.3 Modulator	429
13.3 Power Amplifiers	431
13.3.1 Amplifier Operating Characteristics	431
13.3.2 Efficiency	433
13.3.3 Push-Pull Architecture	434
13.3.4 Classes of Amplifiers	435
13.3.5 Switching Architectures	438
13.3.6 Amplifier Linearization	444
13.3.7 Basic Power Modules	445
13.3.8 Combiners	448
13.3.9 Output Filters	449
13.3.10 Noise-Power Ratio	450
13.4 Concluding Remarks	451
References	451
Chapter 14 Early-Entry Organic Electronic Support	**453**
14.1 Introduction	453
14.2 Target Model	454
14.3 Intercept System Model	455

14.4 Simulation Results	459
14.4.1 Performance Versus the Number of Target Nets	460
14.4.2 Search Bandwidth	462
14.4.3 Noise Factor	464
14.4.4 Postprocessing Time	466
14.4.5 Mission Duration	467
14.5 Concluding Remarks	469
References	471
Chapter 15 Detection and Geolocation of Frequency-Hopping Communication Emitters	473
15.1 Introduction	473
15.2 Analysis	473
15.3 Simulation	476
15.3.1 ES System Operation	478
15.3.2 Results and Analysis	481
15.3.3 Discussion	484
15.4 Concluding Remarks	486
References	486
Chapter 16 Signal Detection Range	487
16.1 Introduction	487
16.2 Noise Limits on Detection Range	490
16.3 Targets	491
16.4 Detection Range with the Reflection Propagation Model	491
16.4.1 Airborne Configurations of ES Systems	492
16.4.2 Ground-Based Detection Ranges	497
16.4.3 Discussion	499
16.5 Concluding Remarks	501
References	503
Chapter 17 Electronic Attack: UAV and Ground-Based	505
17.1 Introduction	505
17.2 Signal Propagation at Long Ranges	506
17.3 Jamming ERP	507
17.4 Targets	508
17.5 RLOS	508
17.6 UAV Jammer	509
17.7 Ground-Based Jammer	511
17.8 Expendable Jammer	513
17.9 Concluding Remarks	514
References	516

Appendix A Probability and Random Variables	517
A.1 Introduction	517
A.2 Means, Expected Values, and Moment Functions of Random Variables	517
A.3 Probability	520
A.3.1 Conditional Probability	521
A.3.2 Random Variable Distribution and Density Functions	522
A.4 Gaussian Density Function	525
A.5 Kurtosis	527
A.6 Skewness	527
A.7 Useful Characteristics of Probabilities, Density Functions, and Distribution Functions	528
A.8 Concluding Remarks	528
Reference	528
Appendix B Simulated Networks	529
B.1 Introduction	529
List of Acronyms	539
About the Author	545
Index	547

Preface

This book grew out of a need for an introductory level technical text on the basic aspects of communication electronic warfare (EW) systems. EW as a topical area consists of many aspects and probably the area treated least in detail is communication EW systems. Traditional books on EW signal collection and processing have concentrated on radar signals. This has been, in part, because of the closeness with which the governments of the world have held such information. After all, EW is a countermeasure technique targeted against, in this case, communication radio systems. The United States and other developed countries have produced radios to thwart such countermeasures as best they could. In the past that has not been very good and open discussions about how to attack radio communications ran counter to good sense, economically at least. That situation has changed of late, however. Effective radio systems to counter EW countermeasure methods have been developed and fielded. Therefore, the basic principles behind communication EW can now be discussed more openly.

When the author was new to the technical area, there was virtually nothing written specifically about communication EW systems. Everything had to be learned from first principles and experience. Needless to say, there were many mistakes made along the way. This book is an attempt to change that.

The book was written with two, not necessarily distinct, target audiences in mind. The first, but not necessarily the more important, consists of engineers first being introduced to the world of communication EW technologies and systems. There is a very real potential overlap of this group with the second group for which it was written: practicing EW military professionals. Some sections of the book contain derivations of equations, but those interested in the communication EW operational issues discussed can skip these sections.

For the technical audience, there is adequate depth in many areas for a concentrated and rewarding reading. The chapters on signal processing (Chapter 10), direction-finding position fixing (Chapter 11), and hyperbolic position fixing (Chapter 12), were written particularly with this audience in mind. The chapters that concentrate on operational issues (Chapters 1, 3, 5, 6, 7, 14, 15, 16, and 17) should appeal to this audience as motivational for studying the field in the first place. These chapters provide an understanding as to why there is an operational interest in EW as well as discuss some of the practical limitations.

For the audience with more of an operational interest, the more technical chapters can be skipped, although it would probably be better to just skip the math and focus on the operational sections of even the more technical chapters. These chapters also contain some important discussion of such issues. Chapters 1, 3, 5–7, and, in particular, 14–17 should appeal to those with operational interests. These chapters focus more on how EW systems are, or can be, employed. There are very real limitations on what such systems can do and many of these are presented.

Acknowledgments

Many colleagues over the years at the government laboratory where I have spent my entire professional career contributed in some way to the contents herein. These colleagues are too numerous to delineate here—they know who they are. Herb Hovey, however, needs to be particularly mentioned because he was largely responsible for my career. Herb was the Director of the Army Signals Warfare Laboratory from before the time I joined the laboratory in 1976 until Herb retired in 1990. He was a widely respected engineer in the government military intelligence and EW circles during the time he was engaged in them. Without his continued support and tolerance, I would not have experienced enough or developed adequate knowledge to write this book.

This book would not have been possible without the unlimited patience and understanding of my wife Debbie, to whom this book is dedicated. The many hours of lost family time necessary to produce it has been, for the most part, her albatross to bear.

Chapter 1

Communication Electronic Warfare Systems

1.1 INTRODUCTION

Just as the Agrarian Age, of necessity, gave way to the Industrial Age in the latter part of the nineteenth century, the Industrial Age gave way to the Information Age in the latter part of the twentieth century. Pundits of modern armies comment about a revolution in military affairs, where attacking information systems (as well as protecting one's own) plays a substantial part in conflicts of the future. Thus is born military *information warfare* (IW), sometimes referred to as command and control warfare, where protecting and attacking information and the systems that process it come into play.

1.2 INFORMATION WARFARE

IW is the appellation applied to conducting warfare-like actions against an adversary's information systems or protecting one's own information systems from such activities [1–3]. Attacking information systems can serve several purposes. It can deny the availability of key information at important junctures. It can provide false information causing an adversary to reach incorrect conclusions. It can degrade the confidence adversarial decision makers have in their information bases also by providing false information. This is just a partial list of the effects of IW-like activities.

It is frequently stated that the war in the Persian Gulf was the first information war [4, 5]. This is because of the extensive use of modern information technologies that affected the way the war was fought. This is not to say, however, that the *use* of information in war is a new concept. Information has always been important [6]. What is new are the methods implemented to use or affect information usage by an adversary. Modern information technologies permit such new affects.

IW can be applied to both nonmilitary and military situations. One commercial firm may conduct electronic espionage against a competitor in order to gain a competitive advantage. This is an example of commercial IW. Jamming the communications of an adversary when hostile activities are occurring is an example of military IW.

IW is generally (and incorrectly) construed to refer to attacks on computer networks. However, IW as a category can be interpreted to include just about everything that has to do with the use of information. This yields a very broad field and a fundamentally problematic taxonomy. Thus it's becoming accepted to refer to the more useful, but limiting, definition of attacks on computer networks. For our purposes, this is too limiting.

Command and control warfare (C2W) is IW applied in a military setting [7]. C2W is comprised of five so-called pillars: (1) physical destruction of information systems, (2) *psychological operations* (PSYOPS), (3) deception, (4) *operational security* (OPSEC), and (5) EW.

Physical destruction of information systems is pretty much self-explanatory. It can be accomplished in several ways, from hand-emplacing explosive devices to missiles that home on radiated energy. It is, obviously, the extreme case in that in general hostilities must be under way, where the other forms of IW are subtler in their application.

PSYOPS has been used for years. It attacks the psyche of an adversary and it also can take many forms. Acoustic attacks can be used to try to convince an adversary to take some action. Pamphlets can be dropped from aircraft over adversarial forces to attempt to accomplish the same thing. In a sense, the simple *unmanned aerial vehicles* (UAVs) used in the Gulf War were a form of PSYOPS. Just upon sight by the soldiers they surrendered *to the UAV.*

Deception involves actions taken to create the appearance of the existence of a situation that is not actually present. It can be as simple as placement of cardboard resemblances of vehicles in an area to give the impression of the presence of more forces than are actually present, to the transmission of radio communications in a region to give the same impression. The movement of coalition forces on the Saudi Arabia side of the Kuwait border created the appearance of an attack against Saddam Hussein's forces in Kuwait in the incorrect region [8].

OPSEC are procedures emplaced to ensure the protection of friendly information. Protecting classified information by proper utilization of security containers is an example. Taking care during telephone conversations not to give away what might be sensitive information is another.

EW, the last pillar of IW, is the subject of the remainder of this book. It is discussed at length in Section 1.3.

1.3 ELECTRONIC WARFARE

Throughout history, military warfare has been about measures to defeat one's enemy and countermeasures to those measures on the other side. In modern times, nowhere is this more evident than in IW, and in particular, communication EW. EW practices are always undergoing changes due to an adversary changing their capabilities based on a countermeasure developed by the other side. Theoretically this measure/countermeasure process could go on unbounded. In practice it does not, however, largely because of economics. Eventually things get too expensive. This is true in EW just as it is true in other areas.

This book is about that part of IW that attacks information systems by withdrawing from, or imparting energy into, communication systems so that the intended transport of information is either intercepted, denied, or both. As indicated above, that part of IW is called communication EW. More specifically, the systems that are intended to attack communication systems are discussed. By simple extension, these systems can sometimes be used to "screen" friendly communications from similar actions taken by an adversary so they can also be used in an information protection role.

Most of the more important salient attributes of these systems are covered herein, although no claim is made that all facets are included. The intent is to give the reader a not-so-technical introduction to such systems, relying on engineering terminology as little as possible. There will be derivations of equations. Such derivations are included in a few places to indicate that most of the aspects of designing EW systems are based on fundamentally sound first principles, and these cases illustrate some of these first principles.

There are several forms of EW that include virtually the entire frequency spectrum. Herein EW will be limited to what is normally referred to as the radio frequency part of the spectrum. This starts at about 500 kHz and extends into the hundreds of gigahertz. This specifically excludes audio-frequency measures at the low end, and infrared and electro-optics at the high end.

It is generally accepted that EW has three distinct components: (1) *electronic support* (ES), (2) *electronic attack* (EA), and (3) *electronic protect* (EP) [9]. ES is comprised of those measures taken to collect information about an adversary by intercepting radiated emissions. EA refers to attempting to deny adversaries access to their information by radiating energy into their receivers. EP involves activities undertaken to prevent an adversary from successfully conducting ES or EA on friendly forces.

1.3.1 Electronic Support

ES attempts to ascertain information about an adversary by intercepting radiated energy. This radiated energy can be emitted from any type of transmitter such as those in communication networks. It could also stem from radars, telemetry transmitters, or unintended radiation as from computer clocks.

Some important information can be gleaned from just the measurement of a few external parameters associated with a transmission. The frequency of operation, the modulation type, the bit rate, and the geolocation of the transmitter are examples of such external parameters. In some cases it is also possible to intercept the internals of a transmission. When this occurs it is normally termed *signals intelligence* (SIGINT), where the goal is to generate intelligence products about an adversary. ES is usually restricted to collection of external parameters.

1.3.2 Electronic Attack

When radiated energy from a friendly source is used to deny adversaries access to their information, it is referred to as EA. Again, the adversarial radiated energy could be from communication transmitters, radars, or telemetry transmitters. It is important to remember, however, that in EA it is the receiver that is the target of the attack. Examples of EA include radio jamming and radar deception.

1.3.3 Electronic Protect

It is frequently desired to take measures to disallow an adversary to conduct ES and EA against friendly forces. As *dominant battlespace knowledge*[1] (DBK) [10] becomes more and more important in the era of IW, more reliance on accurate, timely, and thorough information is necessary. Therefore, it becomes more important to protect friendly information from manipulation by an adversary. It is also important to deny the availability of friendly information to those adversaries.

Many disciplines can be grouped within EP. *Emission control* (EMCON) is perhaps one of the simplest forms. Through EMCON, the use of friendly transmissions is limited or precluded for a certain period of time, usually at critical junctures. Simply the presence of such emissions can provide information to an adversary as to friendly force sizes and, perhaps, intentions. EMCON prevents an adversary from intercepting and identifying the operating frequency of a friendly communication network, thus protecting the frequency information from being available to the adversary. Another form of providing such protection is low probability of intercept (spread spectrum) communications (which is discussed in Chapter 2).

Screen jamming is a form of EP. This is when a jammer is placed between friendly communication nets and an adversary's SIGINT systems to prevent the latter from intercepting the communications. Screen jamming is a useful tool

---

[1] DBK is knowing more accurate information about the battlespace at a particular point in time than the adversary. See [1] for a thorough discussion of DBK.

when attempting to break contact to restructure the nature of hostilities, for example.

Encrypting communication nets is another form of EP. It prevents an adversary from gleaning information from the intercept of communication transmissions. The ready availability of practical and effective encryption algorithms to all countries are causing more communications of all types to be encrypted.

EP is a category in a much larger area called information assurance. In general, assuring the integrity and security of information is a complex task in a complex world. The U.S. military is just beginning to get a handle on the enormity of the problem [11]. This book is primarily about ES and EA against communication targets. Many of the technologies applied to ES and EA discussed, however, are applicable to EP. Little more will be said specifically about EP herein.

1.4 ELECTRONIC SUPPORT

ES refers to those measures taken to gather information from *radio-frequency* (RF) emissions, by noncooperative intercept of those emissions. When the emissions are from a communication system, then it is known as communication ES. Herein, the communication will be assumed and communication ES will be referred to as simply ES. The term noncooperative in this sense means that the communicator and interceptor are not cooperating—the communicator does not want the interceptor to be successful.

ES is the collection of communication signals for the purposes of gleaning information from them. This information can be in the form of intelligence or combat information. The difference between these two is what the information is used for as opposed to the collection means. Another difference might be the type of information collected—signal externals versus internals, for example. Generation of intelligence usually requires access to the internals of a signal, a human analyst, and a relatively extensive amount of time. On the other hand, combat information is information that is readily apparent from the data and does not require extensive analysis—usually no human analysis at all. The externals of a signal are usually all that are used to generate combat information. Externals are signal parameters such as frequency of operation, baud rate, and location. Combat information can be used directly to support ongoing operations, such as real-time targeting. Combat information, of course, can be, and is, also used for intelligence generation.

A ground-based ES scenario is depicted in Figure 1.1. The transmitter is communicating with a receiver, while the ES system is trying to intercept that transmission. The fundamental distinction between ES and the normal reception of communication signals by the intended receiver is that the former is non-cooperative while the latter facilitates communication between the transmitter and the receiver. That is, the transmitting entity is not trying to communicate with the ES system but it is with the receiver. In older forms of communication, analog AM and FM, for example, this distinction was not so important. The reception of signals by either the receiver or the ES system depended mostly on the geographical relationship between the entities involved. With modern digital communications, however, this distinction is significantly more important.

Cooperating communication systems can adjust for imperfections in the communication channel between them. An example of this is evident when computer modems are first interconnected for telephone Internet communications. The tones and rushing noises heard at the beginning are the modems doing just that—determining the quality of the channel that interconnects them (normally, but not always, these days one or more telephone lines) and setting the communication speed depending on these measurements. A noncooperative ES receiver does not share this luxury. The transmitter not only does not try to measure the channel between the transmitter and the ES system, but also normally would not even want the ES system there at all. This simple distinction between cooperative and noncooperative communication situations to a large extent determines the complexity of the ES system.

Classical ES for communication signals usually entails determining the geolocation of the emitting entity; determining its frequency of transmission, if

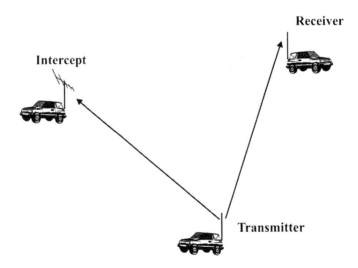

Figure 1.1 Notional scenario for ground-based electronic support.

appropriate; determining the type of signal it is (e.g., analog versus digital); and, if possible, determining the type of weapon system the signal is associated with, such as artillery and rocket forces. Such information is useful in determining the *electronic order of battle* (EOB), which refers to the entities on the battlefield and what kind of RF electronic systems they possess.

1.4.1 Low Probability of Detection/Interception/Exploitation

These terms apply to ways of processing *electromagnetic* (EM) signals so that either it is difficult to know that the signal is present or, if it is detected, that the information contained in it is difficult to extract [12, 13].

Low probability of detection (LPD) attempts to "hide" the presence of signals. One technique puts the signal below the level of the noise, so that the signal cannot be differentiated from the noise upon a casual look at the spectrum. This technique is called *direct-sequence spread spectrum* (DSSS). Another technique jumps the signal around in frequency so that a fixed-tuned receiver will rarely see the signal. This technique is referred to as *frequency-hopping spread spectrum* (FHSS).

If the nature of the communication system is such that it is difficult to establish LPD, then it may be more desirable to yield to the possibility that the signal may be detected, but that once it is, it is difficult to extract the information contained in it. *Low probability of interception* (LPI) and *low probability of exploitation* (LPE) are the terms used in this case. FHSS is an example of this.

Most EM signals with which the reader may be familiar, such as TV stations or FM and AM radio stations, are fixed in frequency. Most practical low-probability schemes take such narrowband signals and spread them out so they occupy much larger portions of the electromagnetic spectrum than they would otherwise. One technique that doesn't spread the signal out in frequency but instead stretches it out in time is called *time hopping*. This technique is not used as much as the other LPI techniques for precluding detection. Time hopping naturally occurs, however, for military tactical communications when it is not known a priori when an entity is going to communicate. Such communications are frequently referred to as *push-to-talk* (PTT).

Frequency hopping changes what is called the *carrier frequency* often, say, 100 or more times per second. This is akin to shifting the station to which your car radio is tuned 100 times per second. Even if you knew where the next frequency was (and one of the features of frequency hopping is that unintended receivers do not) it would be difficult to tune your radio fast enough to keep up. Of course with modern digital receivers, the switching speed is not a problem, but not knowing the next frequency is.

The DSSS technique mentioned above for generating low-probability signals (the focus here is on LPD) is a little more difficult to understand. In this technique a special signal called the *chip sequence* is added (modulo two, $0 + 0 = 0$, $0 + 1 = 1$, $1 + 0 = 1$, $1 + 1 = 0$) to the information signal (which has already been converted to digital) before it is broadcast. The chip sequence is a digital signal that changes states at a much higher rate than the information signal. The net effect is to have the resultant signal much broader in frequency extent, and at any particular frequency much lower in amplitude, than the information signal would otherwise occupy (if located at that frequency). It is much lower in amplitude at any single frequency since the same signal power is present, but it is spread across a much broader frequency range. Such a low-amplitude signal can be difficult to detect if you do not know it is there.

One technique that has nothing to do with technology but that is an effective LPD/LPI/LPE technique is called EMCON mentioned above. This is using the technique of radio silence during certain portions of an operation to preclude radio signals from being detected that, in some cases, could indicate that an operation is occurring. It is normally the case that once an operation starts, much of the a priori planning no longer applies. Since much of tactical command and control is accomplished with RF communications, employing EMCON can be difficult when plans must be changed. To be detected while employing EMCON would require a sensor other than a SIGINT sensor.

Collection of wideband signals such as these requires receiving equipment that is also wideband. Unfortunately, it is a law of physics that the larger the bandwidth of the receiving equipment the more background noise enters the receiver along with any desired signals. Nevertheless, there are some techniques that have been developed, including compressive receivers and digital receivers, that help to minimize this effect. These receiver architectures will be discussed later in more depth in Chapter 9.

1.4.2 Future Communication Environments

As discussed in Poisel and Hogler [14], future communications of concern to the U.S. military forces will be largely based on commercial technology developments. The principal reasons for this are diminishing military budgets and increasing sophistication of commercial communication means around the world. Communication is one of the cornerstones of the Information Age, and it will be vitally important to everyone living in it eventually. Therefore the private sector is expected to invest heavily in the development of advanced communication means, precluding the necessity of military investments in communication technology—other than, of course, to adapt it to military applications and purchase it. Such adaptation will range from no modifications at all to ruggedizing the housing and electronics inside. The underlying signaling technology will not often be modified.

1.4.3 Wired Communications

ES against wired communication systems is considerably more difficult to conduct, since the signals are not radiated into space. Many wired systems, such as traditional telephone systems, do utilize wireless communications at some point in the transmission process, and at these points of course wireless techniques can be used. However, for the case where the communications does not use wireless at any point, one is reduced to some sort of physical access to the transmission media or equipment.

Such physical access can lead to wire taps on phones, for example. It is also sometimes possible to measure the unintentional radiation from the communication equipment, called *Van Eyke radiation* by Schwartau [15]. This, however, utilizes approaches that are the same as those for wireless communications. Lacking physical access for wired communication systems essentially precludes ES operations as considered herein.

Wired communications are difficult to employ on a tactical battlefield, although not at all difficult for stationary employment. It has been said that a great deal of Eastern Europe has been prewired by the former Soviet Union so that echelons of advancing artillery could use wire communications instead of their radios in the event of a conflict there. No doubt this wire is rapidly disappearing.

1.4.4 ES Summary

Interception of communication signals on the battlefield, herein collectively called ES, will continue to be an important source of information. If the communication signals are not encrypted then in some cases the intent of an adversary can be inferred. Even if they are encrypted, however, locating the source of the signals, the volume of signals, and other external parameters can be used to garner information.

Modern communication devices have gone to great lengths to prevent their interception. These techniques are referred to as low probability of intercept, low probability of detection, and low probability of exploitation. Wideband receiving systems must be used against such signals and the ranges of these systems is frequently limited because of the amount of noise associated with wide bandwidths. Compressive and other forms of analog receivers, as well as digital receivers, have been designed to be used against these signals.

The accuracy at which these signals can be geographically located depends strongly on the *signal-to-noise ratio* (SNR) of the signal at the receiver. Low SNR signals are more difficult to locate accurately than those with large SNRs. The accuracy normally increases as the inverse square root of the number of samples of the signals that are taken, as well as inversely to the integration time and measurement bandwidth.

1.5 ELECTRONIC ATTACK

Denying adversaries the effective use of their communication systems is the goal of communication EA. Any electronic means of attacking such targets, short of lethal weapons, falls under the guise of EA. The scenario shown in Figure 1.2 summarizes a situation for EA on communication nets, in a ground-based setting. The transmitter is attempting to transport information to the receiver while the jammer is attempting to deny that transport. This denial is in the form of placing more jamming energy into the receiver than the transmitter's information signal. The particular communication signal involved does not matter; however, the effectiveness of the jammer does depend on the signal type.

Conducting EA from standoff EW systems is complicated by the fact that frequently fratricide occurs, that is, interfering with friendly communication systems. It is therefore sometimes more fruitful to place EW systems closer to the communication nets that are being jammed than to friendly nets. Delivery mechanisms for such systems could include artillery shells—the 155-mm howitzer shell, for example, can go on the order of 20 km. Another delivery mechanism could include emplacing them by hand or flying them in a UAV. In the latter case the EW systems could either be used on the UAV or they could be dispensed. Since the signal propagation range is greater from the UAV, one would expect

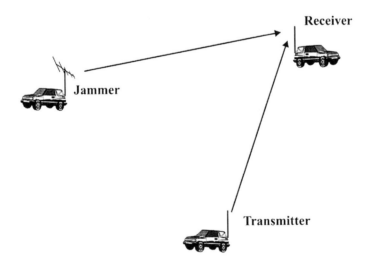

Figure 1.2 Notional scenario for ground-based EA.

better performance from this configuration, although one could encounter the fratricide problem again.

1.5.1 EA Summary

Communication EA is employed in all forms of combat situations: ground-to-ground, ground-to-air, air-to-ground, and air-to-air. It is used to deny or degrade an adversary's ability to command and control forces. The effects of communication EA are temporary—in most cases they last as long as the jamming signal is present. Digital communication signals can be affected more easily than analog signals, because a threshold *bit error rate* (BER) is required in order for digital communications to work. Typically this is about 10^{-2} or better for machine-to-machine communications. Thus causing problems on 1 bit out of 100 or more should be sufficient to deny machine-to-machine digital communications. Encrypted communications requires resynchronization when synchronization is lost due to jamming or other reasons.

1.6 TYPICAL EW SYSTEM CONFIGURATION

Presented in this section is a configuration for a "typical" communications EW system. It will be used throughout this book so that the usage of each component can be illustrated. It certainly is not the only configuration available for RF EW systems, the design of which depends on the particular application. A block diagram for the typical EW system is shown in Figure 1.3. Some of these components may not exist in every system, and some systems may contain others. If the EW system were for ES only, the exciter, *power amplifier* (PA) filters, and EA antenna would not be present for example. In this figure, the dotted lines represent RF or other types of analog signals, whereas the solid lines represent digital control signals.

1.6.1 System Control

One or more computers typically exercise the control of such a system. If there is only one, then the control is centralized. If there is more than one, then the control can be either centralized or distributed, depending on the chosen architecture.

The solid lines indicate a system control bus, which may actually consist of more than one. Typical buses would be Mil Std 1553, VME, VXI, IEEE 1394 (Firewire), IEEE 802.3 (Ethernet), and ANSI X3T9.5 *fiber distributed data interface* (FDDI). The lighter lines are intended to indicate signal paths, whether RF or lower in frequency. If there are operators present that interface with the

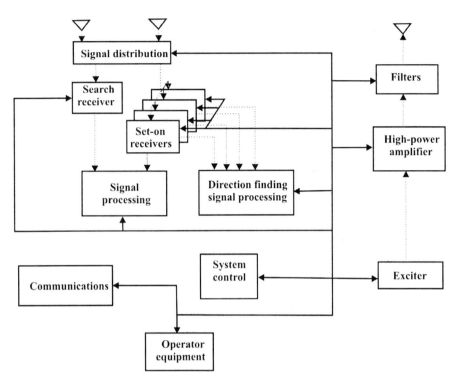

Figure 1.3 Block diagram for a typical EW system.

system, then there are usually workstations connected to the system control for that purpose.

1.6.2 Antennas

Antennas are used to extract the *electromagnetic* (EM) energy from the propagation medium. For most cases of interest herein the propagation medium is the atmosphere surrounding the Earth (the air). Antennas convert the EM energy into electrical signals, which is a form of energy that the rest of the system can use.

Antennas are also used in the reverse fashion for EA applications. They convert electrical energy into EM energy that can be propagated through the atmosphere.

Typically there is more than one antenna performing several tasks associated with an RF EW system. Intercept and direction finding antennas are frequently the same. Intercept and EA antennas also are frequently the same. The specific antenna configuration used depends on the application. Some exhibit gain over other types, thereby increasing the range of operation of the EW system. In some situations, using antennas with gain is troublesome due to mobility requirements, for example. Generally the higher the antenna is above the ground, the further the range of operation. Thus, airborne EW systems typically have a larger range than ground-based systems (for VHF and higher frequencies anyway).

1.6.3 Signal Distribution

Signals from antennas frequently must go to more than one place in an EW system. In this example, the signals go to more than one receiver. In order to maintain the proper impedance the signal splitting must be done carefully. With improper impedance matching, less than maximum power transfer will result, and the signals could experience distortion that they otherwise would not.

Thus, signal distribution is accomplished normally at the output of the antennas and before the receivers. Typically the antenna impedance is converted to the cable impedance, say, 50 ohms. This cable is run from the antenna to the signal distribution and from there to the receivers. A signal splitter in the signal distribution would have an input impedance of 50 ohms and two or more output ports with a 50-ohm output impedance.

1.6.4 Search Receiver

A good deal more will be said about receivers later. Suffice it here to say that there is usually a function to be performed in a RF EW system that searches the frequency spectrum looking for signals of interest. A search receiver accomplishes this function. This receiver typically scans through the spectrum of interest, looking for energy. When it finds energy, measurements are made to characterize it. Those measurements can be performed as part of the search receiver, or by using a set-on receiver.

1.6.5 Set-On Receiver

A set-on receiver is used for relatively long-term analysis of signals detected by other means. There is usually more than one of these. They are turned to a frequency either by an operator or automatically based on energy detected by the search receiver. They may, in fact, be nothing more than channelized filters using the search receiver as the RF portion. The outputs of these receivers are used to measure parameters of signals or, in the case of analog communications, for example, for the operators to listen to.

1.6.6 Signal Processing

Much more will be said about signal processing in Chapter 10. This function comes in many forms, depending on what is trying to be accomplished. Typical functions include detection of the presence of energy at a particular frequency and within a specified bandwidth, determination of the modulation on a signal, and measuring the baud rate of a digital communication signal.

In most communication EW applications the signals being processed are weak compared with the background noise present. Much of the task of signal processing is to extract parameters from such signals regardless of the noise.

1.6.7 Direction-Finding Signal Processing

This signal processing function is broken out because of its importance in many applications. Locating the source of emissions of communication signals is one of the fundamental applications of RF EW systems. This is usually, but not always, accomplished by triangulation where the direction of arrival of a signal at two or more EW systems is measured. The point where the lines corresponding to these arrival angles intersect is the calculated location of the source of the signal. If all of the lines of bearing do not intersect at a single point (and they rarely do for more than two systems), then the emitter location is calculated according to some algorithm to be the centroid of the lines of bearing.

Therefore, accurate measurement of the arrival angle is an important function. Various factors enter this process that make it difficult to measure the angles accurately. Not the least of these is RF noise, caused by the stars in the sky. Interference is another source of noise. This is caused by man-made sources of energy that share the same frequency bandwidth as the signal of interest.

1.6.8 Exciter

If the EW system has an EA function to perform, then it typically will contain an exciter, high-power amplifier, filters, and an antenna as shown in Figure 1.3. An exciter is essentially a RF signal generator, with the capability to modulate the signals generated. The modulation can take many forms, depending on the target signal, but the one used most frequently for communication signals is simply random noise. This signal essentially simply raises the noise level at the target receiver, thus decreasing the SNR. The SNR is a fundamental parameter in communication theory and has been heavily studied. Therefore, accurate prediction of degradation of performance due to jamming a signal can be made. Other forms of modulation that can be used include tone jamming, sometimes effective against *frequency shift key* (FSK) signals.

1.6.9 Power Amplifier

The high-power amplifier for EA purposes amplifies the signal from the exciter. The PA converts the relatively weak (typically 0-dBm, or 1-mW) signal from the exciter into stronger signals for transmission. For communication EA applications, the signal level sent to the antenna is typically 1 kW or so.

Some of the bigger issues associated with power amplifiers include conversion efficiency (fraction of prime power converted to radiated power), frequency coverage (broad bandwidths are difficult to achieve with a single amplifier), heat removal (related to efficiency and reliability), and spectral purity to prevent friendly signal fratricide.

1.6.10 Filters

In order to avoid RF fratricide of friendly receivers, filters on the output of the amplifier are frequently needed, since perfect amplifiers with no spurious responses have not been invented yet. The filters limit the out-of-channel (undesirable) energy that the system emits. These filters must be constructed so they can handle the power levels associated with the output of the PA. Ideally they would introduce no loss of in-band signal power as well.

These filters must be tunable. Typically a jammer will radiate energy at a single frequency and over a limited frequency extent (bandwidth). The filters must be tuned to that frequency if they're to function properly. In some applications, such as EA against frequency hopping targets, where the frequency is changing rapidly, this tuning must be performed even faster than the targets change frequency.

Obviously if there is more than one signal to be jammed by the communication EW system there is more than one filter required, although it may not be necessary to provide more than one PA. The PA itself is wideband, capable of amplification over a broad frequency range. The exciter determines the frequency at which the jamming signal is located, and the filter keeps the signal within a specified bandwidth. Therefore an exciter and filter are required for each frequency, but only one PA is required.

1.6.11 Communications

The communication subsystem can be comprised of several types of capabilities. It is the means for command and control of the system as well as the means to send external data to it (tasking), and the means to deliver a product out of it (reporting). If the system is remotely controlled, then this subsystem is the means to exercise that control.

1.7 CONCLUDING REMARKS

Contained in this chapter are brief descriptions and discussions about most of the more significant issues associated with RF EW systems. Both ES and EA are introduced. The remainder of this book will discuss these subsystems in more detail. This chapter is intended to describe the EW system as a whole before the parts are described in detail.

EW systems constitute the ability to intercept as well as take countermeasures against RF signals. This information can be used to glean *electronic order of battle* (EOB) data about an adversary as well as information on when and how to attack an adversary's command and control and sensor capability. As with any weapon system, the utility of an EW system depends on how it is employed and the situation to which it is applied. In some scenarios, an EW capability may be useless. In other scenarios it might be vital. An example of the latter situation occurred in the Gulf War in 1991. The Iraqi military was so concerned about the allied ability to employ communication EW systems that communication with their forward-deployed troops was almost nonexistent. Whether the EW systems that were employed were effective at performing the EW mission or not is irrelevant. They accomplished their goal simply by reputation.

References

[1] Libicki, M. C., "What Is Information Warfare," The Center for Advanced Concepts and Technology, The National Defense University, Washington, D.C., August 1995.

[2] Schleher, D. C., *Electronic Warfare in the Information Age,* Norwood, MA: Artech House, 1999.

[3] Waltz, E., *Information Warfare Principles and Operations*, Norwood, MA: Artech House, 1998.

[4] Campen, A. D., D. H. Dearth, and T. T. Goodden (eds.), *Cyberwar: Security, Strategy and Conflict in the Information Age,* Fairfax, VA: AFCEA International Press, May 1996.

[5] Campen, A. D. (ed.), *The First Information War*, Fairfax, VA: AFCEA International Press, October 1992.

[6] Von Clausewitz, C., *On War*, London: Penguin Books, 1832 and 1968.

[7] Schleher, D. C., *Electronic Warfare in the Information Age,* Norwood, MA: Artech House, 1999, p. 5.

[8] Scales, R. H., *Certain Victory: The U.S. Army in the Gulf War,* U.S. Army Command and General Staff College, Fort Leavenworth, KS, 1994, pp. 145–147.

[9] Schleher, D. C., *Electronic Warfare in the Information Age,* Norwood, MA: Artech House, 1999, p. 2.

[10] Johnson, S. E., and M. C. Libicki (eds.), "Dominant Battlespace Knowledge," The Center for Advanced Concepts and Technology, National Defense University, Washington, D.C., October 1995.

[11] Gershanoff, H., "Information Assurance: How Do We Win the War to Protect Information?" *The Journal of Electronic Defense,* March 2001, pp. 51–52.

[12] Nicholson, D. L., *Spread Spectrum Signal Design LPE and AJ Systems*, Rockville, MD: Computer Science Press, 1988.

[13] Simon, M. K., et al., *Spread Spectrum Handbook, Revised Edition*, New York: McGraw-Hill, Inc., 1994.
[14] Poisel, R. A., and J. L. Hogler, *Global Communications 2010: Military IEW Challengers,* Technical Report No. IEWD-RT-930001 (U), USA CECOM RDEC IEWD, Vint Hill Farms Station, Warrenton, VA, February 1992.
[15] Schwartau, W., *Information Warfare: Chaos on the Superhighway,* New York: Thunder's Mouth Press, 1994, pp. 137–147.

Chapter 2

Electromagnetic Signal Propagation

2.1 INTRODUCTION

Before describing the characteristics of EW systems it is important to understand some of the fundamental properties of propagating signals, or EM waves. For those familiar with this topic, this chapter can be skipped. For those readers interested in further reading, [1–7] are recommended.

The modes of signal propagation can be dependent on the frequency of the signal, while some of the modes are independent of the frequency. These characteristics are pointed out in the appropriate following discussions.

The design of the RF front ends, especially the antenna, of EW systems depends to a great extent upon how signals propagate. The fundamental equations that govern the movement of signals from a transmitter, through space, and at a receiver are explained in this chapter. In this case, signals refers to RF signals, which for our purposes are assumed to start around 500 kHz.

2.2 SIGNAL PROPAGATION

A signal generated in a transmitter leaves that transmitter at a specified power level and is sent to an associated antenna usually via interconnecting cables. The antenna typically has a gain, which increases the level of the signal in certain preferred directions. As the signal propagates through the atmosphere, it suffers losses due to the spreading of the signal in space and losses due to encountered obstacles. It arrives at a receive antenna at some power level, which usually has some characteristic gain, thus increasing the level of the signal. This signal is then presented to the receiver from the receive antenna.

There are two components that make up an EM wave: the electric field and the magnetic field. One cannot exist without the other. The electric field is designated as **E** and has units of volts per meter. The magnetic field is designated as **H** and has units of amperes per meter. These components are orthogonal to one another and they are both orthogonal to the direction of propagation of the EM wave. This is shown schematically in Figure 2.1.

A signal radiates from an *isotropic* antenna (approximated by a point source) in an ever-expanding sphere, until it encounters something that perturbs that sphere. (An isotropic antenna is one that radiates equally in all directions.) At a significant enough distance from the transmitter, typically taken to be at least 10 times the wavelength, the spherical wavefront is frequently approximated as a plane over the dimensions of most antennas.

There are several modes of signal propagation. The major modes that are of importance for communication EW system design are direct wave, surface wave, reflected wave, refracted wave, diffracted wave, and scatter wave. For ground-to-ground communications, useful VHF, and above, signal propagation is limited to the troposhperic layer of the atmosphere, which ranges in altitude from 9 km at the Earth's poles to about 17 km at the equator [8]. On the other hand, HF signal propagation phenomena takes advantage of the ionosphere for long-distance communication. For ground-to-air or air-to-air communications, the direct wave is the method most frequently used.

2.3 RF BAND DESIGNATIONS

The RF spectrum is divided into designated bands with the common designations shown in Table 2.1. Communication services of some sort are provided in

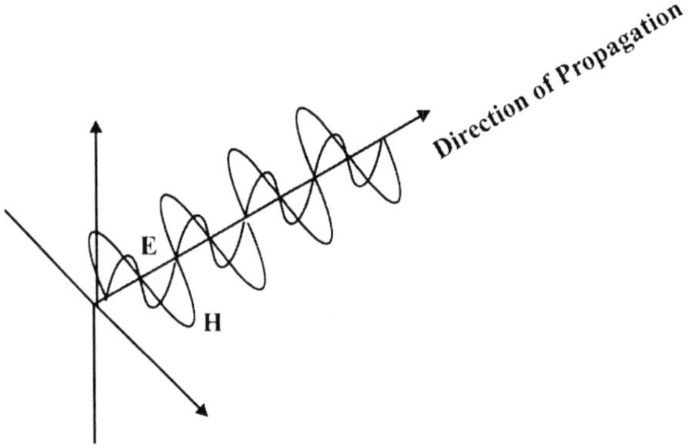

Figure 2.1 E and **H** fields associated with an EM wave.

Electromagnetic Signal Propagation

Table 2.1 Frequency Band Designations

Frequency Band	Name	Designation
3–30 kHz	Very low frequency	VLF
30–300 kHz	Low frequency	LF
300–3,000 kHz	Medium frequency	MF
3–30 MHz	High frequency	HF
30–300 MHz	Very high frequency	VHF
300–3,000 MHz	Ultra high frequency	UHF
3–30 GHz	Super high frequency	SHF
30–300 GHz	Extra high frequency	EHF
300–3,000 GHz	Optical	Optical

virtually all of the RF bands.

The microwave band (500 MHz to 40 GHz) has long been subdivided into subbands. It has recently undergone a change in subband designation, however. The bands, along with their old and new designations, are given in Table 2.2.

Table 2.2 New Designations of the Higher-Frequency Bands

Frequency Band	Old Designation	New Designation
0.5–1 GHz	UHF	C
1–2 GHz	L	D
2–3 GHz	S	E
3–4 GHz	S	F
4–6 GHz	C	G
6–8 GHz	C	H
8–10 GHz	X	J
10–12.4 GHz	X	J
12.4–18 GHz	Ku	J
18–20 GHz	K	J
20–26.6 GHz	K	K
26.6–40 GHz	Ka	K

2.4 POLARIZATION

The *polarization* of an EM wave is the orientation, relative to the Earth, of the electric field component of the wave. Although the polarization of an EM wave can be anything, the most common forms of polarization are vertical, horizontal, circular, and elliptical. In these latter two categories the electric field is intentionally rotated as the signal traverses space. Manipulating the phase of the signal at the transmitter creates this rotation.

VHF and above signals close to the Earth propagate better in the vertical polarization mode than in the horizontal polarization mode. This characteristic occurs because the Earth's magnetic field tends to cancel the magnetic component of EM waves when this component is vertical (which occurs when the electric component is horizontal). Often when an antenna is used both for transmission and reception, orthogonal polarizations are used for the two signals since some degree of isolation is provided by the polarization diversity.

2.5 POWER DENSITY

The *power density* of an EM wave is a measure of the power in the wave at any point in space. Herein it is denoted as P_d and is given in units of watts/meter^2. The *field strength* of an EM wave, E, is a measure of the voltage potential differences in the wave as a function of distance and is given in units of volts/meter. As shown in Figure 2.1, the E and H waves can be considered as vectors, which have both an amplitude and a direction. As vectors they are denoted by **E** and **H**. According to the *Poynting theorem*,

$$P_d = \frac{1}{2}|\text{Re}(\mathbf{E} \otimes \mathbf{H})| \tag{2.1}$$

where $\otimes$ represents the vector cross product. Let

$$\begin{aligned}\mathbf{E} &= E_{\text{peak}} \cos(2\pi f t)\mathbf{e} \\ \mathbf{H} &= H_{\text{peak}} \cos(2\pi f t)\mathbf{h}\end{aligned} \tag{2.2}$$

where **e** and **h** are unit vectors in the direction of **E** and **H**, respectively.

The *characteristic impedance*, Z_0, of the propagation medium is given by the ratio of the magnitudes of the electric wave to the magnetic wave. Thus,

$$Z_0 = \frac{E_{peak}}{H_{peak}} \tag{2.3}$$

In free space, Z_0 is equal to 120π, or 377 ohms. Therefore, since **e** and **h** are orthogonal,

$$P_d = \frac{1}{2}\left|\text{Re}\left(E_{peak} \frac{E_{peak}}{Z_0}\right)\right|$$
$$= \frac{1}{2}\frac{E_{peak}^2}{Z_0} \tag{2.4}$$

However, since for sinusoidal waves, $E_{rms} = E_{peak}/\sqrt{2}$,

$$P_d = \frac{E_{rms}^2}{Z_0}$$
$$= \frac{E_{rms}^2}{120\pi} \tag{2.5}$$

Therefore, the electric field strength is related to the power density by

$$E_{rms} = \sqrt{120\pi P_d} \tag{2.6}$$

From these equations the electric field strength can be calculated versus distance between the transmitter and receiver.

The *sensitivity* of receiving systems is often given in terms of microvolts per meter (μV/m), that is, in terms of field strength. This is a system-level specification in that it reflects the degree to which the system can extract EM wave energy. It best describes the ability of the system antenna to deliver signal power to the remainder of the system. Typical values are 1–5 μV/m. Equivalently, sensitivity is sometimes specified in terms of decibels relative to 1 μV/m. Since a decibel is a measure of power, conversion to power terms is necessary to change such a specification to absolute values. Recall that decibel is defined as

$$dB = 10\log\frac{P_2}{P_1} \tag{2.7}$$

When specifying decibels relative to 1 μV/m, the two power values are calculated assuming the same impedance value, usually the characteristic impedance of free space, Z_0. Thus,

$$dB_{1\mu V/m} = 10\log\frac{(v/m)^2/Z_0}{(1\mu V/m)^2/Z_0} \tag{2.8}$$
$$= 20\log\frac{v/m}{1\mu V/m}$$

So

$$v/m = 10^{-6} \times 10^{dB_{\mu V/m}/20} \tag{2.9}$$

(of course, the 10^{-6} is not used if the results are desired in units of μV/m).

2.6 FREE-SPACE PROPAGATION

Free-space propagation refers to the propagation mode between two antennas where there is no obstacle between the antennas to interfere with the ever-expanding spherical surface of the signal emitted from the transmitting antenna, when the transmitting antenna is isotropic. Free-space propagation is only possible in outer space where there is relatively little matter. Some air-to-air, including satellite communications, approximate free-space situations fairly well, however.

An isotropic antenna is one that radiates equally in all directions simultaneously and is only a model—they do not exist in practice. It is used as a standard against which practical antennas are measured.

The power density at a distance R from the isotropic transmit antenna is the amount of EM wave power that passes through a unit area on the surface of the spherical surface. By simple calculus

$$P_d = \frac{P_{transmitted}}{4\pi R^2} \tag{2.10}$$

where $P_{transmitted}$ is the amount of power emitted by the transmitter.

The amount of power received by the receiver is given by the amount of this power density absorbed by the receive antenna. The effective area, A_{eff}, determines this. Therefore,

$$P_R = P_d A_{\text{eff}} \tag{2.11}$$

The effective area of an isotropic receive antenna is given by

$$A_{\text{eff}} = \frac{\lambda^2}{4\pi} \tag{2.12}$$

Therefore the power received in free-space propagation between two isotropic antennas is given by

$$P_{\text{received}} = \frac{P_{\text{transmitted}}}{4\pi R^2} \frac{\lambda^2}{4\pi}$$
$$= P_{\text{transmitted}} \left(\frac{\lambda}{4\pi R} \right)^2 \tag{2.13}$$

The *gain* of an antenna is a measure of how particular directions are favored for propagation over others. More will be said about gain later. Suffice it here to say that the amount of power transmitted is given by

$$P_{\text{transmitted}} = G_T P_T \tag{2.14}$$

where P_T is the amount of signal power delivered to the antenna, and G_T is the transmit antenna gain in some direction. A similar relationship exists for the receive antenna,

$$P_R = G_R P_{\text{received}} \tag{2.15}$$

where G_R is the receive antenna gain in some direction and P_R is the power available from the antenna.

From above,

$$P_d = \frac{G_T P_T}{4\pi R^2} = \frac{E_{\text{rms}}^2}{120\pi} \tag{2.16}$$

so the free-space rms field strength can be calculated to be

$$E_{rms} = \frac{\sqrt{30 G_T P_T}}{R} \qquad (2.17)$$

Define the *free-space path loss* as the ratio of the power out of the receive antenna to the power input to the transmit antenna

$$L = \frac{P_R}{P_T} = G_T G_R \frac{\lambda^2}{(4\pi R)^2} \qquad (2.18)$$

This is known as *Friis' expression* for free-space path loss. In decibels it is

$$L_{dB} = \begin{Bmatrix} -32.2 \\ -36.6 \end{Bmatrix} - 20\log(f_{MHz}) - 20\log(R) + G_{T,dB} + G_{R,dB}, \begin{Bmatrix} km \\ mile \end{Bmatrix} \qquad (2.19)$$

when G_T and G_R are expressed in decibels. Thus, again in decibels, when P_T is expressed in decibels relative to 1W

$$P_R = P_T - L_{dB} \qquad (2.20)$$

or, when P_T is expressed in decibels relative to 1 mW

$$P_R = P_T - L_{dB} + 30 \qquad (2.21)$$

This expression ignores the cable losses associated with the transmitter and receiver. Denoting these by L_T and L_R, when they are included the expressions become

$$P_R = P_T - L_{dB} - L_T - L_R \qquad (2.22)$$

and

$$P_R = P_T - L_{dB} - L_T - L_R + 30 \qquad (2.23)$$

respectively. Values for L_T and L_R for some common types of cable are given in Table 2.3. These loss values correspond to newly manufactured cable, and the actual loss will depend on such factors as installation parameters, age of the cable, and temperature.

2.7 DIRECT WAVE

If there is line of sight between the transmitter and the receiver then the principal mode of signal propagation is via the *direct wave*. As above, at sufficiently large distances above the Earth's surface, the free-space propagation model may apply to the direct wave. Closer to the Earth, however, that is not the case. In fact, as illustrated in Figure 2.2, the amount of power available from the receive antenna is given by

$$P_R = P_T \frac{G_T G_R \lambda^2}{(4\pi)^2 R^n} \quad (2.24)$$

where P_T = power input to the transmit antenna from the cable connecting the transmitter to the transmit antenna (watts); G_T = gain of the transmit antenna in the direction of the receive antenna (unit-less); G_R = gain of the receive antenna in the direction of the transmit antenna (unit-less); λ = wavelength of the signal (meters); R = distance between the transmitter and the receiver (meters); and n is discussed below.

The electric field strength versus distance for radios with an ERP of 40 and 50 dBm is shown in Figure 2.3. This chart assumes that the propagation exponent n = 2, corresponding to the free-space conditions discussed in Section 2.6. Figure 2.4 shows the electric field strength versus distance when the propagation

Table 2.3 Loss per Foot of Some Common Cable Types

Cable Type	Loss per Foot (dB)		
	At 100 MHz	At 400 MHz	At 1,000 MHz
RG6/U	0.019	0.043	0.065
RG58/U	0.05	0.11	0.2
RG59/U	0.038	0.075	0.11
RG8/U	0.025	0.054	0.092
RG174/U	0.11	0.22	0.32
RG188/U	0.105	0.18	0.3
RG213/U	0.025	0.055	0.095

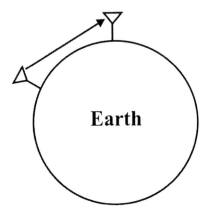

Figure 2.2 A direct wave travels in a straight line from the transmit antenna to the receive antenna.

exponent $n = 4$. Note the change in range on the abscissa. Clearly the signals fall off much more rapidly close to the Earth than high in the air.

Typical specifications for electronic support systems are a sensitivity of 1–5 µV/m. Therefore, the signals from these radios at low altitudes ($n = 4$) can be detected at ranges less than 10 km for ground ES systems. An airborne system, on the other hand, has an intercept range exceeding 150 km for air-to-air intercept with such targets. Thus, airborne systems typically have a substantial advantage for signal intercept.

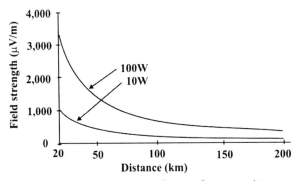

Figure 2.3 Electric field strength as a function of range for two emitter powers, where the propagation exponent $n = 2$.

Electromagnetic Signal Propagation

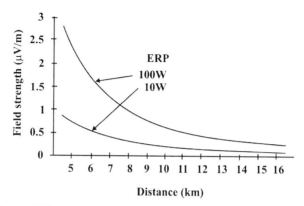

Figure 2.4 Electric field strength as a function of range for two emitter powers, where the propagation exponent $n = 4$.

Unavoidable electronic losses occur at the transmitter and receiver. These losses include cable-resistive loss, frequently expressed as I^2R losses because the loss increases as the current increases. Another source of signal loss at receiving antennas is due to mismatches in the polarization between the signal and the orientation of the antenna. Theoretically no signal is received at all if the signal is vertically polarized and the antenna is horizontally polarized or vice versa. In practice, some signal is received even in this case, however. Herein, these losses are included in the values of G_T and G_R.

There is a propagation model methodology based on the free-space model that is sometimes called the R^n model [9]. The mean path loss at a distance R from the transmitter is calculated as

$$L(R) = L(R_0) + 10n \log_{10}\left(\frac{R}{R_0}\right) \qquad (2.25)$$

where R_0 is a suitably defined reference distance, frequently taken as 1 km for outdoor propagation conditions and 1m for indoor applications. $L(R_0)$ represents the path loss at the reference distance. It is frequently measured, but if it is not otherwise known, it can be estimated by

$$L(R_0) \approx 20 \log_{10}\left(\frac{4\pi R_0}{\lambda}\right) \qquad (2.26)$$

The parameter n, which depends on the environmental conditions, is discussed next. The received power then, at range R, is, as above,

$$P_R(R) = \frac{P_T G_T G_R}{L(R)} \tag{2.27}$$

Examination of (2.27) reveals several items of note about direct-wave signal propagation. First, lower frequencies propagate better than higher ones ($\lambda = 1/f$). Second, the level of the received signal decreases as a power n of the distance between the transmitter and the receiver. Away from the surface of the Earth, $n = 2$. Closer to the Earth's surface, the signal strength decreases faster than $1/R^2$, and $n = 4$ is often used. The actual exponent on R varies with each situation. It ranges from 2 to 16 or more. Therefore, it can be concluded that if an antenna is raised higher into the air, better signal propagation should result. In general this is true, but to obtain a dramatically improved performance, the antennas need to be substantially elevated.

The propagation loss at any range R is a statistical parameter. As such, the loss given above is an average. In many cases, the statistics follow a log-normal distribution. Thus, there is an associated standard deviation of measurements as well. The loss exponent n is given in Table 2.4 for various propagation conditions. For outdoor environments, the standard deviation is in the range of 8 to 14 dB. On the other hand, the propagation loss exponent n and standard deviation σ are given in Table 2.5 for *personal communication systems* (PCSs) in several indoor conditions [9].

All communication paths suffer losses with distance. This is due to the simple fact that the energy or power spreads out with distance, and so the energy and power density decrease. Communication links close to the Earth lose energy approximately as the reciprocal of this distance to the fourth power. Air-to-air links and other free-space direct paths suffer approximately a loss that is the reciprocal of the distance to the second power. Air-to-ground and ground-to-air links suffer loss in between these values. In this section the effects of this propagation loss will be illustrated.

Most lower echelon tactical battlefield radios used for real-time command and control of tactical forces have power levels into their associated antennas of

Table 2.4 Loss Exponent n for Various Conditions

Condition	Loss Exponent, n
Free space	2
Urban area cellular, PCS	2.7–4.0
Shadowed urban cellular, PCS	3–5
In building line of sight	1.6–1.8
Obstructed in building	4–6
Obstructed in factories	2–3

Source: [9], © 1996, Pearson Education, Inc. Reprinted with permission.

Table 2.5 Propagation Loss Exponent n and Associated Standard Deviation σ for Several Indoor Conditions

Conditions	Frequency (MHz)	n	σ (dB)
Indoor: Retail store	914	2.2	8.7
Indoor: Grocery store	914	1.8	5.2
Indoor: Hard partition office	900	3.0	7.0
Indoor: Soft partition office	900	2.4	9.6
Indoor: Soft partition office	1,900	2.6	14.1
Indoor: Factory (LOS)	1,300	1.6–2.0	3.0–5.8
Indoor: Factory (LOS)	4,000	2.1	7.0
Indoor: Suburban home	900	3.0	7.0
Indoor: Factory (obstructed)	1,300	3.3	6.8
Indoor: Factory (obstructed)	4,000	2.1	9.7
Indoor: Office same floor	914	2.76–3.27	5.2–12.9
Indoor: Office entire building	914	3.54–4.33	12.8–13.3
Indoor: Office wing	914	2.68–4.01	4.4–8.1
Indoor: Average	914	3.14	16.3
Indoor: Through one floor	914	4.19	5.1
Indoor: Through two floors	914	5.04	6.5
Indoor: Through three floors	914	5.22	6.7

Source: [9], © 1996, Pearson Education. Reprinted with permission.

10–100W (40–50 dBm). The typical antenna for these configurations is a tuned whip, which, if it's in good condition, has about 2 dB of gain. Most realistic scenarios, however, do not involve perfect hardware implementations. We will assume here that factors such as the net antenna gain and cable loss of these radios is a net 0 dB (no gain or loss).

The so-called *range to horizon* is the distance beyond which one no longer has visual *line of sight* (LOS) between the transmitter and receiver. Assuming that the Earth is a smooth sphere, the range to horizon can be calculated according to the equation obtained from simple geometric principles as

$$r^2 = (R+h)^2 - R^2 \qquad (2.28)$$

or

$$r = \sqrt{2Rh+h^2} \qquad (2.29)$$

where R is the radius of the Earth, h is the elevation of the transmitter, and r is the range to horizon. Frequently h^2 can be neglected relative to R, so

$$r = \sqrt{2Rh} \qquad (2.30)$$

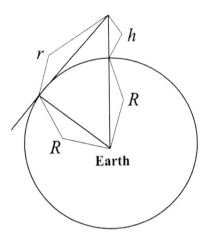

Figure 2.5 Range to horizon.

These distances are illustrated in Figure 2.5. Of course here, the Earth is assumed to be a sphere with an effective radius R.

Assuming that the Earth is a sphere of radius 3,960 miles, then the distance to the horizon for an aircraft at 20,000-feet is 173-miles. At VHF (30–300 MHz) and above frequencies the atmosphere close to the surface of the Earth refracts (bends) radio waves that pass through it as illustrated in Figure 2.6. The net effect is to

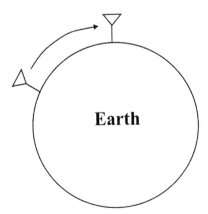

Figure 2.6 The troposphere close to the surface of the Earth refracts radio waves as they traverse through it, causing an extension of the propagation range.

increase the effective radius of the Earth by approximately 1/3. Therefore, a more correct value to use for R in (2.30) is 4/3 R, or 5,280 miles. The resultant calculation is called the *radio line of sight* (RLOS). The 4/3 factor does not always apply, however. In cases of *subrefraction*, the bending of the radio waves in the atmosphere is less than normal and the resultant path length is less than otherwise. In *super-refraction*, the waves are bent more than normal and the path length is longer than otherwise. Using similar arguments, assuming a 4/3-Earth model and a smooth Earth surface, a transmitter and receiver are within RLOS of each other as long as they are within a distance of

$$d = \sqrt{2h_T} + \sqrt{2h_R} \qquad (2.31)$$

of each other, where d is in miles and h_T and h_R are in feet. Thus, if the transmit antenna is at a height of 20 feet and the receive antenna is at 50 feet, the RLOS is 16.3 miles. Caution should be used when applying these equations to transmitters and receivers close to the Earth's surface, however, because the Earth's surface is not smooth and, depending on the frequency, obstacles can (and frequently do) reduce this range or even preclude signal propagation at all. On the other hand, complex phenomena such as edge diffraction and reflections from surfaces can enhance signal propagation close to the Earth.

2.8 WAVE DIFFRACTION

Physical diffraction is caused by a wave impinging on an object. Some energy in the wave appears to be bent by the edge leaving in a direction different from the original direction. Some of the energy in the wave is changed in its propagation direction. At VHF and above, this diffraction accounts for why it is possible to receive signals even though there are significant obstacles between the transmitter and receiver.

When a radio wave encounters an obstacle, a physical phenomenon called *Huygen's principle* explains the wave diffraction that occurs. This principle says that each source on a wavefront generates secondary wavefronts called wavelets, and a new wavefront is built from the vector sum of these wavelets. The amplitude of the wavelets varies as $(1 + \cos \alpha)$, where α is the angle between the wavelet direction and the direction of propagation. Thus, the wavelet with the maximum amplitude is in the direction of propagation and there is zero amplitude in the reverse direction. In schematic form, the waves are generated as shown in Figure 2.7. Thus, the obstacle does not totally block the signal behind it. It is assumed here that the obstacle has small enough thickness that any impacts at the

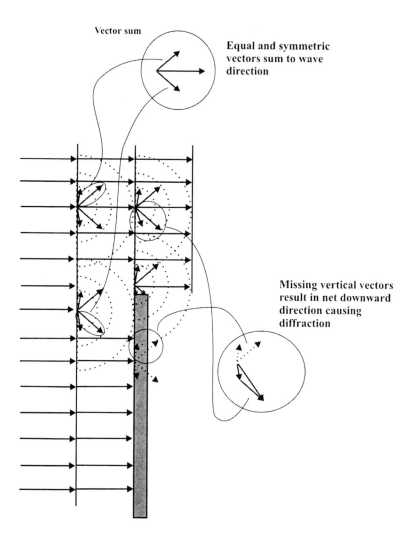

Figure 2.7 Diffraction at an obstacle produces signal energy behind the obstacle.

end of the obstruction are insignificant. In the far field from the obstruction, the E-field at the bottom of the figure has flipped 180°.

When the obstacle is located such that the path-length difference between the direct path and the length of the path made up of the path from the transmitter to the obstacle and the path length from the obstacle to the receiver is a multiple of $\lambda/2$, then the diffraction from the obstacle will tend to cancel the direct path signal at the receiver. The reason for this is that the portion of the signal not blocked by the obstacle will generate signal vectors behind the obstacle but π radians out of phase with the direct path signal. The situation depicted in Figure 2.7 and with the assumption that the thickness of the obstacle has insignificant impact is known as *knife-edge diffraction*. When there is a finite thickness, the attenuation at the edge can be substantially higher than at a knife edge. Diffraction effects also occur behind an obstruction such as a mountain.

2.9 REFLECTED WAVES

The ground and other large surfaces (large relative to the wavelength of the signal) can reflect EM waves. Reflected waves can arrive at a receive antenna out of phase with the direct wave, and, in the case where they are 180° out of phase, can cause considerable fading, depending on the magnitude of the reflected wave compared to the direct wave. These reflections can occur in many ways, one of which is off the ground when the transmitter and/or receiver are close to the Earth's surface. Another form of reflection is off nearby metallic objects, such as cyclone fences close to the transmitter and/or receiver. The ghosts that appear on a television set (when connected to a TV antenna as opposed to a cable or satellite antenna) are a manifestation of such reflections. In that case the reflected wave is received only slightly later than the direct wave, causing the picture to be delayed slightly. The geometrical shapes generated by setting the path difference equal to a constant are ellipsoids (three-dimensional ellipse) with the two antennas at the foci. When the path difference is set equal to $k\lambda/2$ for some integer k, the *Fresnel zones* are generated. For $k = 1$, it is the first Fresnel zone. These notions are depicted in Figure 2.8. Let $\delta_k = d_R - d_D$. Then the Fresnel zones are defined by

$$\delta_k = k\frac{\lambda}{2} \qquad k = 1, 2, \ldots \qquad (2.32)$$

In most cases, it is only necessary to consider the first Fresnel zone.

Signal reflections off objects also affect the signal level received. At low reflection angles of incidence (the case for reflections off the Earth with large path distances and the case of importance here), reflections off the ground impart a π radian phase shift in the reflected signal for both vertical and horizontal polarizations. A rule of thumb useful for determining at what distance such reflections cause propagation changes from $n = 2$ to $n > 2$ in (2.1) is given by

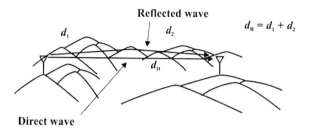

Figure 2.8 An EM wave with a significant reflected component at the receiver. The reflection is off a distant mountain in this case.

$$d_1 = \frac{4h_T h_R}{\lambda} \qquad (2.33)$$

where h_T is the transmit antenna height, h_R is the receive antenna height, and λ is the wavelength, all in consistent units. This corresponds to the distance where the first Fresnel zone first touches a point of reflection. Beyond this distance the signal is grazing the Earth at the reflection point and a constant π radians is destructively added to the phase of the received signal. The amplitudes of the respective signals determine the amount of the destruction, so therefore the characteristics of the reflection point (e.g., smooth or rough Earth and trees) are important.

Reflection of radio waves off the surface of the Earth can be analyzed with the aid of Figure 2.9. For simplicity it is assumed that the Earth is flat in the region

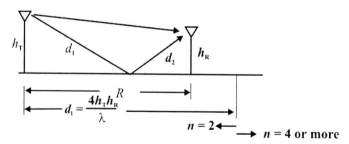

Figure 2.9 Distance d_1 is the distance where the first Fresnel zone touches flat Earth for the first time. Beyond this distance the attenuation exponent n changes for 2 to 4 or more.

between the transmitter and receiver. The power from the receive antenna due to both the direct wave and the reflected wave is given by

$$P_{total} = P_{direct} \left|1 + \rho e^{-j\phi}\right|^2 \tag{2.34}$$

where ρ is the reflection coefficient at the point of reflection and the phase difference between the direct wave and reflected wave, ϕ, is given by

$$\phi = 2\pi \frac{c}{\lambda} \frac{\delta_r}{c} \tag{2.35}$$

where δ_t is the time difference between the two waves and δ_r is the path distance difference. Thus,

$$\phi = 2\pi \frac{\delta_r}{\lambda} \tag{2.36}$$

If $R \gg h_T$ or h_R, then $\theta \approx 0$, $\rho \approx -1$ and

$$\phi \approx \frac{2\pi}{\lambda} \frac{2h_T h_R}{R} \tag{2.37}$$

Now

$$\begin{aligned}\left|1 + \rho e^{-j\phi}\right|^2 &= \left|1 - (\cos\phi - j\sin\phi)\right|^2 \\ &= (1 - \cos\phi)^2 + \sin^2\phi \\ &= \left(1 - 2\cos\phi + \cos^2\phi + \sin^2\phi\right) \\ &= (2 - 2\cos\phi) \\ &= 2 - 2\cos\left(\frac{2\pi}{\lambda} \frac{2h_T h_R}{R}\right)\end{aligned} \tag{2.38}$$

so

$$\frac{P_{total}}{P_{direct}} = 2 - 2\cos\left(\frac{2\pi}{\lambda} \frac{2h_T h_R}{R}\right) \tag{2.39}$$

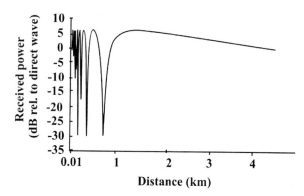

Figure 2.10 Power received due to both the direct wave and the ground reflected wave.

Equation (2.39) is plotted in Figure 2.10 in decibels relative to the direct wave for $h_T = 30$m and $h_R = 2$m at 1,850 MHz. When R is small, the total power oscillates dramatically, but for larger ranges between the transmitter and receiver the power decreases at a rate proportional to $1/R^4$. The reason for this is as follows. For small ϕ, $\cos \phi \approx 1 - \phi^2/2$, so

$$\left|1 + \rho e^{-j\phi}\right|^2 = 2\left(1 - 1 + \frac{\phi^2}{2}\right)$$

$$= \phi^2 \qquad (2.40)$$

$$= \left(\frac{2\pi}{\lambda} \frac{2 h_T h_R}{R}\right)^2$$

therefore

$$\left|1 + \rho e^{-j\phi}\right|^2 \approx 16\left(\frac{\pi}{\lambda} \frac{h_T h_R}{R}\right)^2 \qquad (2.41)$$

and

$$P_{\text{total}} = \frac{P_T}{4\pi R^2} \frac{\lambda^2}{4\pi} 16\left(\frac{\pi}{\lambda} \frac{h_T h_R}{R}\right)^2$$

$$= \frac{P_T}{R^4} (h_T h_R)^2 \qquad (2.42)$$

for isotropic antennas. Note that the total power is independent of frequency. Furthermore, the total power increases as the square of the antenna heights and decreases as the fourth power of the range.

When the antenna gains are considered, both for the transmitter and receiver, then this expression becomes (with G_T and G_R the gain relative to isotropic in the direction of each other)

$$P_{total} = \frac{G_T G_R P_T}{R^4}(h_T h_R)^2 \qquad (2.43)$$

This expression will be referred to herein as the *ground reflection propagation model*. Although in (2.42) the received power is independent of frequency, when the gains are included, the power becomes frequency-dependent because the antenna gains are frequency-dependent.

Using this expression for the received power at all ranges underestimates the power at close ranges. Beyond distance d_1, given in (2.2), it more accurately reflects the total power received, however. At close range the propagation loss increases with $n = 2$ and the antennas are close enough together that the antenna heights do not have an effect. The propagation is effectively free-space.

Reflection amplitude characteristics are shown in Figure 2.11 when the frequency is 100 MHz [10] and the reflections are off the ground. Horizontally polarized waves undergo substantially less amplitude attenuation than vertically polarized signals. Reflection phase characteristics are shown in Figure 2.12. A

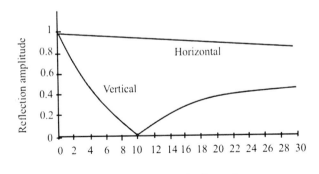

Figure 2.11 Amplitude characteristics of a reflected wave.

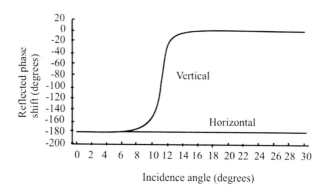

Figure 2.12 Phase shift imparted on a reflected wave.

horizontally polarized wave always undergoes a 180° phase shift, while a vertically polarized one only imparts a significant phase shift for incidence angles less than 12° or so. The region of primary interest for communication EW system design is for low incidence angles—typically less than 10°.

2.10 SURFACE WAVE

At frequencies below approximately 50 MHz, there is a mode of propagation referred to as the *surface wave*, or *ground wave*. This wave propagates along the surface of the Earth out to considerable distances, depending on conditions. At the higher frequencies the attenuation is too high to support propagation. This mode of propagation is illustrated in Figure 2.13.

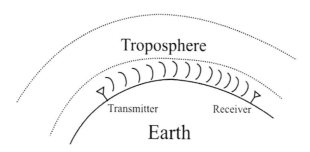

Figure 2.13 Propagation along a trough, one side of which is the Earth, is called surface-wave propagation.

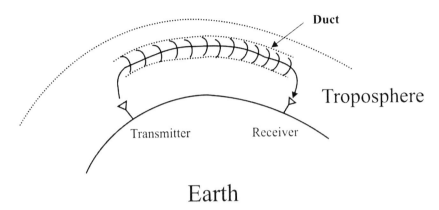

Figure 2.14 Ducting is when signals propagate between two tropospheric layers. Considerable distances can be traversed this way.

2.11 DUCTING

There is a phenomena called *ducting*, wherein VHF and above signals can travel considerably further than RLOS. Essentially the signals are reflected off regions in the troposphere where the refractive index decreases rapidly, forming a duct through which these signals will travel. This is shown in Figure 2.14. Two tropospheric layers or one layer and the Earth's surface can form such ducts.

In EW applications, ducting can be useful for ES from ranges and sites that otherwise would be too far away. On the other hand, determining the geolocation of such signals is difficult. This is because triangulation (defined in Section 11.3) usually will not work since ducts probably will not exist between the transmitter and two receiving sites simultaneously, a requirement for triangulation to work.

2.12 METEOR BURST

This form of communication relies on the thousands of meteors that enter the Earth's atmosphere each day [11]. It also can be classified as a reflected mode. At frequencies around 30–50 MHz, these meteors will reflect radio waves. The phenomenon is illustrated in Figure 2.15.

Signals need to have considerable redundancy in this scheme, and therefore it is only reliable for low data rate communication. Communication is limited to short bursts and the medium only supports digital communications in the low data rate range—up to 600 bps is typical. It is frequently used for relaying remote

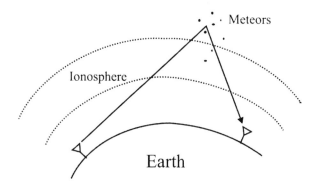

Figure 2.15 Meteor burst propagation reflects the EM wave off meteors. While considerable distance can be covered this way, data rates are quite low.

sensor data where messages are sent infrequently, and each such message consists of only a few hundred bits maximum. Frequently *acknowledgments* (ACKs) and *negative acknowledgments* (NACKs) are used to indicate that a message was received properly. An ACK is sent if the message was received properly and a NACK is replied if not. If not, the message is repeated until it gets through, which means that one or more meteors was in the right place to reflect the signal. Of course, these ACKs and NACKs rely on the same propagation path so they too must be kept short and infrequent since the meteor could very likely no longer be there when the reply needs to be sent.

2.13 SCATTERING

Scatter-wave propagation is caused by nonhomogeneous refractive indices in the troposphere caused by irregular ionization, or by rain. It is also caused by nonhomogeneous refractive indexes on the surface of the Earth. Objects that are smaller than a wavelength will cause scattering for EM waves as well. Typical of this latter category would be street signs and telephone posts scattering UHF signals, as typified by mobile phones in the 900-MHz range ($\lambda = 0.3$m) or PCS systems in the 1,800-MHz range ($\lambda = 0.15$m). At frequencies around 4–5 GHz, there is a reliable propagation phenomena known as *troposphere scattering*. A radio wave with properly oriented transmitting and receiving antennas can communicate over long distances. The configuration is shown in Figure 2.16.

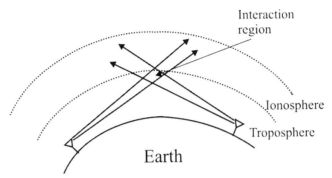

Figure 2.16 Tropospheric scatter propagation relies on the interaction to the two antenna beams at common areas within the troposphere. This type of propagation is possible in the 4–5 GHz range.

The two antenna beams interact in the region where they overlap and the radio wave is essentially reflected from within this region.

2.14 CHARACTERISTICS OF THE MOBILE VHF CHANNEL

Command and control of mobile, tactical military forces frequently are accomplished with RF communications, primarily in the VHF and UHF ranges. This type of communication is subject to slow and fast fading and distortion due to delay spread [12]. As pointed out earlier, $P_d \propto 1/R^n$. The value of n depends on several factors, such as obstacles in the path between the transmitter and receiver. Illustrated at the top of Figure 2.17 is the characteristic path loss for $n = 2, 3$, and 4.

At any given distance from the transmitter, there will be a statistical distribution of the path loss, which therefore imparts the statistical distribution onto the amount of power received. A Gaussian distribution (see Appendix A) as shown often accurately describes this distribution. Objects blocking the direct communication path, called *shadowing,* typically cause large fades. The mean path loss at some distance R due to shadowing produces *slow fades* as shown Figure 2.17. It causes (relatively) long-term variations in signal level at the receiver. This is typical of mobile communications where the receiver or transmitter moves behind large objects (e.g., a mountain). Small fades are caused by locally changing conditions such as changing multipath conditions.

On the other hand, fast fading also occurs in mobile communications. The amplitude of the received signal level is often approximated by a Rayleigh distribution, which is shown in Figure 2.17, middle right panel. The receiver or

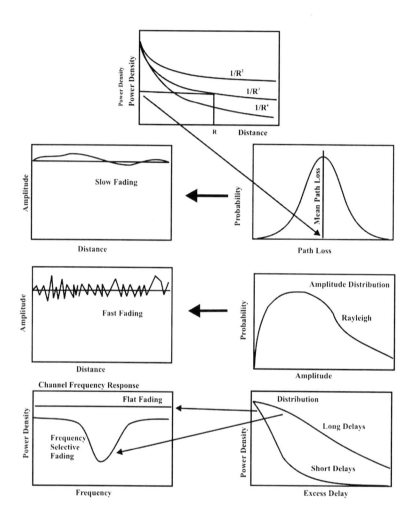

Figure 2.17 The mobile channel is characterized by slow and fast fading and stochastic propagation losses.

transmitter moving behind small objects, causing the signal to be attenuated, produces such fading.

In direct and surface-wave propagation fading can occur in mobile communication paths due to reflections off of large objects or changing diffraction or scattering situations. Even if the transmitter and receiver are not moving, objects moving in the environment surrounding them, such as vehicles or people, can change the propagation conditions. This fading can affect very narrow frequency bands, sometimes affecting only portions of the bandwidth of a signal, attenuating some portions of the frequency band much more than others. The longer-term attenuation characteristics of the ionosphere are attributable to the changing ionosphere, whereas fast fading occurs for other reasons.

The paths taken by reflected waves compared to the direct wave, such as the situation shown in Figure 2.8, are different. These waves add as vectors at the receiver causing the reflected wave to cancel to some degree the direct wave. This path length, of course, varies as the transmitter and/or receiver moves, causing variable fading effects. Even though an EM wave has a well-defined polarization as it leaves the transmitting antenna, the effects of the environment change this polarization as the EM wave propagates. A wave reflected off the Earth or some other surface can change the direction of the polarization and can change its direction of propagation. Typically reflected or refracted waves arrive at a receiver with elliptic orientation. Furthermore, this elliptic orientation varies with time due to the changing propagation conditions. If the receiver antenna is oriented in a particular direction, which it normally is, this changing elliptic polarization will be received as fading phenomena.

The RF channel can also cause time delays due to its impulse response. These time delays can be short or long, depending on the source of the delay. The delay is referred to *excess delay*, where excess is relative to the symbol rate. Long delays are caused by a frequency-selective impulse response. Short delays are due to a relatively flat channel frequency response. The frequency referred to in this case is relative to the bandwidth of the signal.

Reflected waves arrive at a receiver later than the direct or surface wave. This delay time is referred to as the *delay spread* and varies depending on the environmental conditions. In urban environments the delay spread is typically 3 µs; in suburban environments it is typically 0.5 µs, while in rural terrain it can vary considerably, ranging from less than 0.2 µs to 12 µs or more. These values are means since the delay spread is a random variable in most circumstances. The standard deviations of these distributions can vary considerably depending on the specific situation.

This multipath interference tends to smear digital signals and cause one symbol to interfere with others called *intersymbol interference* (ISI). This interference can be quite severe and limits the maximum data rate that the communication channel can support. In an urban channel, for example, when the delay spread is 3 µs, digital signals at 333 Kbps would have symbol n in the direct wave completely overlaid by the $n - 1$ symbol in the delayed signal. If the energy in the delayed signal is strong enough, reliable communication in that case would be impossible.

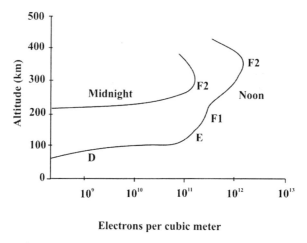

Figure 2.18 Layers in the ionosphere change with time. They determine how HF signals will be refracted.

2.15 PROPAGATION VIA THE IONOSPHERE

Propagation of communication signals via refraction and reflection from the ionosphere is one of the oldest, if not *the* oldest, forms of radio communication [13]. It is certainly the oldest commercial communication method. The signals, which radiate quite a distance by refraction through the ionosphere, are sometimes called *sky-wave* signals. These signals can frequently propagate for thousands of kilometers from the transmitter. Such propagation modes have been useful for communication with ships at sea and were virtually the only way to communicate with ships until communication satellites were invented [7, 13].

It is primarily the radiation from the Sun that generates the ions in the ionosphere. Therefore, there are differences in the propagation characteristics of the ionosphere depending on whether it is day or night. At dawn and dusk the ionosphere is turbulent due to the changing Sun radiation situation. The density of these ions varies with such factors as altitude and time of day.

2.15.1 Ionospheric Layers

The density of the ions in the ionosphere forms layers, designated by D, E, F1, and F2. However, there occasionally occurs a sporadic E layer, referred to as E_S. These layers are actually bands of ions of similar amounts of charge. Their altitude depends on circumstances, but the characteristics shown in Figure 2.18 [7] are representative. The altitudes shown in Figure 2.18 correspond to the height of the average electron density for that layer. Signals reflected off these different layers will have different coverage areas on the ground due to their differences in altitude. The electron density of these layers determines whether a signal is refracted or not. Regions with lower electron density will not refract the signals as well as regions with higher densities. Furthermore the refraction properties are frequency-dependent.

2.15.1.1 D Layer

The D layer is located at approximately 60–90 km above the surface of the Earth. Its half-thickness[1] is typically 10 km. This layer is only present during daylight hours and disappears at night. The ionization is highest at noon when the Sun is at its apogee. The D layer is not very useful for refracting signals, but it does attenuate them as they traverse through to the higher E and F layers.

2.15.1.2 E Layer

Maximum ionization in the E layer is at about 110 km with a typical half-thickness of 20 km. Like the D layer, the E layer only occurs during daylight hours with its maximum ionization occurring around noon.

2.15.1.3 E_s Layer

Occasionally there is an ionospheric layer that occurs somewhat higher than the E layer. It is called *sporadic E*, denoted by E_s, and is located at an altitude of 120 km. It typically is very thin, with a half-thickness ranging from a couple hundred of meters to about 1 km. The ionization, however, is quite intense.

2.15.1.4 F Layer

The F layer is comprised of two sublayers, F1 and F2. The F1 layer is located at an altitude of 170–220 km with a half-thickness of typically 50 km. Like the D and E layers, it only occurs during daylight hours. The F2 layer, on the other hand, is present at nighttime as well. It is located at an altitude of 225–450 km and is typically 100–200 km thick.

The two separate F layers only exist during daylight hours. At night, the two layers combine into one—simply the F layer. Therefore, the lower frequencies

---

[1] Half-thickness is the thickness at which the electron density has dropped to half its maximum.

propagate further at night than during the daylight hours, and frequencies that are usable during the day simply pass on through the ionosphere at night.

The D layer discussed above, as well as the E_s layer, cause changing ionospheric conditions, affecting the attenuation characteristics of the ionosphere. This will cause fading at the receiver.

2.15.2 Refraction

Just as VHF and above signals are refracted by the troposphere as discussed earlier, high-frequency signals are refracted by the ionosphere. At frequencies below about 30 MHz, the ionosphere (altitude 50–500 km) can refract signals, as illustrated in Figure 2.19. Whether it does so depends on several factors. The ionosphere consists of ions (thus, the *ion*osphere), which carry a charge. The density of the charge of these ions is heterogeneous in the ionosphere. This is what the RF signals interact with when they enter the ionosphere, and this interaction is what causes the signals to be refracted or not. Most of the time an equivalent height is used for calculations involving this form of signal propagation. This height is the height of a layer, which would reflect the signal if it were a plane sheet, rather than refract the signal, which is the actual phenomenon involved.

This form of signal propagation is not limited to a single hop. Many hops are possible depending on the conditions of the ionosphere as well as the Earth where the signal returns. The two-hop case is shown in Figure 2.20.

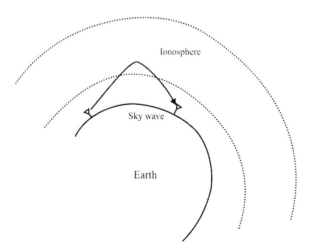

Figure 2.19 Long-range propagation is possible by signal refraction in the ionosphere.

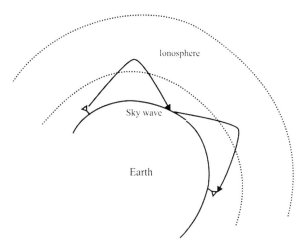

Figure 2.20 Multihop ionospheric refraction is a method to propagate HF signals to considerable distances.

The direct wave emanating from an antenna in the HF range will travel only so far before its energy gets too small to be useful. The sky-wave signal refracted by the ionosphere returns to the Earth beyond a certain distance as shown in

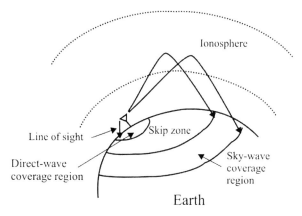

Figure 2.21 Ionospheric propagation creates a skip zone, beyond which the signal returns to the Earth.

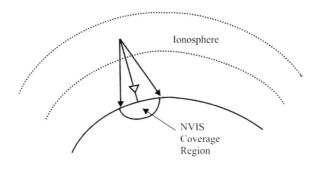

Figure 2.22 NVIS propagation in the HF is for close range. The signal goes straight up and essentially straight back down.

Figure 2.21. In between these ranges no reception is possible. This area is called the *skip zone*.

2.15.3 Near-Vertical Incidence Sky Wave

HF signals can be radiated essentially straight up toward the ionosphere by a properly oriented antenna. Under the right conditions these signals will come virtually straight back down, forming a cone, allowing communication within a range of several hundreds of kilometers from the transmit antenna. This mode of communication is referred to as *near-vertical incident sky wave* (NVIS). The geometry of this mode is illustrated in Figure 2.22, where only half of the ground footprint is shown. To facilitate NVIS communications, the signal must be radiated straight up. This is accomplished by having the antenna arranged horizontally so that the boresight of the antenna is pointed straight up. It must also be at the proper height to maximize the signal component radiated straight up.

2.15.4 HF Fading

Fading of EM waves occurs in the HF range as well. Such fading can be severe, ranging up to 20–30 dB or more. In HF propagation with refraction and reflection via the ionosphere, fading can occur due to changing ionospheric conditions, whether the communication nodes are moving or not. Fading also occurs because of multiple paths taken through the atmosphere by a signal, adding sometimes destructively and sometimes constructively to the signal at the receiver.

Electromagnetic Signal Propagation 51

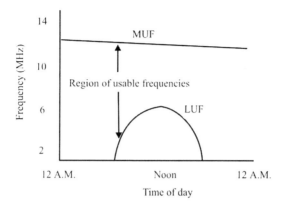

Figure 2.23 Maximum frequency and minimum frequency of ionospheric propagation.

An ionospheric-refracted wave actually is refracted over a region as opposed to a specific point in space. The index of refraction is different over this region, thereby changing the polarization of the wave and giving the effect of making the signal fade at the receiver.

2.15.5 Maximum Usable Frequency and Lowest Usable Frequency

The frequencies supported by the ionosphere at any given time form a band. The lowest frequency is termed the *lowest usable frequency* (LUF), and the highest usable frequency is termed the *maximum usable frequency* (MUF). The lower frequency is determined by the ionization caused by the Sun during the day. Typical characteristics of the MUF and LUF are shown in Figure 2.23.

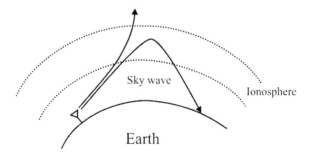

Figure 2.24 An HF signal at the wrong frequency will traverse through the ionosphere without coming back to Earth.

Depending on frequency and incidence angle, a signal impinging on the ionosphere from the bottom will either be refracted or will pass through, not returning to Earth at all, as illustrated in Figure 2.24.

Ionization of the atmosphere on Earth is caused by solar radiation. The solar radiation is not constant, however, and varies on an approximate 10.7-year cycle. The sunspots are dark areas on the Sun, adjacent to which are typically flares of intense radiation. These flares affect the Earth's ionosphere by increasing the degree of ionization. The result is higher critical frequencies, especially in the highest layer, the F2 layer. This, in turn, facilitates longer range communications during peaks of solar activity. In addition, the Sun rotates on an axis, and the radiation activity around the Sun is not constant. A region of high radiation facing the Earth will cause additional ionization, which rotates away. This period of rotation is approximately 27 days.

2.15.6 Automatic Link Establishment

The aforementioned anomalies of communicating in the HF range illustrate some of the difficulties of using the high-frequency range for communication purposes. This is determining the proper frequency to use, which, as just mentioned, depends on the state of the ionosphere. In the past, it has required skilled technicians to make this determination, thus relegating usage of the HF frequency range to technical experts. With *automatic link establishment* (ALE) techniques, the selection of the proper frequency is performed by the radio equipment itself, precluding the requirement for the operator to have extensive technical experience of such propagation and equipment operation.

The equipment does this by probing the ionosphere with signals of varying frequency and determining the frequency that works the best. It determines this by measuring the quality of the signals received at each frequency. Each end of the link performs the signal quality measurement.

2.16 CONCLUDING REMARKS

Propagation of radio waves is facilitated by different mechanisms depending largely on the frequency of the signal. Lower frequencies tend to propagate further than higher frequencies, all else being equal. The propagation characteristics of signals that are normally of interest to military communication were presented in this chapter. These characteristics are of similar importance to designers of communication systems as well as communication EW systems—the propagation effects are on the signals themselves, irrespective of what the signals are being used for.

Some characteristics, however, perturb the operation of communication systems other than communication EW systems. An example of this is meteor burst communications. One of the more important measurable parameters of communication signals of interest to EW systems is the location of the target. This is unobtainable in meteor burst systems since the transmitter sends the signal toward the meteors, not the receiver.

The ionosphere will reflect (refract) HF waves so very long-distance communications are possible. Although there are exceptions, this form of communication is not possible much above the HF range, however. For the higher-frequency range, propagation is normally limited to RLOS (4/3 Earth model), which significantly limits the distance over which two nodes can communicate close to the Earth. For communication EW systems, this normally means that to intercept such signals, it is more efficient to elevate the receiving antenna. Such elevation typically means putting the receiving antenna on some form of aircraft.

References

[1] Hall, M. P. M., *Effects of the Troposphere on Radio Communications,* Stevenege, United Kingdom: Peter Peregrinus, Ltd., 1979.
[2] Shibuya, S., *A Basic Atlas of Radio Wave Propagation*, New York: John Wiley & Sons, 1987.
[3] Parsons, D., *The Mobile Radio Propagation Channel*, New York: John Wiley & Sons, 1992.
[4] Stuber, G. L., *Principles of Mobile Communications*, Boston, MA: Kluwer Academic Publishers, 1996.
[5] Gagliardi, R. M., *Introduction to Communications Engineering*, New York: John Wiley & Sons, 1988.
[6] *Reference Data for Radio Engineers*, Chapter 28, Indianapolis, IN: Howard W. Sams & Co., Inc., 1975.
[7] Davies, K., *Ionospheric Radio*, London: Peter Peregrinus Ltd., 1989.
[8] Hall, M. P. M., *Effects of the Troposphere on Radio Communications,* Stevenege, United Kingdom: Peter Peregrinus, Ltd., 1979, p. 1.
[9] Rappaport, T. S., *Wireless Communications: Principles and Practice,* Upper Saddle River, NJ: Prentice Hall, 1996.
[10] Hall, M. P. M., L. W. Barclay, and M. T. Hewitt (eds.), *Propagation of Radiowaves*, New York: IEEE, 1996.
[11] Schanker, J. Z., *Meteor Burst Communications*, Norwood, MA: Artech House, 1990.
[12] Parsons, J. D., *The Mobile Radio Propagation Channel*, New York: John Wiley & Sons, 1992.
[13] Braun, G., *Planning and Engineering of Shortwave Links*, Siemens Aktiengesellschaft, New York: John Wiley & Sons, 1986, p. 21.

Chapter 3

Noise and Interference

3.1 INTRODUCTION

The signal propagation principles and descriptions in Chapter 2 indicate how signals propagate through space. Theoretically these signals propagate forever to infinite distances. Therefore, a suitably sensitive receiver placed anywhere in the universe would be able to detect these signals eventually. Whereas this is theoretically true, whether it really happens requires more metaphysical arguments than engineering ones. In practice the detection of signals is limited by the amount of noise that accompanies the signal at the receiver relative to the strength of the signal.

While the received signal strength is an important parameter, it is not the only one of significance to determine whether a target is detectable or not. The principle parameter of interest to determine target detectability is the received SNR. This represents the level of the signal, usually its power or energy, compared to the amount of noise accompanying that signal at a receiver. In all but one case of practical interest herein, the level of the signal present must be larger than the noise in order to be properly processed. The one exception is in DSSS, which will be discussed in Section 4.2.7.1. Even in that case, however, the demodulated SNR must be larger than the noise. In some cases the signal must be substantially larger than the noise. Required levels of demodulated SNR can be 30 or 40 dB (signal level 1,000 to 10,000 larger than the noise)—for example, in the case of high-speed digital communications over modems carried by phone lines.

Noise arises from various sources, both external to a system as well as internal to that system. It also takes various forms. It is correct to say that noise is any signal that is not the signal of interest. Thus, noise can be narrowband, wideband, structured, or unstructured, for example. A few of the more important sources of noise that influence the design of communication EW systems are presented in this chapter.

A good discussion of noise sources and how they affect the mobile VHF channel is contained in [1]. Discussion about VHF and above noise sources and other interference effects are contained in [2] and [3]. For a discussion of the effects of noise on HF signal propagation, [4] is a good source.

3.2 THERMAL NOISE

Thermal noise is caused by the random motion of electrons that are excited by being raised above absolute zero temperature measured in degrees Kelvin (K) where 0K = –273°C. It can be radiated or carried along wires just as other EM energy can. It is characterized by its wide bandwidth. What is wide, of course, depends on the situation of interest. Herein it refers to when the noise frequency spectrum is at least as wide as the communication channel width under consideration.

Thermal noise has a Gaussian amplitude distribution in the time domain. See Appendix A for a discussion of probability distributions. As indicated there, such a distribution is characterized by a mean value, denoted as μ, and a standard deviation, denoted as σ. The amount of power in this noise source is given by σ^2.

Another unique characteristic of this type of noise is its frequency distribution. Over bandwidths of interest to the problems considered herein, broadband noise is assumed to have a uniform spectral density. That means that the spectrum has a constant value over the frequency range under consideration—all frequencies are contributing to the noise equally.

3.3 INTERNAL NOISE SOURCES

Thermal noise in circuits is caused by the heating of electrons. Electrical current flow can be thought of as controlling the direction of this random motion in one direction or another. The actual motion of electrons can only be determined statistically, however, and the random motion of these electrons causes noise. Ideal electronic components are only an engineering approximation. Every capacitor, inductor, and semiconductor has an associated resistance. It is within this resistance that the noise generation occurs.

Each circuit in the receiver generates such thermal noise, some of which is added to the signal. When all of these noise sources have their effects reflected to the first signal input stage, the *noise figure* results, denoted by F. That is, F represents all of the internal noise sources, and once these are so reflected, the remainder of the receiver is assumed to be noise-free. When F is expressed in decibels, it is referred to as the *noise factor*.

This noise can be summarized in the equivalent system noise, N, given by $N = kTBF$, where k is Boltzman's constant, $k = 1.38 \times 10^{-23}$ joules per degree Kelvin, T is the absolute temperature measured in degrees Kelvin, B is the noise bandwidth in hertz, and F is the system noise figure with no units. This noise exhibits a Gaussian statistical characteristic.

Other sources of noise internal to communication EW systems are oscillators and clocks and the circuits that distribute them. Oscillators and clocks are radiated and can be unintentionally picked up by other circuits. Fast signals such as these exhibit high frequency noise characteristics but are typically not broadband. The effects of these noise sources are called *EM compatibility* (EMC) and *EM interference* (EMI). More about these areas will be presented in Section 5.2.2.

3.4 EXTERNAL NOISE SOURCES

There are several noise generators external to an EW system. These sources vary depending on the frequency under consideration.

Man-made noise comes from several sources that are generated by machinery or other man-made devices. Examples of these sources are automobile ignitions, welding machines, and microwave ovens. Clearly the amount of this noise will depend on the number of such interference sources present. More noise will be generated in manufacturing settings than on farms, for example. More man-made noise will be found in cities than in the country.

The atmosphere surrounding the Earth contains a certain amount of heat energy at any given time. This heat energy warms up the electrons in air, which in turn radiate a certain amount of thermal noise. This noise is picked up by the antennas in a communication system and is manifest as thermal noise at the system inputs. Compared to other noise sources, this one contributes in a relatively minor way.

Noise is also generated by energy sources (stars) in the universe, to include the Earth's own Sun [3]. If an antenna is pointed toward a star, for example, then that star will cause significant broadband energy to be introduced into the system. These noise sources tend to be wideband, containing noise energy across a significant portion of the spectrum. This form of noise is sometimes referred to as *galactic noise*.

Another principal cause of *atmospheric noise* is lightning strikes from thunderstorms. This source of noise is prevalent in the HF frequency range and therefore can propagate for significant distances. Thousands of thunderstorms occur each year with many more thousands of lightning bolts. Thus, this type of noise is a problem almost everywhere.

The amount of noise generated by these sources is typified by the data shown in Figure 3.1, where the variation with frequency is obvious. The total amount of noise present is given by a multiplicative factor on kTB as above. Thus,

$$N_{total} = N \times N_{external}$$
$$= kTBFN_{external}$$
$$\frac{N_{total}}{kTB} = FN_{external} \tag{3.1}$$
$$N_{total, dBkTB} = 10\log\frac{N_{total}}{kTB} = 10\log F + 10\log N_{external}$$
$$= F_{dB} + N_{external, dBkTB}$$

So the total noise in decibels above kTB is given by the noise figure in decibels plus the external noise in decibels relative to kTB. That is why the ordinate in Figure 3.1 is in the units indicated.

As seen in Figure 3.1 [5], in general the external noise decreases as frequency increases. Also in general, the levels of noise produced by sources such as these are determined by measurements as opposed to experiment. These levels will be dependent on the environment in which the communication system is to operate.

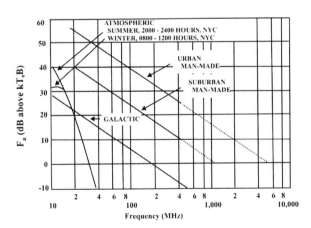

Figure 3.1 Levels of external noise. These levels should be interpreted as typical as the noise in any given circumstance will be unique to that circumstance. (*Source:* [5], © 1975, Howard W. Sams. Reprinted with permission.)

For a more complete discussion of noise, see [1–5].

Just as EMI can be a source of noise internal to a system, it can be a source of unwanted noise external to a system. Systems within close proximity of an EW system can radiate signals that interfere with the proper operation of that system. This can be a particularly significant problem for EW systems because such systems are typically broadband relative to, say, communication systems. Broadband systems are more susceptible to a broader range of EMI than many other systems.

Intentional jamming of an EW system can be considered as noise and can be treated the same. This type of jamming is sometimes used to screen friendly communications from intercept by an adversary's communication EW systems. This is a form of EP, another category of IW.

3.5 COCHANNEL AND MULTIPATH INTERFERENCE

Cochannel interference and multipath interference are two manifestations of the same problem. It is when two or more signals are received by a communication or EW system at the same carrier frequency (channel). The difference between cochannel interference and multipath is whether the interference is correlated to the signal of interest or not. If the interfering signal is from a second transmitter, then, in all likelihood, the signals are not statistically correlated with each other. If the interfering signal is a copy of the same signal reflected by some metallic object, then the signals will be correlated, although the reflected signal is delayed from the other and can be substantially weaker.

The effects are different if the interference is cochannel versus multipath. Multipath interference essentially causes degradations of receive performance due to phase differences between the direct signal and its reflection. Under the worst conditions the reflected signal can totally cancel the direct signal. A familiar case of multipath interference is ghosting on broadcast television mentioned in Section 2.9. The delay shows up as "ghosts" of the figures on the screen, somewhat physically offset from the intended picture.

Cochannel interference shows up as random perturbations of either of the signals present. The effects are more noise-like than multipath interference, although the interference has more structure than random background noise. An example of this type of interference is AM radio signals in car radios. As a car moves from the area of good reception of one signal into that of another, sometimes two signals can be heard at the same time.

Cochannel interference frequently is not as significant in *frequency modulation* (FM) or *phase modulation* (PM) situations because of the *capture effect*. If one signal is 6 dB (typically—can be less) or higher than another interfering signal, then the stronger signal is "captured" by the demodulation circuitry and the weaker signal is rejected.

3.6 CONCLUDING REMARKS

Noise is the nemesis in any and every communication situation. Thermal noise is generated externally and internally to electronic equipment. Interfering signals can also sometimes be considered as noise sources; such signals can be from different transmitters or the same transmitter. The latter is referred to as multipath interference.

References

[1] Parsons, D., *The Mobile Radio Propagation Channel*, New York: John Wiley & Sons, 1992.
[2] Hall, M. P. M., *Effects of the Troposphere on Radio Communications*, Stevenege, United Kingdom: Peter Peregrinus Ltd, 1979, pp. 172–185.
[3] Gagliardi, R. M., *Introduction to Communications Engineering,* 2nd ed., New York: John Wiley & Sons, 1988, pp. 154–163.
[4] Braun, G., *Planning and Engineering of Shortwave Links*, New York: John Wiley & Sons, pp. 47–56.
[5] *Reference Data for Radio Engineers,* 6th ed., Indianapolis, IN: Howard W. Sams & Co., 1975, p. 29-2.

Chapter 4

Radio Communication Technologies

4.1 INTRODUCTION

An understanding of communication technologies is important to understand how to design EW systems. These technologies facilitate the exchange of information between two entities, and therefore are the means by which information is passed. Information not shared is useful, but only to the entity that has possession of it. It is far more useful and much more can be done with information if it is shared. Thus, information transport is a vital part of any information-based society, including the military.

There are many good texts available that discuss communication technologies and systems at length. This chapter is included to provide just an introduction to the subject at a depth necessary to understand what is presented here. It is not intended as a thorough review of the subject. For more in-depth reading, the reader is referred to [1–6].

The generic information transport system model used herein is shown in Figure 4.1. There is a source that generates information that is to be transferred to one or more sinks for that information. There are several ways to make a distinction between the techniques for accomplishing this information transfer, but the fundamental distinction made here is whether the transfer medium is a wire and/or fiber or wireless. Note that information storage, on a floppy disk, for example, is another form of information transport system where the information is to be transferred primarily over the time dimension.

There is usually channel noise associated with any communication system, regardless of the communication medium. Here it is assumed that all noise is additive. Some of those that are more important to communication EW system design were discussed in Chapter 3. Noise is any signal that is other than the intended signal. The communication system in Figure 4.1 will be discussed in this chapter. The remainder of the book will discuss in detail the jammer, the intercept system, and their performance.

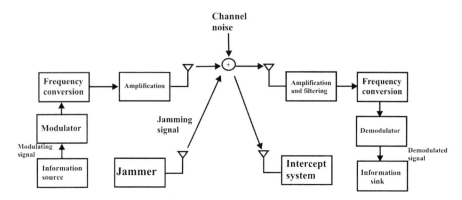

Figure 4.1 Generic communication system model.

Communication systems can be categorized as either broadcast or netted. These appellations, however, are not meant to imply a size. Either can be small or very large. Broadcast communications are usually generated at a single place and sent to many receivers. Examples of this are commercial radio and TV broadcasts. In netted communications, on the other hand, there are several possible sources of information that is to be sent to other locations. Perhaps every communication node on the network can generate such information. The Internet is an example of a large netted network. Military, mobile tactical command and control is normally implemented with networks of VHF and UHF communication radios. Figure 4.2 illustrates the fundamental differences in broadcast versus netted communications.

Either of these communication models can be encountered in military situations. Communication EW systems can be expected to operate against either. An example of a broadcast EW application is when the former Soviet Union jammers denied radio-free Europe to most of Eastern Europe. An example of EW against tactical networks is classical VHF jamming.

Frequently in a tactical mobile network there is one node in the network that is referred to as the *net control station* (NCS). Normally it would be the node associated with the person in charge of the unit using the radios. There is often more communication traffic from the NCS than the other nodes. In some cases that can be used to identify the NCS

Before discussing specific information transport technologies, there are some topics that should be understood that cross the boundaries of the technologies. Section 4.2 briefly introduces these topics.

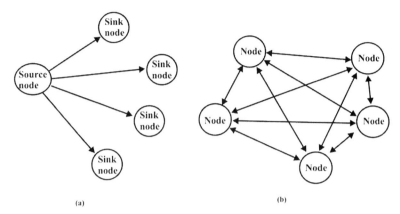

Figure 4.2 Two fundamental types of communications: (a) broadcast and (b) netted.

4.2 MODULATION

Modifying a carrier signal with the information to be transported is called *modulation*. The reason for putting the information-bearing signal on a separate carrier is to move the information signal to a region of the frequency spectrum where transmission is more economical. (As the frequency gets higher, the antenna gets smaller, thereby requiring less real estate. There are other economic reasons as well.)

There are three fundamental ways that a radio signal can be modified to carry information: its amplitude, its frequency, and its instantaneous phase angle. See Figure 4.3. These are referred to as *amplitude modulation* (AM), *frequency modulation* (FM), and *phase modulation* (PM)—the last two are referred to as *angle modulations*.

The signal that is modified by modulation is called the *carrier signal*, while the information-bearing signal, which does the modification, is called the *modulating signal* denoted by $m(t)$. The modulated carrier signal is given by

$$s(t) = A(t)\cos[\varphi(t)] = A(t)\cos[2\pi f_c t + \Psi_c + \phi(t)] \tag{4.1}$$

where $A(t)$ is the amplitude of the signal, f_c is the carrier frequency, Ψ_c is the initial phase of the carrier signal, and $\phi(t)$ is an additional phase term described shortly. AM modifies $A(t)$, while phase and frequency modulation modifies $\varphi(t)$ by modifying $\phi(t)$.

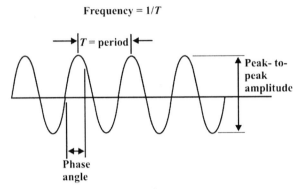

Figure 4.3 Fundamental parameters of an EM sine wave.

The modulation used to convey commercial radio signals in the United States on the carrier frequencies between 550 kHz–1.5 MHz, well-known to all, is AM. In that case the amplitude of the carrier is modified. For the FM radio case, where the commercial broadcast frequencies are 88–108 MHz, the modulation technique is referred to as FM. In this case it is the frequency that is changed by the information-bearing signal. The third parameter that can be changed is the phase of the carrier. The phase is measured relative to a reference at the receiver, so it is relative phase that is important. In addition, there are limits to how much the phase can be changed. Any phase change larger than 2π radians would lead to ambiguities, since $\theta = n2\pi + \theta$ for any integer n.

Work on modulation techniques has long been focused on achieving spectrum efficiency, that is, putting more information through a given amount of bandwidth. One scheme is to reduce the redundancy in human speech and writing. For example, the letter u (almost) always follows the letter q, so why transmit the u? The analog AM and FM communication technologies generally attain much less than the equivalent of 1 bit per second per hertz of bandwidth. Modulating carriers with digital information using these analog schemes are no better. On the other hand, with PM, 4 bps/Hz of RF bandwidth and higher have been achieved, thus explaining the attractiveness of this form of modulation. This, however, does not come freely. Synchronizing sequences and circuitry are required between the transmitter and receiver in order to establish and maintain communications. Furthermore, normally characterization of the communication channel is required in order to prevent one symbol from interfering with the next. Appropriate processing of the communication signal is subsequently required.

Demodulation is the inverse process—removing the modulating signal from the carrier. Demodulation is different from detection, the latter of which is used in two ways. In communication terminology, detection refers to the determination of which of several symbols was transmitted in digital signaling. For communication EW system design, detection most of the time refers to the determination of whether there is a signal present at a channel or not.

There are two fundamental forms of demodulation. The first is noncoherent demodulation. In that case, no carrier phasing information is used in the demodulation process. For coherent demodulation, the carrier phase information is used. The latter generally yields better performance. However, it is often accompanied by more complex hardware implementations. Performance in this case refers to signal detection.

For all of the PM schemes, the receiver needs to know where the reference carrier phase is located. Thus synchronization of the receiver with the incoming signal and coherent detection are required. Recovery of the symbol clock is required in all of the demodulation schemes. Coherent detection is possible with other modulation schemes as well, and when it can be done, it typically provides for 3 dB or more improvement in sensitivity performance.

4.2.1 Amplitude Modulation

The process of amplitude-modulating a carrier signal is illustrated in Figure 4.4. If the modulating signal is a tone signal (single-frequency, zero-bandwidth), then the spectrum of the modulated carrier is as shown in Figure 4.5. For real baseband waveforms that have some finite nonzero bandwidth, the spectrum appears as shown in Figure 4.6.

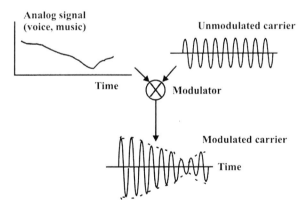

Figure 4.4 AM varies the amplitude of the carrier waveform.

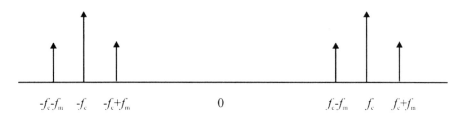

Figure 4.5 Amplitude spectrum of a signal that is AM-modulated with a tone.

Figures 4.5 and 4.6 illustrate the case of *double sideband* (DSB) AM, where substantial portions of the carrier are transmitted with the information part of the signal. If the carrier is suppressed (either completely or partially), it is called DSB *suppressed carrier* (SC). If the upper sideband of the signal is suppressed as well, then the signal is one of the *single sideband* (SSB) forms called *lower sideband* (LSB). Likewise if the lower sideband is suppressed, this form of SSB is called *upper sideband* (USB). One of the sidebands can be suppressed because for AM each of the sidebands contains the same information. Suppression of a sideband reduces the required bandwidth for transmission. As seen in Figure 4.5, the bandwidth of the modulated carrier is either 1 × baseband bandwidth, or 2 × baseband bandwidth. The former is when the modulation is AM SSB and the latter is for AM DSB.

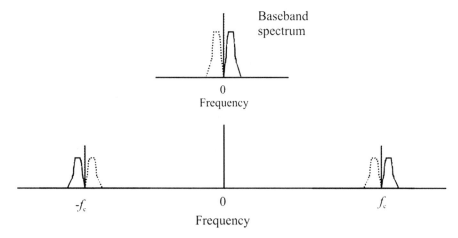

Figure 4.6 Spectrum of AM waveform when the modulating waveform has nonzero bandwidth.

Atmospheric noise affects AM more than other forms of modulation. This is particularly true in the lower-frequency ranges where lightning causes substantial interference to AM signals.

4.2.2 Angle Modulation

The instantaneous phase of the carrier is given by

$$\varphi(t) = 2\pi f_c t + \Psi_c + \phi(t) \tag{4.2}$$

while the instantaneous frequency is

$$f_i = \frac{1}{2\pi}\frac{d\varphi(t)}{dt} = f_c + \frac{1}{2\pi}\frac{d\phi(t)}{dt} \tag{4.3}$$

for PM

$$\phi(t) = k_p m(t) \tag{4.4}$$

where k_p is the PM constant given in radians/volt, or perhaps radians/ampere. For FM

$$\phi(t) = k_m \int_{-\infty}^{t} m(\tau)d\tau \tag{4.5}$$

where k_m is the FM constant given in radians/second/volt or radians/second/ampere.

The modulated carrier waveforms, then, for PM and FM are (ignoring Ψ_c)

$$\text{PM}: \quad s(t) = A\cos[2\pi f_c t + k_p m(t)]$$
$$\text{FM}: \quad s(t) = A\cos[2\pi f_c t + k_f \int_{-\infty}^{t} m(\tau)d\tau] \tag{4.6}$$

Frequency-modulating a signal has the effect shown in Figure 4.7. The instantaneous frequency is changed. The effects in the frequency domain are more complex than in the AM case above because FM is a nonlinear modulation

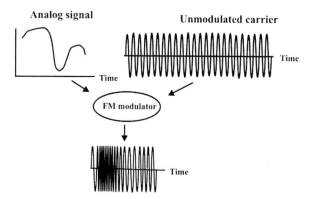

Figure 4.7 FM of a carrier varies its instantaneous frequency around the carrier frequency.

technique. If the modulating signal is a tone, then the spectrum of the modulated carrier is like that shown in Figure 4.8. The spacing of the frequency lines of the modulated signal is equal to f_m. If the modulating signal is not a tone, then the spectrum is very complicated and in most cases has yet to yield to closed form mathematical analysis.

Let $m(t) = M \cos(2\pi f_m t)$ where M is a constant amplitude so that the modulating signal is also a sinusoidal signal such that $f_m \ll f_c$. Then for FM

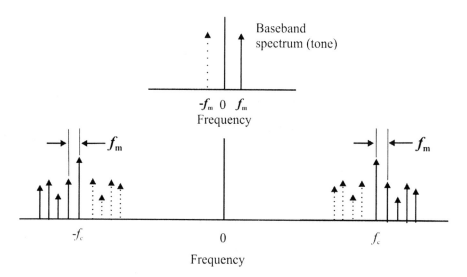

Figure 4.8 Spectrum for FM.

$$s(t) = A\cos[2\pi f_c t + \beta\sin(2\pi f_m t)] \quad (4.7)$$

$\beta = k_m M/2\pi f_m$ is the modulation index. β is the ratio of the maximum frequency deviation of the carrier caused by the modulating signal, to the maximum frequency of the modulating signal.

The cosine term can be expanded as

$$\cos[2\pi f_c t + \beta\sin(2\pi f_m t)] = J_0(\beta)\cos(2\pi f_c t)$$
$$+ \sum_{n=1}^{\infty}(-1)^n J_n(\beta)[\cos(2\pi f_c - n2\pi f_m)t \quad (4.8)$$
$$+ (-1)^n \cos(2\pi f_c + n2\pi f_m)t]$$

where the J_i are Bessel functions of the first kind. For small β,

$$J_0(\beta) \approx 1 - \left(\frac{\beta}{2}\right)^2 \qquad J_n(\beta) \approx \frac{1}{n!}\left(\frac{\beta}{2}\right)^n, \quad n \neq 0 \quad (4.9)$$

Thus,

$$s(t) \approx A\cos(2\pi f_c t) - \frac{A\beta}{2}\cos[2\pi(f_c - f_m)t] + \frac{A\beta}{2}\cos[2\pi(f_c + f_m)t] \quad (4.10)$$

which simplifies to

$$s(t) \approx A\cos(2\pi f_c t) - A\beta\sin(2\pi f_m t)\sin(2\pi f_c t) \quad (4.11)$$

When $\beta < 1$, it is called narrowband FM, and the bandwidth of the signal is about $2 \times f_m$. The spectrum of this signal is shown in Figure 4.9 (taking into consideration the negative frequencies as well).

FM is a nonlinear modulation, as can be ascertained by (4.7)–(4.11). For AM, which is a linear modulation, determination of the bandwidth of the signal was straightforward—it was either 1 × baseband bandwidth or 2 × baseband bandwidth. For FM it is not so straightforward. Carson developed an approximation for the bandwidth of FM signals, which is known as the *Carson rule*. It states that the bandwidth is given by

$$B_{IF} \approx 2(\Delta F + f_m) \quad (4.12)$$

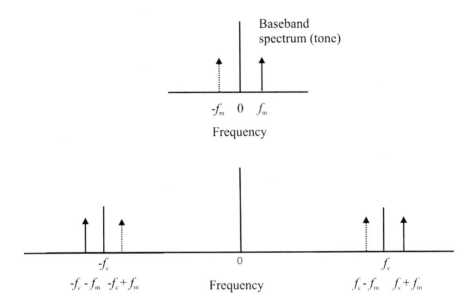

Figure 4.9 Frequency spectrum of narrowband FM.

where β = modulation index = $\Delta F/f_m \gg 1$; B_{IF} = IF bandwidth; ΔF = peak carrier frequency deviation; and f_m = peak-modulating frequency deviation.

An extension to Carson's rule was subsequently developed for smaller modulation indices and is given by

$$B_{IF} = 2(\Delta F + 2f_m) \qquad (4.13)$$

when $2 < \beta < 10$.

An effective and simple FM modulator is made with a *voltage-controlled oscillator* (VCO). Such a configuration is shown in Figure 4.10. The characteristics of the VCO are shown at the top of Figure 4.10. The input voltage (slowly changing relative to the oscillator's frequency) controls the frequency of oscillation. Within its design range, this is a linear relationship.

For most common FM cellular systems $1 < \beta < 3$, so Carson's rule does not apply very well (it is not a very good approximation). For the original North American analog cell phone standard, the *American mobile phone system*

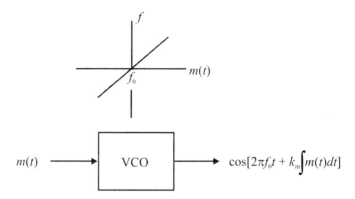

Figure 4.10 An FM modulator constructed with a VCO. The VCO transfer characteristic is shown at the top.

(AMPS), $\beta \approx 2.7$, producing detected SNRs, denoted SNR_d, of about 40 dB from a predetected carrier-to-noise ratio, denoted by CNR_{IF}, of 18 dB.

The derivation of the spectrum for PM is similar to the above for FM and many of the same results apply. Carson's rule for PM is

$$B_{IF} = 2(\beta + 1) f_m \qquad \beta > 10$$
$$B_{IF} \approx 2(\beta + 2) f_m \qquad 2 < \beta < 10 \qquad (4.14)$$

where

β = PM modulation index = $\Delta f/f_m$
f_m = highest modulating frequency.

4.2.3 Orthogonal Signaling

Two baseband signals can be sent at the same time using the same carrier using orthogonal signaling, also called *quadrature modulation*. Two functions are orthogonal if their inner product is zero. In the case of importance here, two periodic functions defined over (0, nT) with period T and integer n given by $f(t)$ and $g(t)$ are orthogonal if and only if

$$\int_0^{nT} f(t)g(t)dt \begin{cases} \neq 0 & \text{if } f(t) = g(t) \\ = 0 & \text{if } f(t) \neq g(t) \end{cases} \qquad (4.15)$$

Specific examples of important orthogonal waveforms in communications are sin $2\pi ft$ and cos $2\pi ft$.

Let $r(t)$ represent the received RF waveform. Furthermore, let $r(t) = A\sin(2\pi ft) + B\cos(2\pi ft)$. Here A and B are two signals to be transmitted that may or may not be related to each other. A and B are treated as constants normally associated with digital communications, either $A, B \in \{0, 1\}$ or, more commonly $A, B \in \{1, -1\}$. Now suppose at the receiver $r(t)$ is divided into two paths as shown in Figure 4.11. In one path is a multiplier where $r(t)$ is multiplied by $\sin(2\pi ft)$, while in the other path it is multiplied by $\cos(2\pi ft)$. These multipliers are followed by integrators where the period of integration is a multiple of the period of the carrier period $T = 1/f$. Then the output of the top integrator will be

$$\int_0^{nT} r(t)\sin(2\pi ft)dt = \int_0^{nT} A\sin^2(2\pi ft)dt + \int_0^{nT} B\cos(2\pi ft)\sin(2\pi ft)dt$$

$$= \frac{A}{2}\int_0^{nT}[1 - \cos(2\pi ft)dt] + \frac{B}{2}\int_0^{nT}[\sin(4\pi ft)dt - \sin(0)]dt \quad (4.16)$$

$$= \frac{A}{2}\int_0^{nT}dt - \frac{A}{2}\int_0^{nT}\cos(2\pi ft)dt + \frac{B}{2}\int_0^{nT}\sin(4\pi ft)dt - \frac{B}{2}\int_0^{nT}0dt$$

$$= \frac{AnT}{2} = KA$$

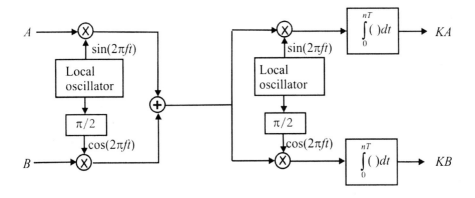

Figure 4.11 Orthogonal modulator and demodulator allow two distinct signals to share the same carrier.

All of the integrals but the first one are zero because any zero-mean periodic function integrated for an integer number of periods is zero, and, of course, the last integral is zero anyway. The second integral in the first line of (4.15) is equal to zero by inspection due to the orthogonality property discussed above. It can similarly be shown that the output of the bottom integrator is equal to $BnT/2$. Therefore, the outputs of the integrators are a constant multiple of the baseband waveforms.

This configuration does impose some requirements on the communication system. For example, the transmitter and receiver local oscillators must be synchronized so they are in phase coherence with one another. If they are not in phase coherence and/or if the integrals are not exactly an integer multiple of periods, the outputs are not exactly equal to a constant times the input, but will be somewhat different. That is, some noise is introduced into the demodulation process. When used for digital signaling, this is not too critical since it only becomes a problem when the transmitted symbol was an A, for example, and the other possible symbol is $-A$. If the demodulated symbol is closer to $-A$ than A, a symbol error will occur. The lower the SNR, the more likely this possibility will be.

If $A = a(t)$ and $B = b(t)$ are not constant, but functions of time, then this analysis and configuration still works. However, if $a(t)$ and $b(t)$ are the information-bearing signals, then it is necessary to take their derivatives at the transmitter before being submitted to the modulation process. That way the output of the integrators at the receiver will be functions of the baseband signals, not their integral as they would be otherwise.

4.2.4 Access Methods

The techniques used to allow a user of a communication channel to use the channel is referred to as the *access method* of the communication system. There are principally four such access methods and they will be introduced here. Combinations of these access schemes are frequently implemented to take advantage of the characteristics of each.

4.2.4.1 Frequency Division Multiple Access

In *frequency division multiple access* (FDMA) the available frequency channels are divided among the users. It is one of the oldest forms of access. Each user has sole use of the channel until it is no longer needed and it is relinquished. FDMA is illustrated notionally in Figure 4.12. Each user does not normally keep the channels forever, as would be implied by Figure 4.12. They do keep it whether they have anything to send or not, however, so it can be an expensive method of access.

4.2.4.2 Time Division Multiple Access

If a frequency channel is to be shared among more than one user at (essentially) the same time, then some method must be devised that allows this sharing. In *time division multiple access* (TDMA), use time in a channel is divided among the users. At one instant one user would be using the channel and at the next, another user has the channel. Usually this happens so fast that the users do not know that the channel is shared. TDMA is notionally illustrated in Figure 4.13. Normally the frequency spectrum would be channelized as well.

ALOHA

ALOHA is a TDMA scheme where a user transmits when desired. This results in collisions when more than one user transmits at the same time. If these collisions cause reception problems, the data will be received in error. This, in turn, causes no ACK to be sent back to the transmitter or a NACK to be sent. If there is no ACK from the receiver, the message is sent again.

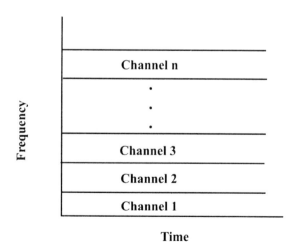

Figure 4.12 In FDMA, the frequency spectrum is divided.

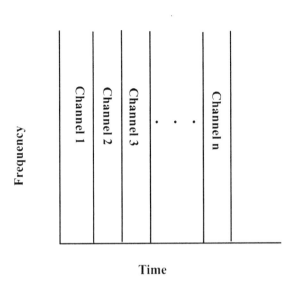

Figure 4.13 In TDMA the frequencies are shared among multiple users who each get a time slot. Normally the frequency spectrum is channelized as well.

If

T = Throughput measured in successful packet transmissions per frame time,
P = Packets offered for transmission per packet time,

then the throughput of pure ALOHA is given by

$$T = Pe^{-2P} \qquad (4.17)$$

As shown in Figure 4.14, the peak throughput of pure ALOHA is 0.18. That means that if the channel rate is 1 Mbps, the maximum throughput through that channel is 180 Kbps.

Slotted ALOHA

This access method, also a TDMA scheme, is similar to ALOHA except that the transmission boundaries are synchronized in time. Using the notation above the throughput of slotted ALOHA is given by

$$T = Pe^{-P} \tag{4.18}$$

Thus the maximum throughput of slotted ALOHA is twice that of pure ALOHA. Its maximum throughput is 0.36 as shown in Figure 4.14, yielding 360 Kbps maximum (on average) through a 1-Mbps channel.

Although the throughput through a channel can be greater for slotted versus pure ALOHA, as shown in Figure 4.14, there is a penalty. This penalty takes the form of system complexity because of the requirement to synchronize all transmitters in the network with each other.

Carrier-Sensed Multiple Access

When the channel is first sensed to see if someone is already using it before a transmission is tried, it is called *carrier-sensed multiple access* (CSMA), which is a form of TDMA. It is possible to detect when a collision occurs, for example, by listening for an acknowledgment. With such collision detection the notation is CSMA/CD.

In wired networks, the maximum slotted ALOHA throughput is 36%, whereas the maximum for CSMA is 50%. Thus, there is considerable overhead with using these schemes. For wireless systems, the capture effect can increase the throughput.

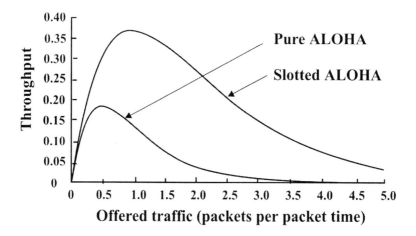

Figure 4.14 Performance comparison of ALOHA and slotted ALOHA protocols.

Demand Assignment Multiple Access

A separate, lower data rate, and therefore a less expensive channel, can be used to request the use of a higher speed, and therefore a more expensive channel. Requesters are assigned usage of the channel according to some preestablished prioritization scheme. Users are notified when the channel is theirs to use. This is referred to as *demand assignment multiple access* (DAMA), and is really a form of TDMA since the channels are being shared on a time basis.

4.2.4.3 Code Division Multiple Access

In *code division multiple access* (CDMA), DSSS or FHSS is used to share the channel [7]. All frequencies (within a band) at all times (that have been allocated) are shared by all users. A different spreading code for each user in the band is used.

4.2.5 Duplexing

Duplexing refers to the method that allows a communication system to facilitate two (or more) users access to the same communication medium. In half-duplex systems, only one user can transmit at the same time without interference. Older forms of military tactical communications facilitated by HF and VHF voice radio are examples of half-duplex systems. In full-duplex systems, either or both of the users of the communication path can transmit at the same time without interfering with one another. Modern mobile digital communication systems, as exemplified by cellular phones, are full-duplex systems. There are two generally used forms of duplexing: time division and frequency division.

4.2.5.1 Time Division Duplexing

In *time division duplexing* (TDD) the same frequency channel is used for the upstream and downstream data but at different times (see Figure 4.15). It is conceptually similar to TDMA described above, but the context within which the phrase is used implies something different. Some form of duplexing is required for any full-duplex communication system, and when the two channels are separated in time but use the same frequency if is referred to as TDD.

The upstream and downstream data are sequenced in time, one after the other. There is generally a guard time allocated between the two time events in order to ensure the signals do not interfere with one another and to allow for system delays, such as those associated with finite propagation times.

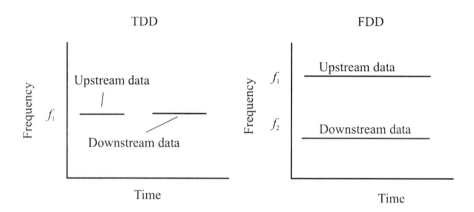

Figure 4.15 TDD and FDD are two forms of access for full-duplex communications.

4.2.5.2 Frequency Division Duplexing

Similarly, in *frequency division duplexing* (FDD), the data streams are sent at the same time but in different frequency channels (see Figure 4.15). This is similar to FDMA, with the difference focusing on the duplexing aspects.

4.2.6 Digital Signaling

Digital signaling has spread considerably in recent years [8]. The reason is simple: Generally signals that are modulated by digital methods are more spectrally efficient that those that are analog-modulated. The carrier for digital signaling can be either amplitude-, frequency-, or phase-modulated, or combinations.

Wireline communication, such as that provided by the *public-switched telephone network* (PSTN), is a fairly benign environment. Induced noise is low and interfering signals can be controlled. Communication by radio is a different story. It is a much harsher environment in that external noise is higher, interfering signals cannot normally be controlled, and knowing the propagation parameters is difficult at best.

4.2.6.1 Pulse Code Modulation

In order to transmit analog information, such as human speech and television images, over digital channels, the signals must first be converted into digital form. This is normally done with analog-to-digital (A/D) converters. A/D converters change an analog signal into digital representations of those signals. These digital representations are referred to as pulse codes in the domain of digital communications. The output of an A/D converter is simply the (discrete) binary approximation of the amplitude of the voltage or current at the input to the A/D at each sample time. When these digital words are modulated onto a carrier, the results is called *pulse code modulation* (PCM). This modulation can take many forms, some of which are discussed next.

Differential PCM

PCM assigns a number in digitized form to the amplitude of an analog signal. If the signal is not changing too rapidly from one sample to the next, there will be considerable redundancy from one sample to the next (for example, there is not too much difference between 1 and 1.01V). An attempt to remove this redundancy was developed, called *differential PCM* (DPCM). In DPCM, instead of encoding every sample with the full number of bits of the A/D converter, there are fewer than the total number of bits used; frequently only 1-bit difference from one sample to the next is used. That bit is +1 if the next sample is larger than the current sample, and it is –1 if the next sample is smaller.

Adaptive Differential PCM

A variant of DPCM is *adaptive differential PCM* (ADPCM). It is similar to DPCM except that if the next sample is substantially larger than the current sample, there are provisions for using more than +1 or –1 to encode the change. Figure 4.16 shows how ADPCM works. Each digitized sample of an analog sample is compared with the last sample. While for DPCM, if the sample is larger, +1 (bit) is used to represent the change and if it is smaller, a –1 (bit) is used to represent the change, as seen in Figure 4.16 more than 1 bit is used in ADPCM. There are only +1 bits and –1 bits, no zero bits, so if there is no change in the signal, +1 and –1 will alternately be coded. This can be seen in Figure 4.16 where the analog signal is reversing directions and therefore the changes are small enough to be less than the difference in one sample. In ADCPM, where, after a certain number of attempts (three in this example), the digital samples fail to catch up with the analog waveform, the size of the step is increased (by one step here) until the analog waveform is matched or exceeded. At that point, the polarity of the step is reversed and the amplitude reduced back to one step size.

Since changes larger than 1 bit are used, more than 1 bit is necessary to encode the difference. For example, up to four levels can be encoded with two binary bits. ADPCM allows the digitized waveform to catch up faster.

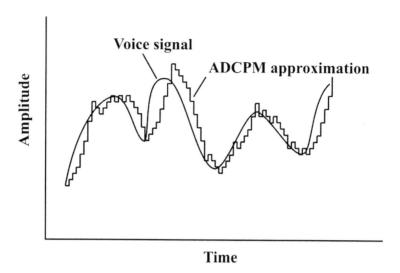

Figure 4.16 ADPCM.

4.2.6.2 Power Efficiency

Power efficiency is determined by the required energy per bit to noise level, as specified, for example, by the SNR in watt-hours per bit per hertz of bandwidth, or E_b/N_0, in order to produce a specified BER. Over a ground-to-ground tactical army communication channel, noise and interference dictates that a BER of 10^{-2} is typical, whereas over a satellite link the BER is typically 10^{-5} or better. The BER is a system design parameter and many techniques can be used to achieve low BER even over noisy, normally high BER channels. Usually an *additive white Gaussian* (AWG) noise channel is assumed, especially for satellite links.

4.2.6.3 Bandwidth Efficiency

Bandwidth efficiency is defined as the number of bits per second that can be transmitted over 1 Hz of bandwidth. The higher the bandwidth efficiency, the more information that can be transmitted over that channel. The amount of frequency spectrum is fixed—there is only a certain amount of it. In order to increase the amount of communication once there are signals traveling over every channel in a given area of the spectrum, it is necessary to increase the bandwidth efficiency. Modulation techniques must be bandwidth-efficient to allow the maximum number of users possible to use the propagation medium.

4.2.6.4 Other Considerations for Digital Communications

As discussed in Section 3.5, multipath is the simultaneous reception of two or more copies of a signal. It is caused by reflections of the signal from potentially one or more objects. Some forms of modulations are more tolerant to multipath effects than others, so in cases where toleration of multipath is required, those methods may be appropriate.

Another form of interference is when other signals are impinging on the receiver at the same time as the intended signal. This is called cochannel interference and some forms of signal processing are more tolerant than others.

Economics always is a consideration in all but the most exceptional cases. Therefore, initial investment in the communication system is usually a concern. In most cases, however, cost of ownership involves much more than the initial investment. *Operation and support* (O&S) costs typically run much higher than the initial investment. Minimizing those can often be of great importance. An example of this is for cellular systems. If there were no rechargeable batteries to reduce the cost of powering the handsets, it is unlikely that modern cellular systems would be economically viable.

When the RF spectrum is tightly packed, as all the lower RF communication bands are today, out-of-band spillover of energy is a consideration. This is accidentally putting the energy from one channel into adjacent channels. Of course, some spillover is inevitable in any practical implementation. Some schemes are more tolerant than others, however.

The narrower the main lobe width in the frequency spectrum, the more spectrally efficient the modulation scheme. This is not the whole story, however, as sidelobes of energy can show up in adjacent channels, as well as channels far away from the intended channel. While each of the modulation schemes exhibits its own unique characteristics, it is certainly a goal of them all to keep the energy of the signal all within the desired channel. Not only do the sidelobes interfere directly with the signals in the other channels, but also the energy spilled over interacts with the signals in the sidelobes. When the system is nonlinear, this interaction can generate intolerable intermodulation interference. Spillover energy is wasteful, as is the energy in the intermodulation products, exacerbating the power inefficiency.

4.2.6.5 Constellations of Amplitude and Phase Modulations

Most of the digital modulation schemes described herein can be visualized with the aid of the *constellation*. An example of a constellation is shown in Figure 4.17. Both the phase and amplitude, in general, can and are modulated (changed)

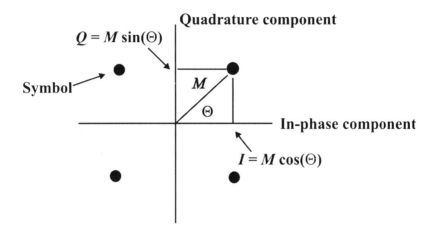

Figure 4.17 AM/ PM constellation.

with these modulations. By convention, the component of the vector that is in phase with the carrier (before modulation) is called the *in-phase component,* while the component on the other axis is called the *quadrature component.* Each combination of amplitude and phase (in phasor notation) or in-phase and quadrature in *x, y* dimensions is called a *symbol.*

4.2.6.6 Amplitude Shift Key

When the amplitude of the carrier signal is changed between two (or more) levels, frequently either A and 0, then the carrier is *amplitude shift key* (ASK)–modulated. The resultant signal is as shown in Figure 4.18 along with a simple notional implementation. This is the form of modulation originally invented by Morse for use with the telegraph.

4.2.6.7 Frequency Shift Key

In *frequency shift key* (FSK) modulation the carrier signal is changed between two (or more) frequencies according to the digital signal as shown in Figure 4.19. It is notionally implemented by switching rapidly between the two frequencies as shown. Alternately, the switching could be between two baseband frequencies and the subsequent signal, consisting of f_{m1} and f_{m2} used to frequency-modulate

Radio Communication Technologies 83

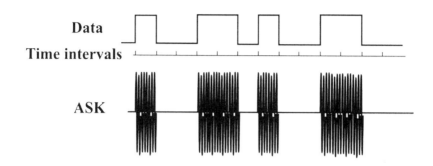

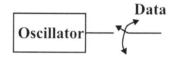

Figure 4.18 ASK modulation.

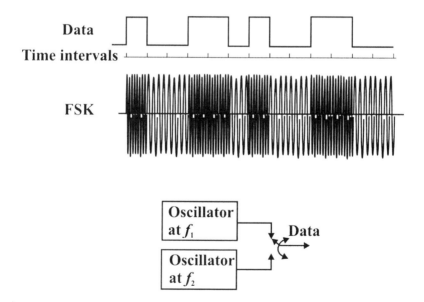

Figure 4.19 FSK changes the transmitted frequency between two (or more) frequencies.

Minimum Shift Key

Minimum shift key (MSK) is a special type of FSK where phase continuity is maintained at the symbol boundaries. It is implemented with a frequency shift that is equal to one-half of the modulating signal bit rate. This frequency shift is the smallest it can be while still maintaining orthogonality. Since phase continuity is maintained, there are no rapid changes in the phase, and MSK maintains the *constant modulus* (amplitude, defined below) property even after modulation.

Gaussian MSK

When the modulating signal is passed through a filter with the transfer characteristics given by

$$H(f) = e^{-\alpha f^2} \tag{4.19}$$

and then used to modulate the carrier via MSK, it is called *Gaussian MSK* (GMSK). Constant α is a filter parameter that controls the slope of the skirts of the filter. Gaussian MSK exhibits good spectral efficiency with sharp cutoff at the band edges. This cutoff characteristic is very important for narrowband applications such as cellular radio. It also has reasonable power efficiency. It is used in second-generation cellular systems such as *global system mobile* (GSM) at 1.35-bps/Hz.

GMSK can be demodulated in several ways. The VCO scheme indicated in Figure 4.20 is one way. The instantaneous phase of the VCO is compared to the

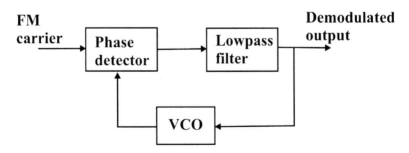

Figure 4.20 VCO in a phase-locked loop for demodulation of MSK, including GMSK and analog FM.

incoming FM signal. The output of the phase detector will increase or decrease depending on this phase difference. The increasing and decreasing function is precisely the modulated data signal. Note that this scheme also works for analog FM and regular FSK.

Alternately the simple clipper/discriminator can be used. Like most other forms of digital modulation, however, better performance can be achieved with coherent demodulation.

When cellular phone networks were first implemented, they used analog modulations—specifically FM was the modulation of choice. The North American version was called AMPS. In Europe, on the other hand, each country installing such systems used their own standard. This, as one would expect, caused these systems to not be interoperable and roaming among countries was impossible. This caused the European community to develop GSM, a second-generation standard based on digital signaling technology. It is a TDMA-based system with TDD. In the United States there are competing standards for digital cellular services, the most popular of which is IS-95, which is a CDMA modulation technique. The other standard is IS-54, which utilizes TDMA technology. Initially by edict of the FCC in the United States both of these digital standards were forced to be dual-mode, capable of both digital and analog operation. IS-136 is the same TDMA standard as IS-54 except the dual-mode operation required with older analog AMPS is removed.

There are digital data mobile services available in many areas as well as digital cellular services. Three of the most popular mobile data services are MOBITEX, *cellular digital packet data* (CDPD), and *advanced radio data information service* (ARDIS). MOBITEX and CDPD use GMSK, while ARDIS uses FSK. These are regional services that can be contracted for typical businesses to communicate within themselves.

4.2.6.8 Phase-Shift Key

The phase of the carrier signal is changed in *phase-shift key* (PSK) modulation, while the amplitude (modulus) is held constant. There are various types of PSK as discussed herein. As the types get more complex, as one would imagine, the implementation gets more complex as well. Due to the fact that the phase of the carrier must be measured, some form of synchronization of the receiver with the carrier is required. After this synchronization is obtained it must be maintained; the latter is called *phase tracking*.

The types of modulations that are most often used for power efficient communications are *binary PSK* (BPSK), *quadrature PSK* (QPSK), and *offset QPSK* (OQPSK). BPSK uses two phase states (or changes from one symbol period to the next), 0 and $\pm\pi$ radians (0 and $\pm 180°$), to carry the information

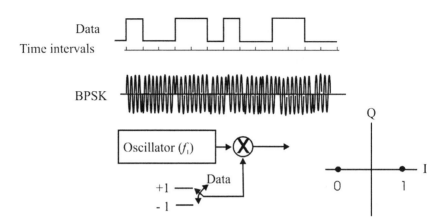

Figure 4.21 BPSK modulation.

signal. It is actually the transition from one state to the other that is usually of importance. QPSK uses four states, 0, ±π/2, and π radians. OQPSK also uses four states, but they are offset from zero by typically π/4 radians. MSK is also a popular modulation scheme. It is OQPSK with half-sinusoidal pulse shaping prior to modulation. As previously indicated, MSK has some special characteristics for spectral efficiency.

BPSK

The simplest form of PSK is BPSK. The phase of the carrier is changed between two states, corresponding to 1 and 0, or 1 and –1. The time waveform is illustrated in Figure 4.21.

Implementation can be as simple as multiplying the carrier by +1 and –1, which, mathematically, is the same as shifting the phase by π radians. Because of this simplicity, it was frequently used in earlier forms of satellite communications and spread spectrum implementations. It is not as efficient, either power or spectrally, however, as other forms of PSK discussed below.

QPSK

In QPSK, two data bits are combined together to form a single symbol as shown in Figure 4.22. As shown on the constellation, each symbol is π/2 radians apart; they all have the same amplitude (usually taken to lie on the unit circle for simplicity). Notice that there are amplitude variations of the carrier where there

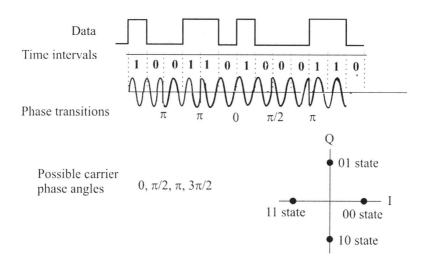

Figure 4.22 QPSK modulation. Two bits at a time are combined to form a symbol.

are phase discontinuities. The amplitude variations are greatest when the phase shifts π radians.

OQPSK

As illustrated in Figure 4.22, normal QPSK exhibits a nonconstant amplitude characteristic. This can be seen at the phase transition from 0 to π where the amplitude can go from the maximum of the carrier to the opposite extreme. Because of this amplitude variation, QPSK is not appropriate for any implementation where the amplifiers are operated close to saturation in a highly nonlinear region, such as satellite communications and digital cell phone systems. Such amplitude variations cause intermodulation products to be generated. To address this deficiency in QPSK, π/4 OQPSK was devised. The constellation for this modulation scheme is shown in Figure 4.23. Only ±π/4 and ±3π/4 phase transitions are allowed. This prevents the large amplitude variations from occurring by not allowing the signal to pass through the origin where the amplitude is zero. Upon each symbol transition, the constellation shifts counterclockwise by π/4 radians, which is where the prefix "offset" comes from. So at $t = t_i$, the symbol at π/4 might correspond to the data bits 00, while at $t = t_{i+1}$, the symbol at π/2 corresponds to the data bits 00. The other symbols shift likewise.

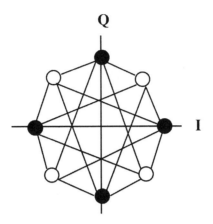

Figure 4.23 π/4 OQPSK constellation.

Intermodulation Distortion Minimization

Satellite communications are one of the chief methods for communications over long distances. The amplification device on satellites is called a *traveling wave tube* (TWT); it generates the power levels that allow communications at the frequencies at which satellites operate. These active devices are operated at or near saturation to achieve maximum efficiency. At saturation, the amplification becomes nonlinear, and nonlinear devices generate amplitude distortions of the signals they are amplifying. One way to minimize such distortions is to use modulation techniques that allow the modulated carrier signal to have constant amplitude (also called the modulus). Constant amplitude modulations are also required in order to reduce the sidelobes in the spectrum generated by operating the *high-power amplifier* (HPA) in a nonlinear mode.

Like GMSK, π/4 QPSK modulation is popular in second-generation cellular phone systems. The U.S. digital AMPS standard, IS-54, uses π/4 OQPSK at 1.62 bps/Hz. The Japanese digital cell implementation uses it as well at 1.68 bps/Hz. Some others are the European Tetra at 1.44 bps/Hz and the new Japanese *personal handy phone* (PHP) system.

In summary, nonlinear amplification (on satellites, in cellular handsets, and elsewhere where power efficiency is at a premium) is the primary motivation for using π/4 OQPSK modulation, where the constellation is shifted by π/4 each successive symbol interval with the result that ± π transitions are not allowed. This allows for a more constant signal envelope, which reduces out-of-band

Table 4.1 Efficiencies and Other Data on Simple Forms of Digital Modulations
(A = best, D = worst)

Modulation	Bandwidth Efficiency		E_b/N_0 (dB) at $P_b = 10^{-5}$	Immunity to Nonlinearity	Implementation Complexity
	Nyquist	Null to Null			
BPSK	1	0.5	9.6	D	A
QPSK	2	1	9.6	C	B
OQPSK	2	1	9.6	B	C
MSK	N/A	0.67	9.6	A	D

Source: [9], © 1994, IEEE. Reprinted with permission.

energy, while facilitating the performance advantages of QPSK. Table 4.1 [9] shows some common modulation techniques and compares them with the above metrics.

2^nPSK

As just illustrated for QPSK, modulating bits can be combined together to form higher types of constellations for transmission. This is referred to as 2^nPSK, where implementations have included n as high as 8 and higher.

4.2.6.9 Quadrature Amplitude Modulation

When both the phase angle and amplitude are used to encode data bits into symbols, it is called *quadrature amplitude modulation* (QAM). The amplitude and phase of the carrier signal both depend on the value of the binary word to be encoded. The phase refers to the change of the phase of the carrier from the last symbol as opposed to an absolute value. Both I and Q amplitudes as well as phase are modulated. There are 2^n discrete symbols in QAM. When $n = 1$, BPSK ensues. When $n = 2$, it is the same as QPSK above. The in-phase component is referred to as I and the other component is referred to as the quadrature component, or Q. I and Q transmit two data streams simultaneously, as discussed in Section 4.2.3.

As an example, suppose $n = 3$ corresponding to 8QAM. Further suppose that the coding process follows Table 4.2. Using Table 4.2, the following digital stream

001011010100101011101010100110101000

separated into octets for clarity becomes

Table 4.2 Phase Shifts for the Example

Binary Word	Amplitude	Phase
000	1	None
001	2	None
010	1	$\pi/2$
011	2	$\pi/2$
100	1	π
101	2	π
110	1	$3\pi/2$
111	2	$3\pi/2$

001 011 010 100 101 011 101 010 100 110 101 000

which is coded as shown in Table 4.3.

The constellation for the above 8QAM is shown in Figure 4.24. This 8QAM coding would facilitate 3 bits per symbol, where a symbol is one of the points in the constellation. For one symbol period, 1/1,200 second for 1,200 baud and 1/2,400 for 2,400 baud, for example, that symbol is transmitted. If the channel is a voice grade channel such as that used for modems over telephone lines, for example, then the resultant data rate would be 3,600 bps for 1,200 baud and 7,200 bps for 2,400 baud. The amount of encoding is a power of two, and 8QAM, 16QAM, 64QAM, and 256QAM are not uncommon. Furthermore, the symbol points need not lie on the axes.

Table 4.3 Phase Changes

Data Word	Amplitude	Phase Change
001	2	None
011	2	$\pi/2$
010	1	$\pi/2$
100	1	π
101	2	π
011	2	$\pi/2$
101	2	π
010	1	$\pi/2$
100	1	π
110	1	$3\pi/2$
101	2	π
000	1	None

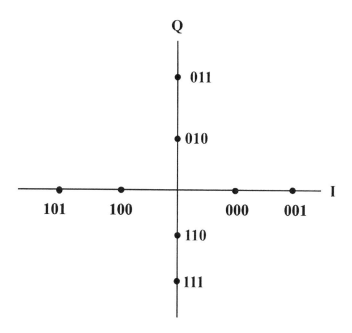

Figure 4.24 Constellation for 8QAM.

Modern modems that are used to interconnect *personal computers* (PCs) to the local Internet service providers use QAM as the modulation technique below 33 Kbps. The advantage is speed. Local telephone lines only provide a bandpass of 300–3,000 Hz, typically, and at 1 bps/Hz (a speed faster than most older encoding schemes) the throughput of these lines could only go to about 3,000 bps. With QAM encoding, higher than 1 bps/Hz can be achieved. Each symbol in 16QAM, for example, represents one of 16 states, which is 2^4, so four bits are sent for each symbol, resulting in a 4:1 increase.

QAM is used extensively for terrestrial radio links. It is not suitable for satellite links, however, because of the amplitude variations inherent in the modulated signal as previously discussed.

4.2.6.10 Spectral Efficiency

Various modulation schemes exhibit different amounts of energy spillover into adjacent channels, as well as channels further away. Figure 4.25 shows a comparison of these modulation techniques. One of the major tradeoffs, as is evident from Figure 4.25, is the amount of energy in the next channel versus the amount of energy in channels further away [10].

4.2.6.11 Linear Modulations

This is a term that broadly refers to the cases where the amplitude is changed to transfer information—the information could be in analog or digital form. The general definition for linearly modulated signals is [11]:

> Let $\varphi[f(t)]$ be the modulated signal and $f(t)$ be the modulating signal. Then the modulation scheme is linear if $\{d\varphi[f(t)]/df(t)\}$ is independent of $f(t)$. Otherwise it is a nonlinear modulation.

Thus, AM is a form of linear modulation since

$$s(t) = f(t)\cos(2\pi f_c t + \Psi_c)$$
$$\frac{ds(t)}{df(t)} = \frac{df(t)\cos(2\pi f_c t + \Psi_c)}{df(t)} \quad (4.20)$$
$$= \cos(2\pi f_c t + \Psi_c)$$

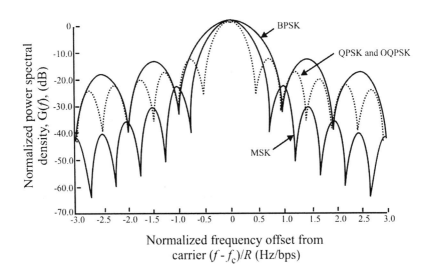

Figure 4.25 Spectral efficiency of MSK, BPSK, QPSK, and OQPSK. (*Source:* [9], © 1994, IEEE. Reprinted with permission.)

and $\cos(2\pi f_c t + \Psi_c)$ is independent of $f(t)$. The phase can also be modulated in such techniques.

The reason to focus on linear modulations is their relative spectral efficiency. FM can be interpreted as a form of spread spectrum modulation, although it is generally not thought of that way. Part of the definition of spread spectrum is that the spreading signal is independent of the information signal and that is not true with FM. The modulated signal is typically much wider in bandwidth than the bandwidth of the information signal, however. That is not true for AM signals. The bandwidth of AM-DSB is only twice the bandwidth of the information signal, while for AM-SSB, the RF bandwidth is the same as the information signal. Therefore, such signals have better spectral efficiency. The downside is, of course, that they are generally more susceptible to common forms of noise and interference. The latter is becoming less of a problem in PSTN applications where fiber is used as the transport medium. Fiber-optic networks are inherently much less noisy than either cable or wireless.

Such narrow bandwidths could be problematic since receiving equipment must know the frequency of operation very specifically. One way to accomplish this is to imbed a tone in the signals. This type of signaling is referred to as *transparent tone in band* (TTIB), discussed next. The tone is placed at a very specific place in the bandwidth, which is measured to tune the receiver precisely.

As the RF spectrum becomes more crowded, spectral efficiency becomes more important. Newer digital modulation schemes are packing higher bit rates into a given bandwidth—8 bps/Hz and higher are not uncommon. Channel allocations in the future in all parts of the RF spectrum can be expected to be narrower. In the military VHF range in the United States, the channel allocations are currently 25 kHz. These will soon be decreased to 12.5 kHz and 6.25 kHz. Using complex digital modulation schemes, these narrower bandwidths can carry the same information rate that is carried today over the 25-kHz channels. Channel coding is used to combat the higher noise susceptibility of linear modulations.

TTIB

It is sometimes convenient to transmit a synchronization tone with a signal. These signals can be used to synchronize a local oscillator for synchronous detection, for example. One way to do this is with a TTIB. As shown in Figure 4.26, the baseband signal is split apart (although literally the tone need not be in the center of the band) and one or more tones are inserted in the band gap. At the receiver, the tones are removed and the signal is put back together. Not only can information be transferred with the tones, but the width of the band gap can be modulated to transmit information as well.

94   Introduction to Communication Electronic Warfare Systems

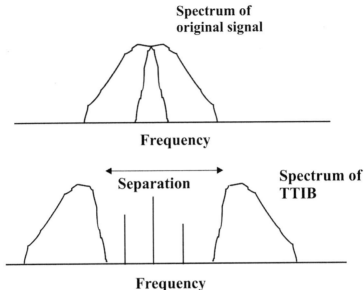

Figure 4.26 Spectrum of TTIB.

4.2.6.12 Orthogonal Frequency Division Multiplexing

Orthogonal frequency division multiplexing (OFDM), also called *discrete multitone* (DMT), is a technique for mitigating the effects of poor channels. Multipath effects, interference and impulsive noise, for example, characterize mobile communication channels. These effects can preclude high-speed communication (several megabits per second). In normal FDM systems, adjacent channels are typically separated by guard spaces—unused frequencies intended to prevent the channels from interfering with each other. This, obviously, is inefficient in terms of spectral utility.

An alternative to this is to make the adjacent channels orthogonal to one another. Each channel then carries a relatively low bit rate, and the channel sidebands can overlap and effective signaling can still occur.

Although in a simplistic implementation, the number of carrier generators and coherent demodulators could get quite large; in fact, when viewed in the Fourier transform domain, the signal can be demodulated by taking the transform at the receiver. A completely digital implementation of the receiver is then possible by taking the inverse transform.

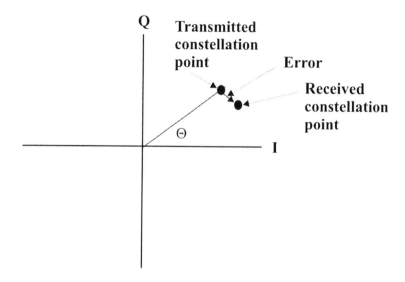

Figure 4.27 Effects of noise on QAM signals.

4.2.6.13 Effects of Noise on Digital Modulations

Either externally induced or internally induced noise, the latter at either the transmitter or the receiver, can cause a received symbol to be in error. The effects are illustrated in Figure 4.27 for QAM, but are similar for any phase-modulated signal. If the noise causes the symbol to appear closer to the expected location of one of the other symbols, the receiver will decide incorrectly which symbol was transmitted. Thus, the BER is a function of the SNR of the signal.

Figure 4.28 shows the required SNRs to produce the indicated BERs for several types of popular digital modulations [12]. These SNRs are postdetection, or demodulated, SNRs, denoted SNR_d herein.

4.2.6.14 Equalization

Equalization is the process of filtering a received signal to account for imperfections in the communication channel. The channel is measured for its frequency response by the transmitter sending tones to the receiver.

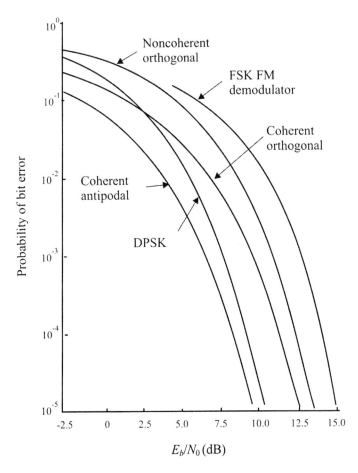

Figure 4.28 Demodulated SNRs required for the specified level of performance (P_e). (*Source:* [4], © John Wiley & Sons, 1988. Reprinted by permission.)

The receiver knows these tones and the time of their transmission. The receiver ascertains how well those tones were received and sets the filter coefficients accordingly. Adaptive equalization for digital communication channels needs to be applied. Without equalization, ISI becomes too large and intolerable interference results, essentially precluding communications. ISI occurs when one information symbol interferes with others. Whereas adaptive equalization is normally applied to wireline communication channels as well, especially for communication by modem discussed in Section 4.4, it is also important for wireless communications. For mobile data communications, an expanding area of digital communications, including the digital overlays of the cellular networks, the modem techniques developed for wireline communications are normally the techniques of choice since it is a well-developed field.

Subsequent symbols in digital channels are delayed and arrive at a receiver later than the symbol that arrived via the direct route. ISI can significantly degrade the channel's performance whenever the delay time (also called the delay spread) is greater than about 10% of the symbol time.

4.2.7 Spread Spectrum

There is a tradeoff that can be made when transmitting information through a channel. It was discovered by Claude Shannon and is described by a theorem given by

$$C = B \log_2 (1 + \delta) \tag{4.21}$$

Equation (4.21) says that the channel capacity C, in symbols per second, can be increased either by increasing the bandwidth B, in hertz, or increasing the SNR δ. Note that this is a theoretical limit and is independent of the type of modulation and other channel parameters.

One method of increasing the bandwidth of a signal is with FM. Another way is to implement spread spectrum modulations. There are two fundamental techniques that fall within the normal definition of spread spectrum: frequency hopping and direct sequence. In the former, the carrier frequency of the signal is changed periodically, or "hopped." In the latter the energy in a relatively narrowband information signal is spread over a much larger bandwidth. Spread spectrum is implemented for a variety of reasons, in addition to increasing the channel capacity. It provides for a degree of covertness, because in frequency hopping the receiver needs to know to what frequency the signal has hopped. This information is normally not known a priori to narrowband ES systems. In direct sequence, since the energy is spread over a wide bandwidth, at any narrowband portion of the spectrum occupied by the signal there is very little energy—frequently below the level of thermal noise present. Thus, the signal is hard to detect.

Another reason to use spread spectrum is for range measurements. This is particularly true for direct sequence, which is the method used in GPS, for example, for determining the range from the satellites to any particular point close to the Earth. Accurate range measurements are possible because of the precise timing in GPS provided by atomic clocks. The correlation between sequences can

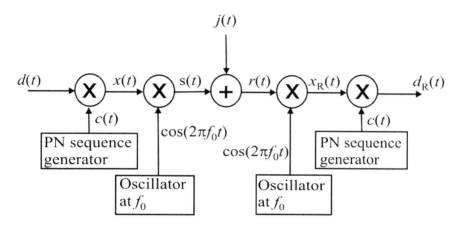

Figure 4.29 BPSK DSSS system.

be measured very accurately and the time offsets can be used to determine range differences.

4.2.7.1 DSSS

Consider the communication system shown in Figure 4.29. The symbol $j(t)$ represents an interfering signal, such as a jammer. A data sequence $d(t) = \{d_i\}$, $d_i \in \{-1, +1\}$ is multiplied by a *chip sequence* $c(t) = \{c_i\}$, $c_i \in \{-1, +1\}$ as illustrated in Figure 4.30. The chip rate, $1/T_c$, is much higher than the data rate, $1/T_b$. Furthermore, the number of chips per data bit is an integer, $N = T_b/T_c$. The product $d(t)c(t)$ multiplies the carrier signal $\cos(2\pi f_0 t)$, forming the transmitted signal

$$s(t) = \sqrt{2S}d(t)c(t)\cos(2\pi f_0 t) \tag{4.22}$$

where S is the power in the signal; the energy per bit is thus $E = ST_b$. This is one form of a BPSK signal. Another equivalent form is given by [13]

$$s(t) = \sqrt{2S}\cos(2\pi f_0 t + d_n c_{nN+k}\frac{\pi}{2}) \tag{4.23}$$

where $k = 0, 1, 2, \ldots, N-1$, n is an integer, and $nT_b + kT_c \leq t < nT_b + (k+1)T_c$. The processing gain of a spread spectrum system is given by the ratio of the bandwidth of the spread signal to that of the nonspread signal. It represents the

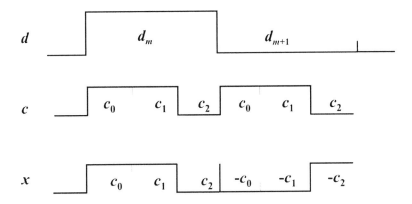

Figure 4.30 Multiplying the data sequence by the chip sequence.

advantage of a spread spectrum system. Thus if the bandwidth of a nonspread signal is B_s and this signal is spread by some means to occupy a bandwidth of B_{ss},

$$PG = \frac{B_{ss}}{B_s} \tag{4.24}$$

then

The received signal $r(t)$ is given by

$$r(t) = s(t) + j(t) = \sqrt{2S}d(t)c(t)\cos(2\pi f_0 t) + j(t) \tag{4.25}$$

This signal is downconverted (filters, amplifiers, and other components are left off the diagram for clarity) by multiplying $r(t)$ by a replica of the signal from the transmitter oscillator. The resulting signal contains components at $2f_0$ and around $f = 0$. It is assumed that the $2f_0$ term is filtered away, leaving only the component around $f = 0$. It is further assumed that these oscillators are at the exact same frequency and are synchronized. The signal is thus

$$x_R(t) = \sqrt{2S}d(t)c(t)\cos(2\pi f_0 t)\cos(2\pi f_0 t) + j(t)\cos(2\pi f_0 t) \tag{4.26}$$

Suppose $j(t) = 0$ for now. Then

$$x_R(t) = \sqrt{2S}\,d(t)c(t)\cos^2(2\pi f_0 t)$$
$$= \sqrt{2S}\,d(t)c(t)\left[\frac{1}{2} + \frac{1}{2}\cos(2\pi 2 f_0 t)\right] \quad (4.27)$$
$$= \frac{1}{2}\sqrt{2S}\,d(t)c(t) + \frac{1}{2}\sqrt{2S}\,d(t)c(t)\cos(2\pi 2 f_0 t)$$

The second term is filtered out, leaving only the first term.
This signal is then multiplied by $c(t)$, yielding

$$d_R(t) = \frac{1}{2}\sqrt{2S}d(t)c(t)c(t) = \frac{1}{2}\sqrt{2S}d(t)c^2(t) \quad (4.28)$$

where it is assumed that the two PN sequence generators are synchronized. Now $c^2(t) = 1$ for all t, so

$$d_R(t) = \frac{1}{2}\sqrt{2S}d(t) \quad (4.29)$$

which is within a constant amplitude of the original data sequence.

The question remaining is why use this modulation format at all, since simpler techniques are available to move a data sequence from one point to another. The advantage becomes clear when $j(t) \neq 0$.

It is assumed that $j(t)$ is centered at f_0, just as the signal is. (See Figure 4.31.) In this case, after multiplying by the local oscillator and filtering off the double frequency term, the remaining signal is given by

$$x_R(t) = \frac{1}{2}\sqrt{2S}d(t)c(t) + \frac{1}{2}j(t) \quad (4.30)$$

Multiplying this signal by $c(t)$ yields

$$d_R(t) = \frac{1}{2}\sqrt{2S}d(t) + \frac{1}{2}j(t)c(t) \quad (4.31)$$

Now, just as the spectrum of $d(t)$ was spread at the transmitter by multiplying by $c(t)$, the jammer signal is spread at the receiver by this multiplication by $c(t)$. Thus, while the multiplication collapses the desired signal to a multiple of $d(t)$, the jammer signal is spread at the receiver, so that the energy per unit bandwidth is

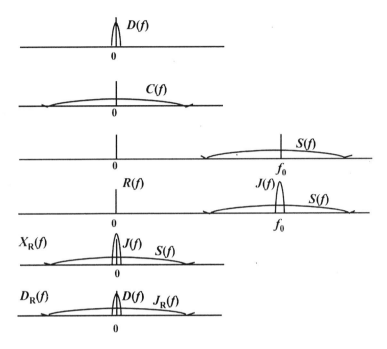

Figure 4.31 Signal spectra of the DSSS system.

quite low. This is illustrated in Figure 4.31. Therefore, spread spectrum exhibits an antijam characteristic. In fact, it was this property that, to a large extent, motivated its development by the military.

Spreading Codes

The codes used to generate the spreading signal, $c(t)$, are selected with specific properties. In particular, the mean value of the chips is approximately zero:

$$\bar{c} = \sum_{i=0}^{N-1} c_i \approx 0 \qquad (4.32)$$

In order for this to be true, the number of -1s must approximately equal the number of $+1$s.

The second significant property is their autocorrelation, which is approximately an impulse function:

$$R_c(k) = \sum_{i=0}^{N-1} c_i c_{i+k} \approx \begin{cases} N & k = 0 \\ 0 & k \neq 0 \end{cases} \quad (4.33)$$

The last property that will be noted here is that their cross-correlation functions are as close to zero as possible. In spread spectrum communication systems, the signals are correlated against a locally generated sequence. If two signals correlate (their cross-correlations are nonzero) then incorrect signals could properly be demodulated at the receiver. If their cross-correlations are zero this will not occur.

In order to achieve a mean value of approximately 0, the PN sequence must have approximately the same number of –1s as 1s in any substantial length of the sequence. This difference should be 0 or only 1. This is called the *balance property*. The *run property* of a PN sequence dictates the following:

- 1/2 ($1/2^1$) of runs are of length 1;
- 1/4 ($1/2^2$) of runs are of length 2;
- 1/8 ($1/2^3$) of runs are of length 3.

There are three types of codes that are frequently used for spread spectrum signaling: m-sequences, Gold codes, and Kasami sequences. They all exhibit the above properties to differing extents. Codes for spread spectrum are generated with *linear feedback shift registers* (LFSRs). Associated with a LFSR is a generating polynomial that governs the code sequence generated by the LFSR. The five-stage LFSR shown in Figure 4.32, for example, has the generating polynomial

$$g(c, a) = 1 + c_1 a_1 + c_2 a_2 + c_3 a_3 + c_4 a_4 + c_5 a_5 \quad (4.34)$$

where all the additions are carried out modulo two, also called the exclusive OR. Also, $a_i, c_i \in \{0, 1\}$. The c coefficients determine whether that stage in the shift register is used in the feedback path, and therefore used in determining which code is generated. By setting $c_i = 0$, then stage i of the shift register is disabled from the feedback path, while setting $c_i = 1$ enables that stage. Since $0 + 0 = 0$, $0 + 1 = 1$, the exclusive OR simply passes the correct bit value on.

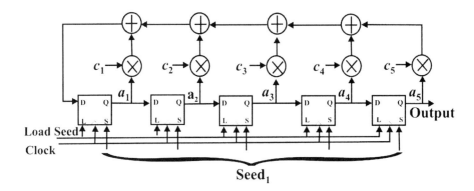

Figure 4.32 Five-stage shift register used for generating codes. The multiplying constants c_i determine whether that stage is used in the feedback loop.

The exponents of the generating polynomial correspond to the state of the LFSR, which is nonzero. Starting from the left is stage 0, then stage 1, and so on. The first and last stages are always used while the intervening stages determine which specific code sequence is generated.

The characteristic of LFSRs that is exploited in code generation for spread spectrum communications is the output sequence that they exhibit. A register is preloaded with a seed and the clock takes the register through its sequence. Registers can be configured to generate output sequences that are as long as they possibly can be. Such sequences are called maximal length sequences, or *m-sequences*. For a register with r stages in it, an m-sequence has a period of $2^r - 1$. That is, the sequence repeats after $2^r - 1$ bits have been generated. The autocorrelation function of an m-sequence is periodic and is given by

$$R(k) = \begin{cases} 1 & k = nN \\ -\dfrac{1}{N} & k \neq nN \end{cases} \quad (4.35)$$

where n is any integer and N is the period of the sequence. An example of this is shown in Figure 4.33 for an $r = 5$-bit register. Although the autocorrelation property of m-sequences is good, their cross-correlation properties are not as good as other orthogonal codes.

Using two equal length m-sequences and combining the results together generate *Gold codes*. An example of this is shown in Figure 4.34. Not all pairs of m-sequences can be used to generate Gold codes. Those that can are referred to as

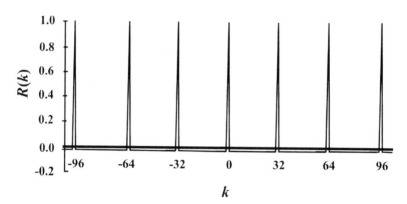

Figure 4.33 Autocorrelation function for an $r = 5$-bit m-sequence.

preferred pairs. The autocorrelation and cross-correlation functions of Gold codes take on three values: $\{-1, -t(m), t(m) - 2\}$ where

$$t(m) = \begin{cases} 2^{(m+1)/2} + 1 & m \text{ odd} \\ 2^{(m+2)/2} + 1 & m \text{ even} \end{cases} \quad (4.36)$$

The advantages of Gold codes are that their cross-correlation functions are uniform and bounded, and a given LFSR configuration can generate a wide family of sequences.

Kasami sequences exhibit very low values of cross-correlation and therefore are useful in asynchronous spread spectrum systems. A similar process used to obtain Gold codes generates them, but various m-sequences are decimated and combined with the original sequence.

Walsh-Hadamard codes are orthogonal and therefore have zero cross-correlations, which makes them highly desirable for spread spectrum communications since the signals are correlated at the receiver during the reception process. The *Walsh functions* are simply determined from a set of matrices called *Hadamard matrices*, which are recursively defined as

$$\mathbf{H}_0 = [0], \quad \mathbf{H}_2 = \begin{bmatrix} 0 & 0 \\ 0 & 1 \end{bmatrix}, \quad \mathbf{H}_4 = \begin{bmatrix} 0 & 0 & 0 & 0 \\ 0 & 1 & 0 & 1 \\ 0 & 0 & 1 & 1 \\ 0 & 1 & 1 & 0 \end{bmatrix}, \quad (4.37)$$

$$\mathbf{H}_{2N} = \begin{bmatrix} \mathbf{H}_N & \mathbf{H}_N \\ \mathbf{H}_N & \overline{\mathbf{H}}_N \end{bmatrix}$$

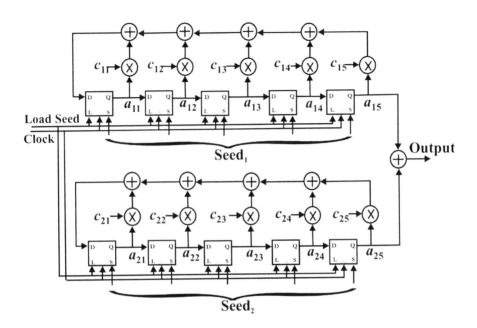

Figure 4.34 Two five-stage linear shift registers connected in parallel with their outputs added (modulo-2) are used to generate Gold codes.

where $\overline{H}_N$ is the element-by-element binary inverse of H_N. All but the all-zero row of a Hadamard matrix forms an orthogonal code word and can be used as a generating polynomial for an LFSR. Each user in the communication system is given a different one of these code sequences and interference is minimized. Actually the all-zero row is orthogonal to the others, too, but it does not generate an interesting output.

The significant downside for Walsh functions in the spread spectrum role is that, while the whole code words are orthogonal, partial code words are not. In fact, partial code words can exhibit significant cross-correlation. In those cases where the CDMA radios are not synchronized with each other, partial correlations will occur and remove much of the advantages of orthogonal codes. For example, while the forward channel (base station to mobile handsets) in IS-95 is synchronous, the reverse channels are not so they would exhibit excessive partial correlations. For that reason Walsh codes are used in IS-95 for the forward channels but not the reverse channels.

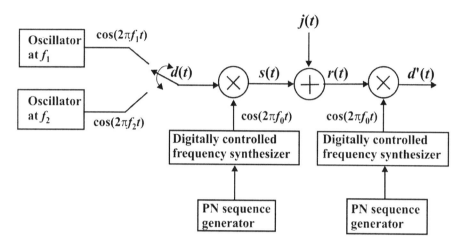

Figure 4.35 Block diagram of an FHSS communication system.

4.2.7.2 FHSS

Consider the spread spectrum communication system shown in Figure 4.35. At the transmitter, the modulating signal is imposed by some modulation method—typically FM—on a carrier frequency that is varied on a regular, frequent basis. Again, amplifiers, filters, and other components are omitted from the diagram for clarity. A digitally controlled frequency synthesizer generates the carrier frequency. The PN sequence generator determines the particular frequency at any moment. This is known as an FHSS system.

The information signal $d(t)$ can be digital or analog but here it is shown as *binary FSK* (BFSK). BFSK is a popular form of modulation for frequency hopping. Although for FSK the two complementary frequencies f_1 and f_2 need not be close together, typically in the VHF frequency range they are two frequencies in the same 25-kHz channel. In many realistic applications of FHSS, noncoherent detection is employed for simplicity and economics. Therefore, even though the PN sequences must be synchronized for proper operation, the local oscillators are not phase-synchronized. Like other noncoherent communication systems, this typically results in about a 3-dB loss in performance.

At the receiver there is an equivalent frequency generator controlled by an equivalent PN sequence generator. When synchronized, the PN sequence generators change frequency control words simultaneously and in synchronization

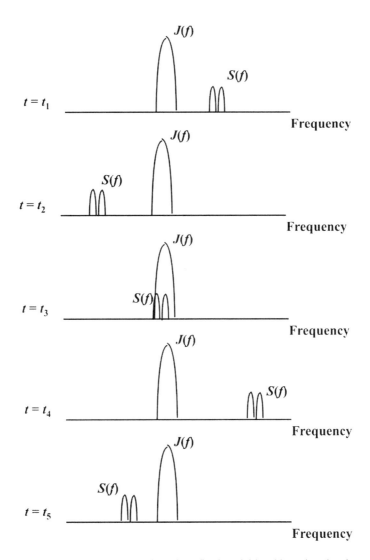

Figure 4.36 Spectra of a frequency hopping with a fixed partial-band jamming signal.

with each other. Synchronizing the PN sequence generators is typically accomplished by using a subset of the available frequency channels to which all of the communication devices tune when synchronization is required (such as when a new member joins the communication net). Synchronization information is always available on these channels.

The functions $j(t)$ and $J(f)$ represent an interfering signal, such as a jammer. The advantage of FHSS systems in the presence of jammers and other interference is illustrated in Figure 4.36. Here BFSK signals are transmitted and the frequency of transmission is varied at the hop rate. For illustrative purposes a narrowband jammer is shown, although jammer waveforms targeted against frequency hoppers would typically cover many more than just the one channel shown here. The hopping signal can hop into the jammer for the duration of the hop. In that case the information during that hop would typically be lost to the receiver. Normally coding, interleaving, and other forms of redundancy are built into the signaling scheme so that if the hopping signal hops into the jamming signal, the data is not lost. Of course, if the signal is digitized voice, some tolerance for lost data bits is inherent anyway. In any case, since the jammer is fixed in frequency, any lost information is only lost for the duration of that hop, and for prior and subsequent hop intervals, the information is transferred unimpeded.

If there is more than one data bit sent per hop frequency, then the system is said to be *slow frequency hopping*. On the other hand, if a single data bit is sent over several hops in sequence, then the system is a *fast frequency hopping* system. The hop rate is denoted as R_h and the data rate is denoted as R_d. Therefore, in a slow frequency hopping system $T_b = 1/R_d \ll T_h = 1/R_h$. R_h for a slow frequency hopping system of 100 hops per second (hps) with a data rate, R_d, of 20 Kbps are typical values. So in this case $T_h = 10$ ms and $T_b = 50$ μs. This data rate would correspond to an approximate rate for digitized speech. On the other hand, for a fast frequency hopping system, for this same data rate, the hop rate might be 40 Khps or faster.

Let the fraction of the total bandwidth, W_{ss}, of an FHSS system that is occupied by a partial-band jammer be denoted as α. Furthermore, let γ_b denote the energy per bit, E_b, to jammer spectral density, J_0, ratio. Thus,

$$\gamma_b = \frac{E_b}{J_0} = \frac{W_{ss}/R_b}{J_{av}/P_{av}} \qquad (4.38)$$

where $E_b = P_{av}/R_b$ and $J_0 = J_{av}/W_{ss}$. For a slow frequency, incoherent BFSK/FHSS system, the performance in the presence of a jammer will be as shown in Figure 4.37 [14]. As a specific example, an airborne, partial-band jammer with an ERP of 1 kW (60 dBm) is standing off a distance of 10 km from a VHF target receiver. The scenario is illustrated in Figure 4.38. Thus, the jammer waveform arrives at the receiver attenuated by approximately 80 dB, yielding $J_{av} = -20$ dBm → $J_{av} = 10^{-5}$ W at the receiver. Suppose $W_{ss} = 10$ MHz; thus, $J_0 = 10^{-12}$ W/Hz. Suppose

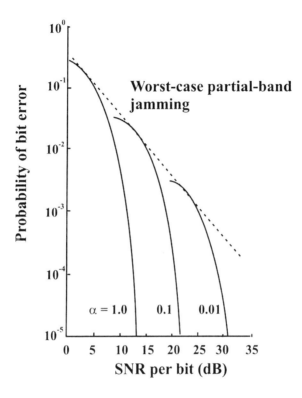

Figure 4.37 Performance of a BFSK/FHSS system in the presence of partial-band jamming. (*Source:* [14], © 1995, John Wiley & Sons. Reprinted with permission.)

the communication link is air-to-ground, with the receiver on the ground, and is 1-km-long. Suppose the target transmitter emits 2W (33 dBm). Over this range in the VHF frequency range, the free-space path loss is about 60 dB, so $P_{av} = -27$ dBm $\rightarrow P_{av} = 10^{-5.7}$W. Suppose $R_b = 20$ Kbps. Thus, $E_b = 10^{-5.7} \div 2 \times 10^4 = 0.5 \times 10^{-9.7}$W·sec and $\gamma_b = 0.5 \times 10^{-9.7}/10^{-12} = 0.5 \times 10^{2.3} \rightarrow 18$ dB. The probability of bit error in this case is approximately 2×10^{-3}.

Code selection for frequency-hopping communication systems is similar to that described above for direct sequence. The codes should have minimum cross-correlation and well-behaved autocorrelations. The receive process, however, is different in frequency hopping systems and therefore the effects of high values of cross-correlations are different. In DS systems, when two codes had a high cross-correlation, the improper signal could be properly decorrelated at the receiver, in

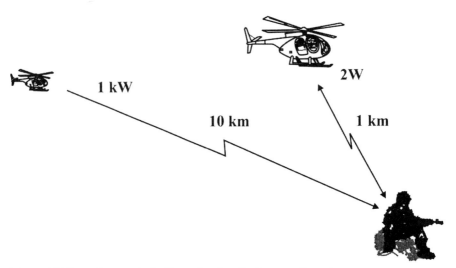

Figure 4.38 Example scenario for illustrating the effectiveness of jamming.

effect causing unwanted interference. In FH systems, two codes with a high cross-correlation will cause the same frequencies to occur in the two sequences more than desired. Thus, the two transmitters will hop to the same frequency more often than otherwise. This also results in unwanted interference, called cochannel interference, even though the details of the cause are different.

4.3 CODING OF COMMUNICATION SIGNALS

Coding of communication signals is generally implemented for one of two purposes. The first is to remove redundancy in the source information. Many sources exhibit redundant characteristics, for example, speech. Coding can remove some of this redundancy to improve efficiency. The second reason for coding is to improve the reliability of the communication. Adding information to the communicated message can increase the probability that the correct message is received.

4.3.1 Source Coding for Data Compression

Many sources of information that are common to humans contain redundant information. Human speech is a particularly bad offender in this regard. In order to maximize the efficiency of using communication resources, it may be desirable to eliminate as much of this redundancy as possible before transmitting the information. The same can be said about storing information, incidentally. The techniques for reducing this redundancy are called schemes for data compression, and some of them are discussed here. Modern day information compression techniques can facilitate the transmission of images at a 300:1 or more compression ratio in some cases without sacrificing image quality [15].

4.3.1.1 Speech Coding Techniques

An *order 0 model* or encoding scheme uses only the current value of the data to be encoded. An example is "u" in English. There is 1/100 chance of this letter occurring in normal English. The *entropy* of this character results in 6.6 bits to encode. Entropy is a term that refers to the amount of randomness in a piece of data. It is a term borrowed from thermodynamics, where it refers to the amount of randomness in gasses. The higher the randomness, the less probable the character is, the higher the entropy is, and the more bits are taken to encode it. Taking this in reverse, the more probable a character is, in order to minimize the number of bits it takes to transmit such common characters, fewer bits are assigned.

For an *order 1 model*, the previous character is included in the encoding process. For example, if a q is the previous character, then the probability that the current character is a u is 95%. In this case it takes only 0.074 bits (on average) to encode the current character. Source coding technology is important to understand for EW system design because source coding is used to get as much information as possible through narrow transmission channels.

Run-length encoding, used extensively in facsimile communications, encodes strings of the same character (such as blank characters) as the character followed by the number of times it is repeated. Huffman encoding allocates the number of bits representing a character in an alphabet (such as the English language) according to the number of times it appears in a message. Those characters that appear the most receive the fewest number of bits allocated, thus minimizing the number of bits necessary to send the message. Arithmetic coding accomplishes similar objectives using similar notions. There is considerable redundancy in higher levels of communications, such as English text. Phrases are often repeated. The Trie-based codes, LZ-77 and LZ-78, address this by using pointers to previously sent phrases, indicating where the phrase began and how long it was. All of these coding techniques are called *lossless* since no information in the messages communicated is lost. They can typically achieve a compression ratio in the range of 10:1.

4.3.1.2 Compression Techniques for Images

In both the commercial world as well as that of the military forces worldwide, the requirement to view and move images around is becoming widespread. In the commercial world, still pictures over the Internet and motion pictures over cable TV are two mainstream applications. A TV picture is comprised of 4.5 MHz of analog bandwidth. If that signal is digitized at the minimum rate, the Nyquist rate, of 9 MHz with 10 bit samples, for example, the resulting bit stream would be 90 Mbps. This is the data rate required to move this signal around as well as to store it on DVD disks. Fortunately, there is considerable redundancy in most video signals and source coding techniques are applied to reduce or eliminate that redundancy. The fundamental tradeoff in implementing these techniques is picture quality versus implementation complexity. The redundancy in images can be represented by the amount of correlation there is between the basic data in the image. This basic data element is called a *pel*.

Transforms are frequently applied to reduce the transmission and storage requirement for images. Perhaps the most common one is the *discrete cosine transform* (DCT), but others that have been used include the Harr transform and the Walsh-Hadarmard transform. The most efficient transform in a mean-squared sense is the *Karhunen-Loeve transform* (KLT); however, it is extremely complex to implement compared to some others. The DCT is popular because it is simple and its performance is close to that of the KLT.

The redundancy in video signals comes in two forms—the spatial redundancy within a single image and the temporal redundancy from one image to the next. To address the former, the *Joint Pictures Expert Group* (JPEG) established a standard for video compression for still pictures. The DCT is used to transform 8 × 8–pixel segments of the picture. The DCT coefficients are normally largest around zero and typically fall off rather rapidly. Therefore, only a few need to be retained to represent the image. In addition, Huffman encoding of the coefficients is applied which assigns shorter code words to the most frequently occurring coefficients, and run-length encoding is applied to those results. The results are extremely efficient coding of single images. Such coding of single images is referred to as *intraframe* coding.

To remove the redundancy from one image to the next, *interframe* coding is used. Just as there is considerable redundancy in the spatial context within a single image, there is significant redundancy from one image to the next. At the speeds of motion pictures, objects typically don't move much from one image to the next.

The H.261 standard uses transmission rates that are multiples of 64 Kbps so compatibility with the *integrated services digital network* (ISDN) could be maintained. It matches a 16 × 16 block from one image to the next and the difference is encoded as opposed to the image itself. Data rates of up to 1.5 Mbps can be implemented with this protocol.

The *Motion Pictures Expert Group* (MPEG) developed three standards for transmission and storage of motion video called MPEG 1, MPEG 2, and MPEG 4. MPEG 1 was developed to support video on CD-ROMs. Data rates of up to about 1.5 Mbps are supported as part of the standard. Forward prediction only as well as both forward and backward (bidirectional) prediction for motion compensation are part of the standard. Forward prediction is similar to that described for H.262, where motion prediction is computed. Backward prediction implements a similar concept except the current image uses the next image to reduce redundancy.

MPEG 1 allows data rates up to about 1.5 Mbps, so a higher-speed standard was needed, particularly for transmission of video over cable TV systems. Thus, the MPEG 2 standard was developed. MPEG 2 allows transmission up to about 15 Mbps. As part of the MPEG 2 standard, it is possible to specify SNR requirements and the coding adapts to those requirements. Thus, it is possible to specify the quality of the coding to apply ahead of time and the images are coded accordingly. The 90-Mbps TV data rate mentioned above can be encoded with MPEG 2 at around 5 Mbps.

Lossy coding can be applied in cases where a certain amount of loss is allowed in coding an image. A still picture, for example, can be adequately represented with some loss in the data in the picture. *Lossless* coding is when losses are not desired or allowed. X-ray coding might be an example of this, where it is critical to transmit an image with the maximum resolution possible. In lossless situations, the resource investment in moving or storing images is less important than keeping all the data possible.

At the other end of the speed spectrum, MPEG 4 was developed. It was intended to apply to videophone applications, such as videoconferencing, where rapid movement was not likely, some loss could be tolerated in the transmission, and the regular narrowband PSTN is the transport medium (at 2,700-Hz guaranteed bandwidth). This standard implements speeds of 5 to 64 Kbps.

4.3.1.3 Chrominance Subsampling

All of the above coding techniques for compressing video images implement what is called *chrominance subsampling*. Each pel is represented by three 8-bit values—one for luminance, or brightness, and two for chrominance, or color. If two adjacent horizontal pels are the same color, determined by averaging the original colors and transmitting or storing just the one color value, then it is called 4:2:2 subsampling. This coding reduces the data rate requirements by one-third and there is little perceptual difference in the image.

Likewise, downsampling the colors can be applied in the vertical direction in the image. This is called 4:2:0 subsampling, but normally the changes in the image are detectable. Compression of one-half is accomplished this way.

If the information in the image can be transferred adequately in black and white, then all of the color information can be removed. Videoconferencing, for example, may not need the color information to be effective. The result is a black, white, and gray image. The data rate in this case has been reduced by two-thirds, however.

4.3.1.4 Steganography

Steganography is the technique of hiding a message within another message such that *detection of the presence* of the first message is prevented. One way to accomplish this is to embed messages in image files. Such files are typically comprised of many pixels and therefore bytes. Depending on the information message to be sent, it may be necessary to change only a few such bits. The image within which the message is embedded is called the *cover image*. An algorithm of some sort inserts the information message into the cover image. The intended receiver of the message knows both the original cover image as well as the coding algorithm so the information-bearing message can be recovered. The encoding algorithm could take many forms. Calculating the exclusive OR on a few of the pixel bits would be one way. Changing 1 bit in every pixel or every n pixels would be another.

Images are either color or gray scale, and either can be used for such purposes. The more random the image, the easier it is to hide the presence of the information message. Changing the least significant bit of the pixel information, for example, one that describes a color image changes the color only slightly—typically imperceptible to the human eye. If a pixel is represented by 3 bytes, or 24 bits, changing 1 bit per pixel in a large image would go unnoticed.

4.3.1.5 Communication EW Implications

Except for MPEG 4, the video standards described here are too wideband to be of much concern to forward-deployed tactical forces. The data rates involved in transporting H.261, MPEG 1, and MPEG 2 preclude the use of most tactical radios. The bandwidth requirements dictate that the higher frequency ranges are required, where adequate bandwidth is available. Communication devices at the division echelon and above during conflict and at all levels otherwise have adequate capacity to carry such communications. Tactical radio systems that are typical of forward deployments, in general, do not have an adequate capacity. The

MPEG 4 standard can be used with these radios to transport video in the near term.

This is not to say, however, that the higher data rate signals are not of concern. Video teleconferencing has become a popular way for military forces to coordinate and plan their activities, and therefore the communications over which this transpires is of considerable interest. Systems and devices to attack such communications are much different from those for tactical communications, however. In those cases where wired and/or fiber cables can be utilized, bandwidth is usually not a problem.

Steganography presents a particularly challenging problem for EW systems. A human observing an intercepted image will rarely be able to tell that a message is hidden in the image. A machine will only be able to tell the presence of the message if it has access to the original cover image and the encoding algorithm.

4.3.2 Channel Coding for Error Control

Channel coding is normally performed in communication and data storage systems (which in reality are a form of communication system where the goal is communication in the fourth dimension—time) for error control. No realistic communication channel is noise-free—noise in this case being defined as anything other than the intended signal. Because of this bit errors inevitably occur.

Some types of communications will tolerate bit errors better than others. Digitized speech, for example, will tolerate a considerable level of bit errors before the speech becomes unintelligible. On the other hand, computer-to-computer digital exchanges, such as bank transactions, must be accomplished almost error-free. In the latter case, great lengths are taken to insure that information is exchanged correctly.

Thus, coding is used to perform this error control. In some cases simply detecting errors is sufficient. In that case either the data within which the bits in error is simply discarded—possible in speech, for example, or the data can be discarded and the source can be requested to resend the data. In other cases it is possible to encode data so that, up to some limit, errors in the data can be corrected at the receiver. There are tradeoffs with both approaches.

One of the most significant advantages of digital communications is the ability to detect and/or correct errors in the data streams. This is done using coding techniques. Although a complete discussion of error detecting and correcting coding is beyond the scope herein, some of the more important topics for IW will be discussed.

The channels over which communication takes place are herein assumed to be *binary symmetric channels* (BSCs). What this means is that the probability of an error occurring is independent of the bit that was transmitted. If the probability of

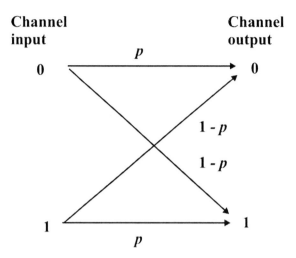

Figure 4.39 Graphic depiction of a binary symmetric channel.

no error occurring is denoted by p, also called the *reliability* of the channel, then the probability of an error is given by $1 - p$. This is represented by the graph in Figure 4.39. Note that if there is a modem involved (discussed later), it is normally considered as part of the channel.

4.3.2.1 Error Detection

It is possible in some system implementations to allow the loss of some data but it is undesirable to let data in error persist. Also, some implementations perform error correction on large portions of a data stream as opposed to the individual pieces of data in it, but the way the accuracy of these large portions is checked is to check each piece of data for correctness. These are examples of when error detection can be used.

Parity

One or more bits can be added to a data word to facilitate error detection. Perhaps the simplest of these schemes is adding a parity bit. A bit can be added so that the total number of 1s in a word (a word is defined here as any set of bits greater than two) is even, in which case it is called even parity, or odd, in which case it is called odd parity. As an example, suppose there are four words to be sent, given by 00, 01, 10, and 11. Further suppose that it is known (to a reasonable degree of assurance) that errors can only occur in one bit in any word. Then these words,

Radio Communication Technologies

Table 4.4 Parity Bit Attached to Data Words

Data Word	Parity Bit	Bit in Error		
		1	2	3
00	0	100	010	001
01	1	111	010	010
10	1	001	111	100
11	0	010	100	111

encoded with even parity, are given in Table 4.4. The data words to be encoded are listed in the left column, and the even parity bit is given in the second column. The possible received encoded words consist of the correct encoded word, or one of those in the last three columns. By inspection it can be seen that none of the encoded words received with errors in them are legitimate so an error occurred in the transmission and it can be detected.

Digital data can be organized into a matrix, such as that shown in Table 4.5 (even parity is used). These 8-bit bytes might represent data bytes from some file to be transferred. A parity bit can be attached for each row and one for each column, shown as the bottom row and rightmost column. If a single error occurs, it will not only be detected but it can also be corrected since the row and column in error will identify a specific bit.

Cyclic Redundancy Check

The *cyclic redundancy check* (CRC) is a very popular coding technique for error detection and is frequently used in *automatic repeat request* (ARQ) schemes. It is popular because it is easy to implement, it has very good error detection

Table 4.5 Examples of Adding (Even) Parity Bits to Rows and Columns

1	0	0	1	1	0	0	1	0
0	1	0	1	1	0	0	1	0
0	1	1	1	1	1	0	0	1
1	0	0	1	0	0	0	0	0
0	1	1	1	0	1	1	1	0
0	0	1	0	1	1	0	0	1
1	0	0	0	0	0	1	0	0
0	0	0	0	1	1	1	0	1
1	1	1	1	1	0	1	1	

capabilities, and there is little extra that needs to be done to check for errors. There are integrated circuits available that compute the CRC.

It is based on the notion that binary data streams can be treated as *binary polynomials*, and it is possible to add, subtract, multiply, and divide these streams of data just as if they were numbers. The binary numbers are the coefficients of the polynomial. For example, if the binary word is given by 1010011, then the corresponding binary polynomial is given by

$$g(x) = 1x^6 + 0x^5 + 1x^4 + 0x^3 + 0x^2 + 1x^1 + 1x^0 \qquad (4.39)$$

or simply,

$$g(x) = x^6 + x^4 + x + 1. \qquad (4.40)$$

The data stream that CRC operates on is called a *frame* and a *frame check sequence* (FCS) is calculated for every frame. In CRC, the FCS is calculated so that the data sequence resulting from concatenating the original data frame with the FCS is exactly (no remainder) divisible by a polynomial. The particular polynomial is selected depending on the type of errors expected to be encountered. Some common polynomials are

$$\begin{aligned}
\text{CRC-12} &: g(x) = x^{12} + x^{11} + x^3 + x^2 + x + 1 \\
\text{CRC-16} &: g(x) = x^{16} + x^{15} + x^2 + 1 \\
\text{CRC-ITU} &: g(x) = x^{16} + x^{12} + x^5 + 1 \\
\text{CRC-32} &: g(x) = x^{32} + x^{26} + x^{23} + x^{22} + x^{16} + x^{12} + x^{11} + x^{10} \\
&\quad + x^8 + x^7 + x^5 + x^4 + x^2 + x + 1
\end{aligned} \qquad (4.41)$$

If $d(x)$ is the binary polynomial corresponding to the data frame to be encoded, then first $r(x)$ is obtained by dividing $d(x)$ by $g(x)$:

$$\frac{d(x)}{g(x)} = q(x) + r(x) \qquad (4.42)$$

The resultant data frame that is sent is then the original frame with the FCS appended to it. At the receiver, this sequence is divided by $g(x)$ and if there were no errors in the transmission, then the remainder of this division is zero.

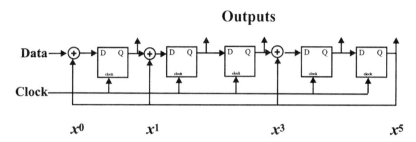

Figure 4.40 Linear shift register for generating and computing a CRC, in this case based on the polynomial $g(x) = x^5 + x^3 + x + 1$.

This code detects error bursts of length less than or equal to $(n - k)$ with probability 1, bursts of length equal to $(n - k + 1)$ with probability $(1 - 2^{-(n-k-1)})$, and bursts of length greater than $(n - k + 1)$ with probability $(1 - 2^{-(n-k)})$ [16]. It has been estimated that when using a 17-bit polynomial, the probability of allowing a 1-bit error go undetected is one in 10^{14}.

Implementation of the CRC algorithm is easy in either hardware or software. The hardware implementation using shift registers is shown in Figure 4.40, where the polynomial $x^5 + x^3 + x + 1$ is implemented. The exclusive OR gates (the circles with the + in them) correspond 1 to 1 with the existence of a nonzero coefficient in the generating polynomial, with the exception of the highest-order coefficient. These gates have as their inputs the prior stage and the highest-order bit as feedback. To use this circuit, when generating the FCS, the data frame is first shifted into the shift register from the left with n zeros appended to the end. After the register has been clocked $n + k$ times, it will be holding the FCS. To check a received frame it is shifted into the register from the left. After $n + k$ clock cycles, the register will be holding all zeros if the frame is error-free.

The CRC procedure works because of the properties of prime numbers. A prime number is a number that can only be divided by itself and one, without generating a remainder. Within certain reasonable assumptions, if an integer of a given number of digits is divided by a prime number then the remainder is unique among all of the numbers with the same number of digits. The shift register in Figure 4.40, for example, effectively divides a data stream by the prime number 41 ($2^5 + 2^3 + 2^1 + 2^0$).

Checksum

A checksum is computed for a set of data words by summing the values of data, without worrying about overflow, as if the data were always numbers. Thus, the

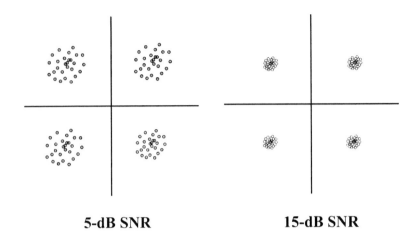

Figure 4.41 Effects of noise on a QAM constellation.

addition is performed modulo 2^n, where n is the number of data bits in a word. The checksum is computed on the data set to be sent (or stored) and is attached to the data as if it were part of the data set. When the data is checked, the checksum is added along with the data and the result is zero unless there is an error. It accomplishes a reasonable degree of error detection. If data is completely random and each data word is 16 bits long, then there is about a 1 in 64,000 chance that the checksum would be computed accurately for an inaccurate data set.

4.3.2.2 Error Correction

There are two generally accepted techniques for error correction: *forward error correction* (FEC) and *automatic repeat request* (ARQ). In the former, bits are added to the outgoing data stream so that an algorithm at the receiver can correct bits that are received in error. In ARQ, extra bits are also added to the outgoing data stream, but at the receiver the received data stream is checked for errors, and if any are present, a request to resend the data is made.

A constellation for OQPSK, shown in Figure 4.41, exhibits some variations in the location of the data points. In the figure, a constellation with a low SNR is shown along with one with a relatively higher SNR. At a sufficiently low SNR, the receiver will declare some of the symbols wrong. Furthermore, all communication systems are limited in signal power at some level. The lower this level is, the closer together the symbols are in the constellation. Therefore, many digital communication systems employ error control, which can be either error detection or error correction; of course, the latter implies the former but not vice versa.

ARQ

ARQ is a relatively simple form of error correction, but it is also one of the least efficient. Using one of the error detection techniques described above or others, a received data word is checked for correctness. If it is determined that the data is in error, a NACK is sent back to the transmitter indicating that the data should be sent again. When it is resent, the whole data sequence is resent. Thus in low SNR environments, the channels can quickly get clogged resending messages received in error. FEC schemes were devised to address this inefficiency issue inherent with ARQ.

FEC

When bits are added appropriately to each digital word prior to conversion to symbols in order to correct errors, it is called FEC. It does, of course, reduce the number of data bits that can be sent in a given bandwidth but in many cases this is a favorable tradeoff. Of course, as the bandwidth is increased, more data bits can be sent but this means that more noise enters the receiver, thereby increasing the BER. The goal, then, in FEC code design is to the reach an optimum tradeoff between the number of bits added versus the bandwidth necessary to transmit the code words.

There are three major types of error correcting coding techniques: (1) block codes, (2) cyclic codes, and (3) convolutional codes. Block coding usually encodes large blocks of data at a time, and the encoding depends only on the current data to be encoded. Convolutional coding encodes a block of data, and the encoding depends on the current block of data to be encoded, as well as prior blocks of data. To use the best features of both block coding and convolutional coding, they are frequently combined.

For a simple example of error correction, suppose that there are only two possible data words that can be sent given by 01 and 10. Furthermore, suppose that it is known (to a reasonable degree of assurance) that only one bit error can occur in any symbol. One possible encoding is shown in Table 4.6 for this case. The data word to be sent is in the leftmost column and the error bits attached are in the second column. These two columns concatenated could be received when there is no transmission error. The remainder of the columns contain possible received encoded words depending on which encoded word was sent and which bit is in error. Comparing the two rows clearly shows that whichever encoded symbol is received it can be mapped back into the correct data word, under the assumption that only 1 bit (or none) is in error, thus forming an error correcting code.

This example points out the overhead necessary to implement error correction; 2 bits were added to correct 2-bit data words. This is an extreme case, however, and as the size of the data word increases, the number of check bits relative to the number of data bits decreases.

Burst error protection. Suppose each column of the following matrix is a data word to be sent and they are encoded so that up to 2 bits in error in each data word would be corrected at the receiver. Transmission of these data words would normally be $x_{1,1}$, $x_{1,2}$, $x_{1,3}$, ..., that is, vertically down the column. Suppose the order were changed so that the data bits were sent horizontally, that is, $x_{1,1}$, $x_{2,1}$, $x_{3,1}$, ... Then a burst of errors up to 14 bits long could occur and the data words would be delivered correctly, because a burst this long or shorter guarantees that no more than 2 bits in any original data word are in error. Normally a burst of 14 bits would destroy two or three data words (in this case word = byte), depending on where the burst started.

$$x_{1,1}\ x_{2,1}\ x_{3,1}\ x_{4,1}\ x_{5,1}\ x_{6,1}\ x_{7,1}$$
$$x_{1,2}\ x_{2,2}\ x_{3,2}\ x_{4,2}\ x_{5,2}\ x_{6,2}\ x_{7,2}$$
$$x_{1,3}\ x_{2,3}\ x_{3,3}\ x_{4,3}\ x_{5,3}\ x_{6,3}\ x_{7,3}$$
$$x_{1,4}\ x_{2,4}\ x_{3,4}\ x_{4,4}\ x_{5,4}\ x_{6,4}\ x_{7,4}$$
$$x_{1,5}\ x_{2,5}\ x_{3,5}\ x_{4,5}\ x_{5,5}\ x_{6,5}\ x_{7,5}$$
$$x_{1,6}\ x_{2,6}\ x_{3,6}\ x_{4,6}\ x_{5,6}\ x_{6,6}\ x_{7,6}$$
$$x_{1,7}\ x_{2,7}\ x_{3,7}\ x_{4,7}\ x_{5,7}\ x_{6,7}\ x_{7,7}$$
$$x_{1,8}\ x_{2,8}\ x_{3,8}\ x_{4,8}\ x_{5,8}\ x_{6,8}\ x_{7,8}$$

Hamming weight and distance. The *Hamming weight* of a binary word, denoted by $w(.)$, is the number of nonzero elements in the word. Thus, if (1, 0, 1, 1, 1, 0) is such a word, then $w(1, 0, 1, 1, 1, 0) = 4$.

The *Hamming distance, d,* is the minimum number of bit positions in encoded words by which any two code words are different. For example, if the code

Table 4.6 Attaching Error Correction Bits to Data Words

Data Word	Error Correction Bits	Bit in Error			
		1	2	3	4
01	00	1100	0000	0110	0101
10	11	0011	1111	1001	1010

consists of the two words (1, 0, 1, 1, 1, 0) and (0, 0, 1, 1, 0), then $d = 3$ since the two words are different in the first and last two bit positions. The performance of a code is partially specified by d as follows:

1. The number of errors that can be detected is less than or equal to d.
2. The number of errors that can be corrected is less than or equal to $\lfloor (d - 1)/2 \rfloor$ where $\lfloor x \rfloor$ denotes the greatest integer less than x.

Block codes. When there are k bits in a data word to be sent, and n bits are sent, then $n - k$ bits are added to each of the words. These $n - k$ bits are check bits selected in a particular way so that error control is maintained. There are 2^k possible messages to be sent and there are 2^n possible encoded words that are transmitted. The 2^k messages are called a block code and each of the possible 2^n transmitted symbols is called a code word.

While there are many kinds of codes available, the example included here is from the linear, systematic block codes. In these types of codes, parity bits are obtained from modulo-two addition of the data bits (linear), and the resultant parity bits are appended to the end of the data bits (systematic).

Hamming codes. As noted above, it is possible to organize data into a matrix and assign parity bits to the rows and columns of this matrix. Such an arrangement facilitates some error detection and error correction. This notion can be generalized where parity bits are applied to various combinations of the data bits (not just the rows and columns). If the number of data bits is given by d and the number of parity bits is given by p, then $d + p + 1 \leq 2^p$.

The Hamming code word, c, is the p bits appended to the d bits for each word. Such a code is identified as a (c, d) Hamming code. A Hamming code is identified by a generator matrix $\mathbf{G}$, given by

$$\mathbf{G} = [\mathbf{I} : \mathbf{P}] \tag{4.43}$$

where $\mathbf{I}$ is the identity matrix and $\mathbf{P}$ specifies the particular Hamming code. The data words given by $\mathbf{d} = [d_1, d_2, d_3, d_4]$ are multiplied by $\mathbf{G}$ to form the code words $\mathbf{c} = [d_1, d_2, d_3, d_4, p_1, p_2, p_3]$, that is, $\mathbf{c} = \mathbf{d}^\mathrm{T} \mathbf{G}$ (all math is modulo 2). The columns of $\mathbf{P}$ are selected so that each column is unique, thereby forcing the parity calculation to be performed over different sets of the data bits, and no resultant row is zero.

An example of a (7, 4) Hamming code is given by the generating matrix

$$G = \begin{bmatrix} 1 & 0 & 0 & 0 & 1 & 1 & 1 \\ 0 & 1 & 0 & 0 & 0 & 1 & 1 \\ 0 & 0 & 1 & 0 & 1 & 0 & 1 \\ 0 & 0 & 0 & 1 & 1 & 1 & 0 \end{bmatrix} \qquad (4.44)$$

and suppose that the following data word is to be encoded

$$\mathbf{d} = \begin{bmatrix} 1 \\ 1 \\ 0 \\ 1 \end{bmatrix} \qquad (4.45)$$

then the following code word is generated

$$\mathbf{c} = \begin{bmatrix} 1 & 1 & 0 & 1 & 0 & 1 & 0 \end{bmatrix} \qquad (4.46)$$

Clearly the first 4 bits are the original data word, while the last 3 bits form the appended parity bits.

At the receiver, the incoming data stream is checked by a matrix given by

$$\mathbf{H} = \begin{bmatrix} \mathbf{P}^T & \mathbf{I} \end{bmatrix} \qquad (4.47)$$

to form a *syndrome*, $\mathbf{s}$, for the data word, $\mathbf{s} = \mathbf{H}\,\mathbf{c}$. If $\mathbf{s} = \mathbf{0}$, then the data was received correctly, and if $\mathbf{s} \neq \mathbf{0}$, then an error has occurred.

In this example,

$$\mathbf{s} = \mathbf{Hc} = \begin{bmatrix} 1 & 0 & 1 & 1 & 1 & 0 & 0 \\ 1 & 1 & 0 & 1 & 0 & 1 & 0 \\ 1 & 1 & 1 & 0 & 0 & 0 & 1 \end{bmatrix} \begin{bmatrix} 1 \\ 1 \\ 0 \\ 1 \\ 0 \\ 1 \\ 0 \end{bmatrix} = \begin{bmatrix} 0 \\ 0 \\ 0 \end{bmatrix} \qquad (4.48)$$

(Remember that the arithmetic is carried out modulo two.) On the other hand, if [1001010] were received, then

$$\mathbf{s} = \mathbf{Hc} = \begin{bmatrix} 1 & 0 & 1 & 1 & 1 & 0 & 0 \\ 1 & 1 & 0 & 1 & 0 & 1 & 0 \\ 1 & 1 & 1 & 0 & 0 & 0 & 1 \end{bmatrix} \begin{bmatrix} 1 \\ 0 \\ 0 \\ 1 \\ 0 \\ 1 \\ 0 \end{bmatrix} = \begin{bmatrix} 0 \\ 1 \\ 1 \end{bmatrix} \qquad (4.49)$$

indicating an error has occurred in transmission.

Cyclic codes. A code is called *cyclic* when $(x_0, x_1, \ldots, x_n)$ is a code word whenever $(x_1, x_2, \ldots, x_n, x_0)$ is a code word. The latter is called a permutation of the former. BCH codes, developed by Bose, Chaudhuri, and Hocquenghem, are cyclic codes. Reed-Solomon codes are one form of BCH codes. They are used in CD players.

Convolutional codes. Convolutional code words are generated by combining sequences of bits together. Thus, the current code word depends on the current bit to be encoded as well as some number of previous bits. The bits are sequentially loaded into a shift register and the stages of the shift register are combined modulo-two in such a way to generate the convolutional code. The length of the shift register is called the *constraint length,* denoted as K, while the ratio of the number of bits in the input data word to the number of bits in the output, denoted by v, is called the *rate*. For example, shown in Figure 4.42 is such a convolutional encoder. Here the constraint length $K = 3$ while the rate $= 1/v = 1/2$ since two output bits are generated for each input bit. In this example

$$c_1 = a_1 + a_2 + a_3 \qquad (4.50)$$
$$c_2 = a_1 + a_3$$

The initial seed for this shift register is zero.

Convolutional coders such as these are finite state machines, where the next state is determined by the input data and the current state. The state machine for

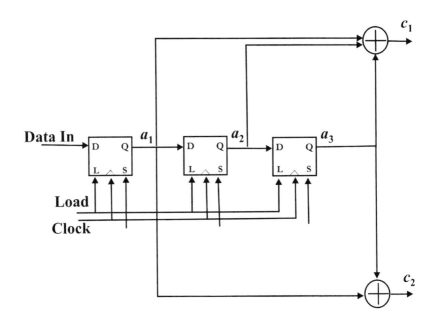

Figure 4.42 Convolutional encoder where $K = 3$ and rate = 1/2.

this example is shown in Figure 4.43. The state of this machine is defined as (a_1, a_2) while the output is (c_1, c_2).

A *trellis diagram* is a way of displaying the operation of a convolutional encoder. The trellis diagram for the encoder shown in Figure 4.42 is shown in Figure 4.44. In this diagram, x/yy refers to x = input and yy = output after the next clock pulse loads the input into the first stage and all the other stages are shifted to the right one place.

Viterbi decoding. Viterbi decoding is a form of maximum-likelihood decoding. It is implemented with a convolutional decoder. The concept of maximum-likelihood decoding arises because of the Hamming distance notion mentioned above. If a noncode word is received, the most likely code word that was sent is the one with the minimum Hamming distance from the received word. This is perhaps best illustrated with an example. Suppose that the code under consideration is of length three. The probability there are no errors is given by probability bit 1 is error-free × probability bit 2 is error-free × probability bit 3 is error-free = $p \times p \times p = p^3$ since by the binary symmetric channel assumption bit errors are independent of each other. Likewise, there are three ways of having 1 bit in error and the probability of this is given by

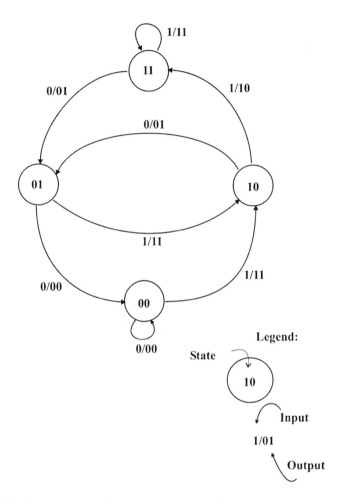

Figure 4.43 Finite state machine for the encoder shown in Figure 4.42.

$$P(1 \text{ bit in error}) = \frac{\begin{array}{l} P(\text{bit 1 wrong and bits 2 and 3 error-free}) \\ +P(\text{bit 2 wrong and bits 1 and 3 error-free}) \\ +P(\text{bit 3 wrong and bits 1 and 2 error-free}) \end{array}}{3} \qquad (4.51)$$

$$= \frac{(1-p) \times p \times p + (1-p) \times p \times p + (1-p) \times p \times p}{3}$$

$$= (1-p)p^2$$

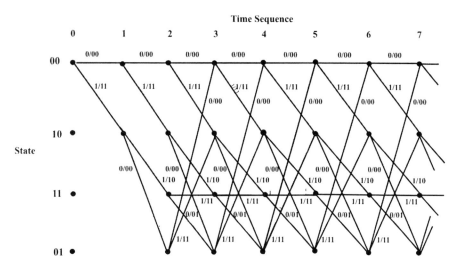

Figure 4.44 Trellis diagram for the encoder in Figure 4.42.

Lastly, the probability of there being 2 bits in error is given by

$$P(\text{2 bits in error}) = \frac{\begin{array}{l}P(\text{bit 1 and bit 2 wrong and bit 3 error-free})\\+P(\text{bit 1 and bit 3 wrong and bit 2 error-free})\\+P(\text{bit 1 error-free and bit 2 wrong and bit 3 wrong})\end{array}}{3} \quad (4.52)$$

$$= \frac{(1-p)\times(1-p)\times p + (1-p)\times(1-p)\times p + (1-p)\times(1-p)\times p}{3}$$

$$= (1-p)^2 p$$

Suppose $p = 0.9$. Then $P(\text{no errors}) = 0.73$, $P(\text{1 bit in error}) = 0.08$ and $P(\text{2 bits in error}) = 0.009$. Therefore, the maximum probability, if there are errors present, corresponds to 1 bit in error—that is, the code word that is a distance of one away from the received word. Thus, the most likely code word is the one with minimum Hamming distance from the code word received.

If the code words are short, it is common practice to compare the received coded words, possibly with errors, to a lookup table to find the word that is the closest to the received word. For long code words such a lookup technique is impractical. The Viterbi decoding algorithm was devised as a more efficient way of doing this comparison. What the Viterbi decoding process does is prune the list over which the search is performed upon receipt of a code word. The most likely code words are kept for a period of time. When a new coded symbol is received, the Hamming distance between the sequence made up of the concatenation of the new symbol with the received symbols and each surviving path through the trellis is calculated. Those sequences with the smallest Hamming distance are kept for the next symbol period, and the process is repeated. After a certain specified number of symbol periods, a decision is made about the received sequence of symbols. The operation of the algorithm is best understood by considering an example. Suppose in the above encoder the depth of code words at the receiver is set at five symbols and the maximum Hamming distance is set at $d = 4$ bits. Then through the first five symbols received, all of the possible sequences are kept. These are shown in Table 4.7 for the example received sequence shown. When the sixth symbol is received, the first symbol is dropped and only those symbols with a Hamming distance less than five are kept. These are shown in Table 4.8. Note that the Hamming distance is the total distance, even though the symbols received older than five are dropped. The total distance is easily calculated since the difference in the last symbol received and the possible symbols from the trellis is simply added to the running total.

In this example, the maximum Hamming distance was set at $d = 4$ in order to have the sequence retained. It may make more sense to adopt the strategy of keeping the sequences with the three smallest Hamming distances, for example, than to set d at a fixed number.

Per-survivor processing. A recent addition to the field of maximum-likelihood decoding convolutional codes is called *per-survivor processing* (PSP) [17]. Depending on the decoding scheme used, most decoders require as inputs some or all of the parameters associated with a received signal. These parameters could include the SNR and the carrier phase. Therefore, they are known, assumed, or estimated in a global sense and then input to the decoding algorithm, such as the aforementioned Viterbi algorithm, for all the paths through the trellis.

In PSP this is not the case. The channel is assumed to be unknown. In fact, the channel can be time varying. Each surviving path through the trellis retains its own estimate of these parameters and they are updated with each new symbol received. It is claimed that significantly better performance ensues from PSP than other conventional types of decoders.

Soft decoding. In the decoders described so far, at some point a decision is declared that a decoded bit is a one or a zero or that a specific sequence has been received. This is called *hard decision decoding*. In maximum-likelihood

Table 4.7 First Five Time Intervals for the Decoder Corresponding to the Encoder Shown in Figure 4.42

Time interval	00	1	2	3	4	5	6	7
Transmit data		1	0	1	1	0	1	0
Encoder state	00	10	01	10	11	01	10	01
Error-free coded sequence		11	01	11	01	01	11	01
Received sequence		11	01	10	01	11		
Surviving code words							Previous state	Current state	Hamming distance

						Previous state	Current state	Hamming distance
	00	00	00	00	00	00	00	7
	00	00	00	00	11	00	10	5
	00	00	00	11	00	10	01	7
	00	00	00	11	10	10	11	6
	00	00	11	00	00	01	00	7
	00	00	11	00	11	01	10	5
	00	00	11	10	01	11	01	7
	00	00	11	10	11	11	11	6
	00	11	10	01	00	01	00	5
	00	11	10	01	11	01	10	3
	00	11	10	11	01	11	01	5
	00	11	10	11	11	11	11	4
	00	11	00	00	00	00	00	7
	00	11	00	00	11	00	10	5
	00	11	00	11	00	10	01	7
	00	11	00	11	10	10	11	6
	11	00	00	00	00	00	00	5
	11	00	00	00	11	00	10	3
	11	00	00	11	00	10	01	5
	11	00	00	11	10	10	11	4
	11	00	11	00	00	01	00	5
	11	00	11	00	11	01	10	3
	11	00	11	10	01	11	01	5
	11	00	11	10	11	11	11	4
	11	10	01	00	00	01	00	7
	11	10	01	00	11	01	10	5
	11	10	01	11	00	10	01	7
	11	01	01	11	10	10	11	4
	11	10	11	01	00	01	00	5
	11	10	11	01	11	01	10	3
	11	10	11	11	01	11	01	5
	11	10	11	11	11	11	11	4

decoding, however, likelihood ratios are computed based on the received bits. These ratios are probabilities with values between zero and one. If these values are output from the decoder and used to make decisions about the received bit as

Table 4.8 Sixth Time Interval for the Encoder Shown in Figure 4.42

Time interval			1	2	3	4	5	6	7		
Transmit data				0	1	1	0	1	0		
Encoder state	00	10	01	10	11	01	10	01			
Error-free coded sequence			01	11	01	01	11	01			
Received sequence			01	10	01	11	11				
									Previous state	Current state	Hamming distance
Surviving code words			11	10	01	11	00		10	01	5
			11	10	01	11	10		10	11	4
			11	10	11	11	01		11	01	5
			11	10	11	11	11		11	11	4
			00	00	00	11	00		10	01	5
			00	00	00	11	10		10	11	4
			00	00	11	10	01		11	01	5
			00	00	11	10	11		11	11	4
			00	11	00	11	00		10	01	5
			00	11	00	11	10		10	11	4
			00	11	10	11	01		11	01	5
			00	11	10	11	11		11	11	4
			01	01	11	10	01		11	01	5
			01	01	11	10	11		11	11	4
			10	11	01	11	00		10	01	5
			10	11	01	11	10		10	11	4
			10	11	11	11	01		11	01	5
			10	11	11	11	11		11	11	4

opposed to having the decoder output a one or zero, it is called *soft decision decoding*.

In hard decoding some of the information available within the decoding process is not made available to the decision processes that decide which bit was received. This information loss results in degraded performance. Soft decision decoding typically improves error performance by about 2 dB. That is, the same BER performance is obtained with 2 dB less SNR.

Concatenated codes. These coding techniques use one code followed by another, called outer and inner coders/decoders. Interleaving is also usually employed to increase the bit error performance. The error detecting and correcting effects are different.

Turbo codes. One of the classic problems with employing coding techniques has been the complexity of the decoder. Code generators are normally simple devices, but as codes get more complex, the decoders become complicated as well.

Turbo codes do not have this difficulty. They use what is referred to as *maximum a posteriori* (MAP) processing. If $P(A)$ is the a priori probability of event A occurring, an event in this case being the transmission of code word A, that is $P(A)$ is the probability of A occurring at all, then Bayes' theorem says

$$P(A|B) = P(B|A)P(A)/P(B) \tag{4.53}$$

$P(A|B)$ is the a posteriori probability of event A occurring given that event B was observed. $P(B|A)$ is called the *likelihood function* and $P(B)$ is called the *evidence*. The MAP criteria selects the A for which the a posteriori probability is maximum.

These codes are also called *parallel-concatenated convolution codes*. The encoder is made up of two or more convolutional encoders, each of which implements a *constituent code*. Interleavers, denoted here by π, interconnect the constituent encoders. The interleavers approximate what is known as a uniform interleaver that perfectly scrambles the input sequence. The code words generated by the encoder consist of the original data input to the first encoder followed by the bits from the constituent encoders, in the appropriate order. Normally the systematic bits from any but the first constituent encoder are not transmitted. An example of a Turbo encoder is shown in Figure 4.45 [18].

The decoder is comprised of one constituent decoder module for each encoder module. The decoders perform soft decision decoding of their input sequences, and the same interleavers used in the encoder interconnect them. It has been reported that BERs of 10^{-6} with $E_b/N_0 = -0.06$ with code rates of 1/15 have been achieved with Turbo code implementations through simulations [19].

Several error-correcting techniques are compared to the Shannon limit in Figure 4.46 [20]. White Gaussian noise is assumed for this data. The *turbo-coded channel* (TPC) is seen to exceed the performance of all the other coding techniques considered and is only about 1 dB or so away from the Shannon limit for this example. It is about 1.5 dB better in performance for most BER $< 10^{-4}$ as compared to the next best code, the *Reed Solomon/Viterbi* (RSV) decoder. The uncoded channel case is also shown as the binary channel (PSK).

The advantages of coding as described above can be seen in Figure 4.47 [21] where the probability of a bit error, or equivalently, the BER, is plotted versus the SNR as specified by the E_b/N_0 for several coding schemes. As shown, very good performance is possible for low SNRs over uncoded data streams.

Radio Communication Technologies 133

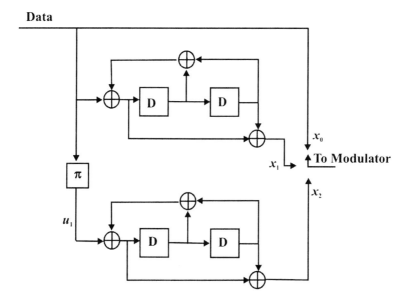

Figure 4.45 An example of a Turbo code encoder.

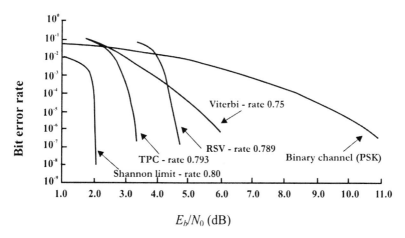

Figure 4.46 Comparison of several coding techniques with the Shannon limit. (*Source:* [20], © 1998, EE Times. Reprinted with permission.)

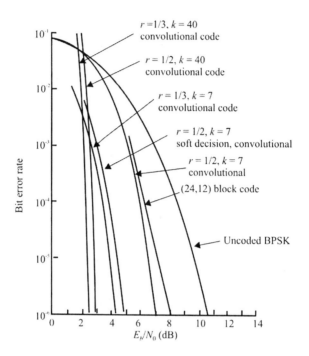

Figure 4.47 BER performance of some codes with BPSK. The BER drops off faster as the SNR is increased if the data is encoded. (*Source:* [21], © 1988, John Wiley & Sons. Reprinted by permission.)

4.4 MODEMS

A modem, a term that is formed from the concatenation of modulator and demodulator, is a device that converts information from the digital domain into the analog domain and back again. Data must be in analog form if it is to be transmitted over much of the existing worldwide telecommunications infrastructure.

Modems are primarily used to communicate between digital devices—principally computers—over analog networks. They are key components of wide area networks, but the data rate they can transmit is relatively slow. Although initially used for wireline communications, they can also be used with radios in wireless networks. Since wireless environments are typically much more error-prone, some form of error detection and correction is necessary. Historically these capabilities have not normally been part of the modem, but are inserted into the data stream before the stream is sent to the modem for transmission. With

Table 4.9 Characteristics of Many of the ITU Modem Recommendations

Recommendation	Bit Rate (bps)	Baud Rate (sps)	Modulation	Coding
V.17	7,200	1,200	QAM	None
V.17	9,600	1,200	QAM	None
V.17	12,000	1,200	QAM	None
V.17	14,400	1,200	QAM	None
V.21	300	300	FSK	None
V.22	1,200	600	PSK	None
V.22bis	1,200	600	QAM	None
V.22bis	2,400	600	2^4QAM	None
V.23	600	600	FSK	None
V23	1,200	1,200	FSK	None
V26	1,200	1,200	BPSK	None
V26	2,400	1,200	2^2PSK	None
V.27bis	2,400	1,200	2^2PSK	None
V.27bis	4,800	1,200	2^3PSK	None
V.27ter	2,400	1,200	2^2PSK	None
V.27ter	4,800	1,600	2^3PSK	None
V.29	4,800	2,400	2^2PSK	None
V.29	7,200	2,400	2^3QAM	None
V.29	9,600	2,400	2^4QAM	None
V32	2,400	2,400	BPSK	None
V.32	4,800	2,400	QPSK	None
V.32	9,600	2,400	2^4QAM	None
V.32	9,600	2,400	2^5QAM	Trellis (2 dim)
V.32bis	4,800	2,400	QAM	Trellis (2 dim)
V.32bis	7,200	2,400	2^5QAM	Trellis (2 dim)
V.32bis	9,600	2,400	QAM	Trellis (2 dim)
V.32bis	12,000	2,400	QAM	Trellis (2 dim)
V.32bis	14,400	2,400	2^6QAM	Trellis (2 dim)
V.32terbo	16,800	2,400	QAM	Trellis (2 dim)
V.32terbo	19,200	2,400	2^9QAM	Trellis (2 dim)
V.33	14,400	2,400	2^7QAM	Trellis (2 dim)
V.34	28,800	2,400–3,249	2^8–2^9QAM	Trellis (4 dim)
V.34bis	33,600	2,400–3,249	2^8–2^9QAM	Trellis (4 dim)
V.90	56,000	N/A	PAM	None

sps = samples per second, bps = bits per second

convolutional coding as represented by Trellis code modulation and Viterbi decoding, that picture is changing.

Standards, or specifications, have been developed so that interoperability between modems from different manufacturers can be accomplished. The *International Telecommunication Union-Telecommunications Standardization Sector* (ITU-TSS), or ITU-T for short, is an agency of the United Nations that has traditionally been the governing body for such standards, by mutual agreement. Such specifications are called recommendations by the ITU. Table 4.9 lists the

Table 4.10 Channel Capacity of a 2,700-Hz Channel

SNR (dB)	Channel Capacity (bps)
5	5,590
10	9,340
20	17,977
30	26,912
40	35,877

recommendations that apply to modem communications. All but the last recommendation transmit digital data with analog techniques over modems. V.90, on the other hand, does not convert the digital data to analog forms. The digital data is sent over the channel in digital form.

Modems built according to the higher data rate specifications are backward-compatible with the lower data rates, and frequently modems can handle several other specifications as well. These specifications provide for training for the channel, which is a form of equalization. Training sequences are used for this, where the data being sent is known to the receiving modem. The received sequence is examined for accuracy and the ISI is determined. If there is too much interference or other degradation, such as noise, the sending modem is notified to lower the data rate. In this way, handshakes between the sender and receiver establishes the highest rate that the channel can support and that is the rate used for the message transfer.

According to Shannon's theorem given earlier (4.1), the maximum capacity of a memoryless voice grade channel corrupted by only Gaussian noise of (guaranteed) 2,700-Hz bandwidth is given in Table 4.10 for various SNRs. The phone system in the United States has a specified (theoretical) noise-limited dynamic range of 39.5 dB, but more typical values are in the 30–35-dB range. Since the guaranteed bandwidth is only 2,700 Hz, data rates should be expected to be 27 Kbps or less. SNRs are typically substantially less than this for RF links, especially in the VHF range.

Because of this, frequently coding schemes are employed in modems to increase the data rate. *Trellis coding* (TCM) is a popular choice. Using TCM, modems have been able to achieve 8 bits per symbol to provide a data rate of 19.2 Kbps at a channel signaling rate of 2.4 Ksymbols per second. This is four times better than GMSK or $\pi/4$ QPSK.

Unencoded QAM at 9,600 bps over a 2,400-baud voice grade channel yields a 14-dB advantage as compared to TCM at the same rate. TCM signals have about the same BER performance at 19,200 bps over this channel.

V.32 at 9,600 bps specifies a convolution code for its TCM which facilitates full-duplex communication over a two-wire circuit. A coding gain of 4 dB is

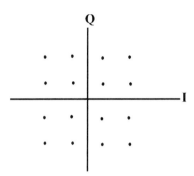

Figure 4.48 Constellation for V.32.

achieved in V.32. Details on the V.32 are presented here to illustrate a typical encoding scheme. The two principal modes in V.32 facilitate either 4,800 bps or 9,600-bps data rates, both at a 2,400 full-duplex baud rate over two wires. The former is for unencoded transmission, while the latter implements TCM. The 16QAM constellation for the 4,800-bps rate is shown in Figure 4.48, while in Figure 4.49 the 2^5QAM constellation for 9,600 bps is shown.

For 2^4QAM, 4 data bits are combined into each one of the constellation points for transmission. For 2^5QAM, obviously 5 bits are needed, but these 5 bits are constructed from the same 4 data bits. The fifth bit is a result of the TCM added

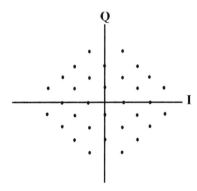

Figure 4.49 Constellation for V.34.

to the data stream for error control. The first 2 data bits, denoted $Q1$ and $Q2$, are first differentially encoded via the formulas

$$Y1_n = Q1_n \oplus Y1_{n-1} \qquad (4.54)$$

$$Y2_n = (Q1_n + Y1_{n-1}) \oplus Y2_{n-1} \oplus Q2_n \qquad (4.55)$$

(n here refers to the bit interval number).

This encoding guards against a 180° phase ambiguity in the channel. The last two data bits, $Q3$ and $Q4$, are not encoded. $Y1$ and $Y2$ are then used to form $Y0$, which carries the FEC information. The formation of $Y0$ is illustrated in Figure 4.50 [22], as is the method for constructing the symbol mapping. $Y0$ is the product of a $r = 2/3$ $K = 3$ convolutional encoder.

The combinational logic in this encoder, as well as other TCM encoders, is designed according to rules defined by set composition on the constellation for decoding. The Euclidean distance between constellation members in any given set is maximized at each step of the decomposition process.

At the demodulator a Viterbi decoding algorithm is used with V.32 as mentioned earlier. This is a convolutional code decoder that minimizes the mean squared error.

In V.34, data rates higher than 28,800 bps are possible. Rates of 33,600 bps and 38,400 bps are theoretically possible, but ultraclean and balanced lines are necessary. When V.42 is combined with V.34, rates up to 115,000 bps are theoretically possible. Normal telephone lines have a bandwidth that is nominally 300–3,000 Hz. Small changes in these frequencies can have significantly deleterious effects on the data rate possible over these lines. The necessary bandwidths for transmitting these high data rates are given in Table 4.11. Modern

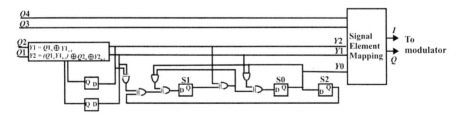

Figure 4.50 One implementation of a circuit to implement the coding for V.32. (Courtesy Texas Instruments. Reprinted with permission.)

Table 4.11 Bit Rates Possible Through Band-Limited PSTN Channels

Symbol Rate	Carrier Frequency	Bandwidth Requirements	Maximum Bit Rate
2,400 sps	1,600 Hz	400–2,800 Hz	21,600 bps
	1,800 Hz	600–3,000 Hz	21,600 bps
2,743 sps	1646 Hz	274–3,018 Hz	24,000 bps
	1829 Hz	457–3,200 Hz	24,000 bps
2,800 sps	1,680 Hz	280–3,080 Hz	24,000 bps
	1,867 Hz	467–3,267 Hz	24,000 bps
3,000 sps	1,800 Hz	300–3,300 Hz	26,400 bps
		500–3,500 Hz	26,400 bps
		375–3,376 Hz	26,400 bps
3,200 sps	1,829 Hz	229–3,429 Hz	28,800 bps
	1,920 Hz	320–3,520 Hz	28,800 bps
3,429 sps	1,959 Hz	244–3,674 Hz	28,800 bps

sps = samples per second, bps = bits per second

modems can adjust these bandwidths to be compatible with the actual PSTN lines being used by exchanging training sequences that have a known bit pattern to be measured as well as by *probing* the channel. Probing is transmitting and measuring the response of the channel at known frequencies and levels. ISDN is expected to change that picture dramatically since no modems are required—the communication is digital in origin and remains that way throughout the path. Interface devices will be required, however. While the modem recommendations discussed above are not inextricably connected to the Internet, the V.90 recommendation was inspired by it. A modem modulates and demodulates low-frequency carriers with digital data using some form of QAM. This inherently limits their speed to less than about 30 Kbps due to quantization noise in the A/D conversions involved. The downstream leg of the V.90 is not modulated this way. It is sent in digital form, as if the PSTN were a digital network. The information is *pulse amplitude modulated* (PAM) onto the line in the downstream direction, from the service provider to the customer's premises. In this way, depending on the quality of the phone line, rates up to 56 Kbps are possible. Since the majority of high-speed traffic on the Internet is downstream, with the upstream normally consisting of slow keystrokes, for example, substantially higher throughput over the Internet is possible with V.90. The analog upstream leg in V.90 is V.34.

In fact, the vast majority of the PSTN in the United States is digital. In most populated places the only remaining analog portions are on the local loop. This is the part of the network that connects the local office with the residences and businesses. The remainder of the PSTN is digital. This statement does not necessarily apply in countries other than the United States, however.

4.5 FACSIMILE

Facsimile (fax) communication is the transfer of images—typically typed pages—over analog communication systems. Group I fax is an obsolete standard where either AM or FM was used as the modulation scheme. In the former, the blacker the pixel, the higher the amplitude of the carrier, while in the latter, the blacker the pixel, the higher the modulating tone. Group II fax is also an old standard that is almost obsolete. It uses vestigial sideband AM as its modulation scheme, and the whiter the pixel, the higher the tone that was sent. These standards have been almost totally replaced by Group III fax.

Run-length encoding is used in the Group III fax protocol. It takes each line of pixels and run-length encodes them before transmission. This is one-dimensional encoding in that each line is encoded, as opposed to the whole page. Two-dimensional encoding encodes the first line with run-length encoding as just described, but subsequent lines are encoded as differences from the previous line. Significant compression can result since there is frequently quite a bit of correlation between lines of text on pages. The specification is error-detecting but not error-correcting; therefore, errors can propagate down the page this way. Because of that, a limit of two lines is encoded this way for standard-resolution faxes and four for high-resolution faxes.

The Group IV fax standard covers facsimile transmissions over ISDN at 64 Kbps. In this case the compression is the same as just described for two-dimensional compression except that there is no limit on the number of lines that are encoded based on the first line.

Within each of these groups there are classes defined, which essentially describes the amount of processing that a modem must do to transmit and receive facsimile messages. In class 1 most of the work must be done by the computer, and the fax modem work is limited to modulation of the carrier and asynchronous-to-synchronous conversions. In the class 2 standard, the modem is smarter and must do more of the processing. In particular the modem is responsible for modulation and asynchronous-to-synchronous conversion, as well as taking care of the fax protocol. The computer is responsible only for the compression of the image (as well as overall management, of course).

4.6 COMMUNICATION SECURITY

One of the most significant issues associated with information transport is the security, or privacy, of the information. Current, unclassified techniques to deal with this problem are discussed in this section.

4.6.1 Data Encryption

Encryption is perhaps the most common way to protect information that is to be transported from one place to another. Information is encrypted when a mathematical operation is performed on it to make it unreadable by unauthorized persons. Cryptography, the study of encryption techniques, is an electronic protect measure for the benefit of whoever uses it. It refers to the process of transforming information from its current state to another where the content is hidden, and from which, with the proper knowledge of such components as encryption keys, the original content can be retrieved and without which the data cannot be retrieved. Any information in any state can be encrypted. The most well known perhaps is the encryption of written or radio communication. Other types include encrypting data on a disk of a PC and parents spelling words that their 2-year-old child cannot understand. Sir Arthur Conan Doyle enabled Sherlock Holmes with a special skill in dealing with cryptography.

Historically, this is an area that has been dominated by the government, but recently the private sector has become more interested. Techniques have been devised to protect information, while making it relatively easy to move the encryption keys among legitimate communication nodes.

Authentication is the process of verifying the identity of the generator of some data and, therefore, the integrity of that data. A *principal* is the party whose identity has been identified. A *verifier* is the person who wants the verification. *Data integrity* means that there is assurance that the data received is the same as the data generated. *Confidentiality* refers to protection from disclosure of information to those unauthorized to receive it. *Authorization* is the process used to ensure that a principal may perform an operation.

User nonrepudiation refers to the inability of the source of the data to deny that the data was generated by that source. That is, if the process is correct and it identifies a particular source of the data, then certainly that is the source of the data.

4.6.2 Public Key Encryption

In public key encryption, each entity i has a public key U_i, known to everyone, and a private key R_i, known only to the entity. A message to i is encrypted by using U_i while R_i is used by i to decrypt the message and, if encrypted with a user's public key, it can only be decrypted with the user's private key.

Public key encryption is *asymmetric encryption* since each end of a link uses a different cryptologic variable to encrypt and decrypt data. This is as opposed to *symmetric encryption* where the same cryptologic variable is used at both ends.

The excerpt in Figure 4.51, taken from [23], describes by example how this encryption technique works.

Because of the extensive (and therefore slow) computations involved in using public key cryptography, most of the time it is used to exchange session cryptologic keys so that communications can proceed in a symmetric manner, which is much faster. It is also used for exchanging digital signatures.

Session keys are the equivalent to what is called the one-time-pad. Such a scheme uses a crypto key only once (for the current session). After that, it is discarded. Such techniques are virtually impossible to decrypt since the keys are only used once.

4.6.3 Digital Signatures

A security issue that has arisen because of the mass movement toward electronic commerce (for example, retail sales over the Internet) is how one knows that the person at the other end of a transaction is who he or she claims to be. In the old analog world of plastic credit cards and bank checks, it was a matter of checking the signature on the card or check and the picture on an identification. These are not as yet practical for use over the Internet.

Public key cryptology can be used to facilitate such checking, however. The approach takes the appellation *digital signatures*. In order to understand how they work it is necessary to understand the functioning of hashing, which is a process of taking one number (A) and converting it into another (B), where the inverse (knowing B, find A) is next to impossible. The two most popular hashing algorithms are ones invented at MIT by Ron Rivest, called MD5, and one invented by NIST and NSA called the *secure hash algorithm* (SHA). The former generates a 128-bit number and the latter a 160-bit number. The larger the number of bits, the more secure the data.

A message that is to get through without being changed—or if it is changed, the change is detected at the receiving end—can be sent by hashing the contents of the message, encrypting this hashed message with a public key, and sending the encrypted hashed value along with the message. The receiver decrypts the hashed value with the sender's private key, hashes the message with the same algorithm that the sender used, and compares the hash value sent with the one calculated at the receiver's location. If they match, then the receiver is assured that the message arrived unchanged, while if they do not match, it can be virtually assured that the message did not come from the sender because of the uniqueness of the hashed value. The hash value is the digital signature since it is the only number that can be generated from the document; and furthermore, the encryption/decryption process only works accurately if the public and private keys are correct—if they are not, the hash value will decrypt incorrectly.

Encryption 101
Public-key cryptography uses different keys for encrypting and decrypting, which are created as shown below, and then assigned to each user. Whenever someone wants to send a message, he looks up the recipient's public key and uses it to encrypt the message. The recipient then uses his private secret key to decrypt the message. For simplicity's sake the example below uses small prime numbers; in practice, the keys are very large numbers, making it difficult for outsiders to factor N and decipher the message. Both sender and receiver must have the same encrypting algorithm in hardware or software in their computer. In this example, adapted from the US Office of Technology Assessment's 1987 report Defending Secrets, Sharing Data, the algorithm is based on the public-key RSA algorithm, named after its three Massachusetts Institute of Technology inventors—Ronald Rivest, Adi Shamir, and Leonard Adelman.

Creating the public key
1. Pick an odd number, E. $E = 5$
2. Pick two prime numbers, P and Q,
where (P-1)(Q-1)-1 is evenly
divisible by E. $P = 7, Q = 17$
3. Multiply P and Q to get N. $N = P \times Q = 7 \times 17 = 119$
4. Concatenate N and E to get the
encrypting or public key.   Public Key = NE = 1195

Creating the private key
1. Subtract 1 from P, Q, and E,
multiply the results, and add 1 $(P-1)(Q-1)(E-1)+1 = 6 \times 16 \times 4 + 1 = 385$
2. Divide result by E to get D. $D = 385/5 = 77$
3. Concatenate N and D to get the
decrypting or private key.   Private Key = ND = 11977

Encrypting the message with the public key
1. The message is converted
to numerical equivalents. The letter
S, for example, may be
represented by 19.   Plain Text = 19
2. The algorithm:
   a) raise plain text to power of E. $19^5 = 2476099$
   b) divide by N $2476099/119 = 20807$ with
   a remainder of 66
3. The remainder is the encrypted value
or cipher text.   Cipher Text = 66

Decrypting the cipher text with the private key
1. The algorithm:
   a) raise cipher text to power of D. $66^{77} = 1.27 \ldots E140$
   b) Divide by N. $1.27 \ldots E140 / 119 = 1.069 \ldots E138$
   with a remainder of 19
2. The remainder is the decrypted
value or plain text.   Plain Text = 19

Figure 4.51 Example of RSA public key encryption algorithm. (*Source:* [23], © 1989, IEEE. Reprinted with permission.)

4.6.4 Data Encryption Standard

One of the ways to attack an encryption system is by brute force. Encryption algorithms are not normally kept secret and the reason for this is simple. If it is necessary to keep the algorithm secret then the algorithm itself is vulnerable—there is something about the algorithm that must be kept from would-be interlopers. Such weaknesses are characteristic of poor encryption systems. It is more normal to make the encryption key secret, but not the algorithm that uses the key. Brute force attacks on an encryption system are conducted using computing power, and every possible key combination is tried until one is found that yields the original message, or at least a message that makes sense.

One might ask how it is known if the original message is obtained or some other message, unless one already knows the original message. While this is an interesting paradox, it is normally the case that anything but the original message will be garbled and senseless, so it is obvious that one has obtained the original message.

The longer a key is, the safer the encryption is against a brute force attack. To see why this is true, if only eight bits are used for encryption, then $2^8 = 256$ different values of the encryption key would have to be tried in order to find the correct key. If one value took 1 ms to try, then it would only take 256 ms (half this value on average) to break the code. On the other hand, if a key is 100 bits long, then $2^{100}/2$ different values would have to be tried. On the same computer, taking 1-ms per try, it would take on the order of 10^{19} years to break the cipher.

Parallel processing can be implemented that substantially reduces the equivalent of the 1-ms per try, however. Also, the 1 ms used in this example is much slower than what high-performance computers can do. Using realistic times and large numbers of parallel computers, cipher systems based on 100 bits can probably be broken by brute force in reasonable times, if the end result is judged to be worth the expense.

The *data encryption standard* (DES) was adopted in 1977 as the standard technique in the U.S. government for encrypting sensitive but unclassified information. [The U.S. *Department of Defense* (DoD), for example, does not use DES for classified data. It was intended to keep an individual's information private, including corporations.] It was made available to the private sector as a means for providing a reasonably secure method for protecting sensitive information. It uses a 56-bit key, which, at the time of its adoption, was a considerable amount of protection against the then-existing and projected computer speeds. This provided 72,057,594,037,927,936 possible different keys.

DES uses a *block encryption* technique as opposed to a *stream cipher*. This means that blocks of data 64 bits in size are encrypted at the same time. In a stream cipher 1 bit at a time is enciphered. *Confusion* and *diffusion* are the techniques at the heart of the algorithm, and the data is operated on 16 times by these two, which is determined by the key. The notion of diffusion is to spread the plaintext over the ciphertext, so that the inherent redundancy in the plaintext (assumed to be there, and if the plaintext is English text, for example, there is considerable redundancy) is hidden. A simple example of this is transposition of the characters in the plaintext. Confusion is the process of disestablishing the relationship of the plaintext from the ciphertext. A simple way to do this is by substitution. It is said that this technique was used by Julius Caesar, by shifting the alphabet by three places and substituting the characters accordingly. As mentioned, at each round through the encipherment process, confusions and diffusions are performed

Another important part of the algorithm is permutations of the data, so that the *avalanche criteria* is met. This ensures that 1 bit of data affects at least two substitutions. In this way, if the ciphertext is changed in only one place, the decrypted plaintext is dramatically different from the preencrypted plaintext. The actual security of the algorithm is accomplished at a nonlinear processing stage imbedded in the algorithm.

The way such keys are used is typically to exclusive OR data bits with the bits in the encryption key. In fact, the processing is more complex than this. A series of shifts and permutations are used that, being known to both the sender and legitimate recipient, can be undone properly. Since the key is unknown to a third party, the resultant data stream looks random.

DES was recertified in 1993 by NIST, but its days were limited since it was vulnerable to attack from modern parallel computers. The coding was broken in 1998. Such brute force attacks are feasible if the end results are worth the effort. International banking is an example of this. A single success out of many tries can yield huge sums of money transferred to a wrong account.

Encryption is frequently viewed as too difficult to depend on for secure, tactical communications. If it has not already, this will change in the future primarily because much of the communication—tactical and otherwise—is becoming digital. Signals that start out as digital are easier to encrypt—modern encryption techniques for electronic signaling normally involve converting analog signals to digital before encryption. The advent of widespread dependencies in the commercial world on private data communications will cause the shortcomings in encryption to be quickly overcome.

4.6.5 Pretty Good Privacy

In 1991, Phillip Zimmerman developed *pretty good privacy* (PGP). It is an implementation of several of the above concepts for encryption as an alternative to DES. Public and private asymmetric keys are used to set up sessions, where a new session key is exchanged for each new conversation. After establishing the session key, encryption proceeds symmetrically. PGP uses encryption techniques based on the RSA algorithms, although some latest algorithms also support Diffie-Hellman key formats. It also implements digital signatures. The keys can be very long, with current implantations in the 4,000-bit range. There is no reason to limit the key size, however, and if they are needed, longer keys will be implemented.

4.6.6 Fortezza Encryption System

It has long been recognized that certain of the information within the DoD was not classified but sensitive in that it is undesirable for nonprivileged users to have access to it. Logistics information is an example. The amount of ammunition moving into a theater of operations on a daily basis could be used to indicate plans for an impending offensive operation, for example. This type of information is normally treated as unclassified but it would be beneficial to not have it known to the adversary. Commerce and industry have equivalent requirements for protecting information from disclosure to unauthorized sources.

To address this requirement the *National Security Agency* (NSA) established the Fortezza program. Like most cryptologic programs, there are two fundamental parts—(1) the hardware and associated software to perform the function and (2) the management hierarchy that deals with the cryptologic keys and how they are managed and accounted for.

The hardware development resulted in a PCMCIA configuration for the implementation, although there is no particular hardware constraint in general that limits this. The PCMCIA architecture has been an industry standard, particularly for small computers, and implementation of the Fortezza system in this configuration would allow for better migration to the maximum possible number of users.

The key management structure has been implemented so that any authorized person or organization can obtain a key for designated authorities. Lists are maintained that are readily available for the holders of these keys.

The Fortezza system implements a two-tiered cryptologic system. Both an asymmetric and symmetric approach are possible within the Fortezza system. In the former, different cryptologic keys are used at either end of the link whereas in the latter the same key is used. The cryptologic keys, which are used to encrypt the message traffic, are first transferred between the two ends of the link using the asymmetric public key encryption scheme described above. Once the cryptologic keys are exchanged, this key is used to encrypt the message traffic symmetrically because of the extensive overhead to perform the calculations for the public key encryption.

The *digital signature standard* (DSS) is based on the *digital signature algorithm* (DSA). It is built into the Fortezza PCMCIA chips.

4.6.7 Escrow Encryption System

A notion for telephone security is the key escrow system proposed in the United States. One implementation of this puts a device unique key, which is part of a chip, called the *clipper chip*, into telephones after manufacture, but prior to going into operation. When put into operation, the key is split into two parts. Each part of the key is given to two different escrow holders. This approach was initially developed to encrypt voice as might be transmitted over telephones, but it is extendable to data communications as well.

Key escrow systems were proposed to allow the government access to private communications as allowed by wiretapping under the law. Without the ability to perform wiretaps, it is said that illegal activity could be hidden that is now accessible to law enforcement. There is considerable controversy about whether the government should be allowed control over encrypted private communications. In this standard there is a *law enforcement access field* (LEAF), which allows legal wiretaps of such encrypted communication. Key escrow systems proposed in the United States use the *escrow encryption standard* (EES).

The purpose of EES is to preclude unauthorized parties from listening to telephone conversations. Each clipper chip is assigned a unique identification number and key. This information is retained by the government, with the encryption key split into two halves. One government agency holds half the key and another holds the other half.

In reality there are three encryption keys involved with every interaction. The session key is unique to each interaction and is generated and used one time in a symmetric mode. The manufacturers of the phones are free to choose the method of generating the session key—it is not part of the EES. The second key is the aforementioned key unique to each clipper chip. It is used to encrypt the session key. The encrypted session key along with the unique identification number of the phone and a computed checksum are then encrypted with the family key, which is common to all clipper chips. This is then sent to the recipient.

The LEAF is sent at the beginning of an EES conversation. If a valid LEAF is not received by the recipient, then conversation is not allowed. The LEAF is 128 bits long. Each chip identification is 32 bits long, and the session key is 80 bits long, while the checksum is 16 bits.

The purpose of the EES is to facilitate legitimate wiretaps, authorized by suitable legal officials (court judges, as wiretaps in the United States are authorized today). Such a wiretap would begin by recording the conversation, including the LEAF. Next the LEAF would be decrypted using the family key, which yields the clipper chip identification, the checksum, and the encrypted session key. Based on the chip identification thus obtained, the two halves of the chip key would be obtained from the holding organizations. Using these two halves the session key would be decrypted leading to the ability to decrypt the conversation.

The *National Institute of Science and Technology* (NIST), is proposed as one of the two escrow agents. The other is proposed to be the Department of Treasury Automated Systems Division.

4.6.8 Over-the-Air Rekeying

One of the difficult problems with any encryption system is rekeying the encryption devices. Large management structures have been established in the past to perform this function. Rekeying is typically performed by personnel that hand carry the new keying material to the encryption device and load the new key. This, of course, is time-consuming and expensive. It has been tolerated by the military because of its necessity to ensure the necessary security.

It is possible, however, to rekey remote devices over the air from a central location, but there are problems associated with that as well, one of which is how one insures oneself that the remote encryption device is in the hands of whom it is supposed to be. Therefore protections must be built into the keying system in the event of unusual occurrences.

4.7 CONCLUDING REMARKS

Several of the more common techniques for communication systems are presented in this chapter. Most of the technologies used by military forces throughout the world are included.

The older schemes of analog communications are rapidly being replaced with digital communication systems. This is because the latter are more spectrally efficient than analog forms and therefore the precious RF spectrum can be used to support more communications. In situations where fiber can be emplaced to facilitate communications, such as in infrastructure-PSTN systems, that is the media of choice since many gigabits per second can be transported over fiber. In many cases fiber or cable cannot be used. Mobile military forces are an example of this as is the situation where the landscape is very hilly and wire cannot be economically put in place. In those situations RF communications over the air are necessary.

In an information-based society, which most of the developed countries in the world are, the respective military forces become information-based as well. In these situations communication of information is critical. Even in tactical situations, data must be exchanged in increasing volumes that require higher data rates.

The demand for secure communications, primarily driven by e-commerce over the Internet, is forcing encryption technology to become more reliable and simple to use. Modern encryption technology is becoming widely available worldwide. This may force communication EW systems to change their modes of operation in order to provide information from such systems.

References

[1] Torrieri, D. J., *Principles of Secure Communication Systems*, 2nd ed., Norwood, MA: Artech House, 1992.
[2] Stuber, G. L., *Principles of Mobile Communication,* Boston, MA: Kluwer Academic Publishers, 1996.
[3] Simon, M. K., S. M. Hinedi, and W. C. Lindsey, *Digital Communication Techniques: Signal Design and Detection,* Upper Saddle River, NJ: Prentice Hall, 1995.
[4] Gagliardi, R. M., *Introduction to Communication Engineering,* 2nd ed., New York: John Wiley & Sons, 1988.
[5] Proakis, J. G., *Digital Communications,* 2nd ed., New York: McGraw-Hill, 1989.
[6] Das, J., S. K. Mullick, and P. K. Chatterlee, *Principles of Digital Communication: Signal Representation, Detection, Estimation, and Information Coding,* New York: John Wiley & Sons, 1986.
[7] Kohno, R., R. Meidan, and L. B. Milstein, "Spread Spectrum Access Methods for Wireless Communications," *IEEE Communications Magazine,* January 1995, pp. 58–67.
[8] Xiong, F., "Modern Techniques in Satellite Communications," *IEEE Communications Magazine,* August 1994, pp. 84–98.
[9] Xiong, F., "Modern Techniques in Satellite Communications," *IEEE Communications Magazine,* August 1994, p. 89.
[10] Xiong, F., "Modern Techniques in Satellite Communications," *IEEE Communications Magazine,* August 1994, p. 88.
[11] Lathi, B. P., *Communication Systems,* New York: John Wiley & Sons, 1968, p. 228.
[12] Gagliardi, R. M., *Introduction to Communication Engineering,* 2nd ed., New York: John Wiley & Sons, 1988, p. 336.
[13] Simon, M. K., et al., *Spread Spectrum Communications, Volume 1,* Rockville, MD: Computer Science Press, p. 143.
[14] Proakis, J. G., *Digital Communications,* 3rd ed., New York: McGraw-Hill, 1995, p. 735.
[15] Anderson, T., "HARC: The Newest Wave in Image Compression," *Technology Transfer Business,* Fall 1995, p. 73
[16] Das, J., S. K. Mullick, and P. K. Chatterlee, *Principles of Digital Communication: Signal Representation, Detection, Estimation, and Information Coding,* New York: John Wiley & Sons, 1986, p. 505.
[17] Polydoros, A., and G. Paparisto, "Per-Survivor Processing for Joint Data/Channel Estimation in Multipath Fading and Co-Channel Interference Channels," Communication Sciences Institute, University of Southern California, Electrical Engineering Systems, Los Angeles, CA, 1996.

[18] Divsalar, D., and F. Pollara, "Multiple Turbo Codes," Jet Propulsion Laboratory, California Institute of Technology, Pasadena, CA, 1995, Figure 1.

[19] Divsalar, D., and F. Pollara, "Turbo Trellis Coded Modulation with Interative Decoding for Mobile Satellite Communications," Jet Propulsion Laboratory, California Institute of Technology, Pasadena, CA, 1998.

[20] Thompson, B., "Error Codes Widen Design Window," accessed July 2001, http://www.eetimes.com/story/OEG19981231S0003.

[21] Gagliardi, R. M., *Introduction to Communication Engineering,* 2nd ed., New York: John Wiley & Sons, 1988, p. 370.

[22] Chishtie, M. S., "Viterbi Implementation on the TMS320C5x for V.32 Modems," accessed July 2001, http://www-s.ti.com/sc/paheets/spra000/spra099/pdf.

[23] Fitzgerald, P., "The Quest for Intruder-Proof Computer Systems," *IEEE Spectrum,* July 1989, p. 25.

Chapter 5

System Engineering

5.1 INTRODUCTION

This chapter introduces the notion of system engineering, in the context of treating the system as a whole. "System" in this case is more than just the hardware manifestation normally thought of when the term is used. This will be discussed more fully in Section 5.2. Suffice it to say that the term system includes all aspects of designing, building, and sustaining the system.

The design of any system, including EW systems, involves many disciplines. Introductions to some of the more important of these disciplines, including system reliability, fault tolerance, environmental concerns, *EM interference* (EMI) and *EM compatibility* (EMC), human factors, and safety, are presented in this chapter. Each of these topics could require a book in themselves to cover adequately, so what is included here are introductions to acquaint the reader with the topics at a very high level—sufficient to be able to address the topics at a cursory level if required.

5.2 SYSTEM ENGINEERING

Kayton defines system engineering as consisting of the following elements [1]:

1. Translate an operational need into a system;
2. Integrate all technical disciplines;
3. Ensure functional and physical interfaces;
4. Identify and abate risks;
5. Verify that the design meets the need.

The aspects involved with every system are shown in Figure 5.1 [2]. Note that only two of the first level blocks refer to the system itself—the hardware, for

152   Introduction to Communication Electronic Warfare Systems

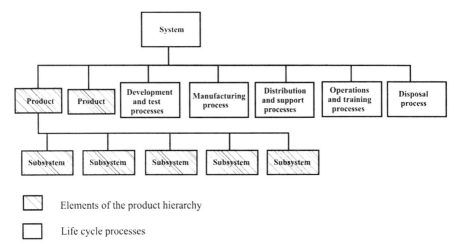

Figure 5.1 Functional areas that need to be considered in the design and use of any system. (*Source:* [2], © 1998, IEEE. Reprinted with permission.)

example. The remainder refers to auxiliary activities affecting the life cycle of the system. The life cycle of a system starts with the research and development phase and ends with the disposal phase, where it is removed from useful service. Manufacturing, sales, distribution, training, and operation are all phases between these. The system engineering process must address all these phases and how they interrelate with one another for any useful system.

The life cycle of a system is notionally illustrated in Figure 5.2 from IEEE Std. 1220-1998. It begins with the system definition, based on satisfying some requirement. The life cycle proceeds into subsystem definition, which is followed by system production and customer support stages, which overlap. The subsystem definition consists of the three steps shown in Figure 5.2: preliminary design followed by detailed design, which is then followed by the *first article integration and test* (FAIT) phase.

The *system engineering process* (SEP), according to IEEE Standard 1220-1998, follows the flow diagram shown in Figure 5.3. The steps indicated in this

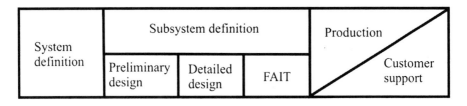

Figure 5.2 Life cycle of a system. (*Source:* [2], © 1998, IEEE. Reprinted with permission.)

process are repeated for each and every action associated with system engineering—all the way from designing the overall system itself, to analyzing incremental changes to the system.

The first step is to analyze the requirements that the system is to address. Tradeoff analyses are performed on the requirements, as there are frequently conflicts in these requirements—they cannot all be met. A requirements baseline is thus established. These requirements are then verified and validated, usually with a customer or user of the system. The functions of the system are then defined with further trade studies and with subsequent verification and validation. Functions are allocated to subsystems, to include software versus hardware allocations. Based on these allocations, the synthesis process begins to actually design and build the physical system, again to include hardware and software. These designs and physical manifestations are then verified, usually to include substantial testing. Once the functions are verified in the physical system, the process is either iterated for design refinement, where feedback can go to any of the previous steps, or the process terminates with a completed system.

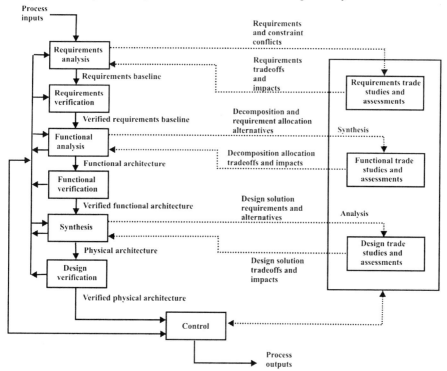

Figure 5.3 Flow diagram illustrating the system engineering process. (*Source:* [2], © 1998, IEEE. Reprinted with permission.)

The early stages in the development and deployment of any system involve system design, which primarily consists of performing tradeoff analyses to ascertain the best system design. Again, Kayton indicates that these tradeoffs almost always boil down to deciding between the characteristics of "performance, reliability and availability, human convenience, and cost." System engineering considerations described below will be described following these basic categories.

Perhaps the most important output of the initial system engineering process is the system specification. This specification, according to IEEE Standard 1220-1998, consists of the elements shown in Figure 5.4. The overall system specification consists of the three parts shown: the product specification, the operational procedures used with the system, and the human aspects consisting of the manpower, personnel, and training specification. The product specification is subdivided as shown in Figure 5.4, eventually consisting of specifications at the component level.

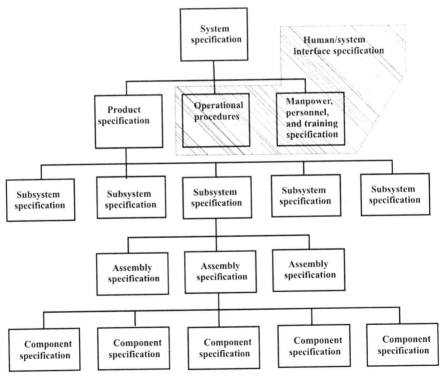

Figure 5.4 The specification tree. (*Source:* [2], © 1998, IEEE. Reprinted with permission.)

With the advent of very highly integrated semiconductor chips, the system engineering process is changing. There is and will be a great deal of emphasis on simulation and modeling in order to establish the system specification. Codesign is the cornerstone to this new philosophy, where all aspects of system design occur simultaneously.

5.2.1 Performance Characteristics

The other chapters of this book discuss the performance characteristics more or less unique to communication EW systems. This chapter discusses some of the system design considerations that apply to systems in general, not just communication EW systems. They thus deal with performance characteristics at the whole system level, as opposed to specific subsystems or subcapabilities.

5.2.2 Environmental Characteristics

Shown in Table 5.1 are the environmental specifications for computer and electronic equipment from IEEE Standard 1156.2-1996 for the case when the equipment is nonoperating. Environmental conditions for tactical military equipment are typically more severe than those for commercial equipment, although most of the areas of concern are the same. The conditions shown in Table 5.1 are commercial specifications so they will apply to those communication EW systems that are built for this environment. PL1 refers to equipment subject to moderate amounts of the condition in question. PL2 refers to the same equipment subjected to minimal amounts. If the unit passes the test, then it can be expected to be able to survive the environmental conditions specified. The specific conditions are shown in the left column of Table 5.1. Table 5.1 contains the conditions for testing for the condition indicated.

Low- and high-temperature environments are concerned with the shipment and storage conditions, after which the equipment will still work. Frequently equipment is shipped and/or stored at temperature extremes because the shipping/storage containers are not temperature controlled. The test is of such duration to simulate a cold or hot soaking, so that the equipment is settled at that temperature.

Electronic equipment can be put into service at a moment's notice from extremely cold conditions. The purpose of the thermal shock test is to verify that the equipment will operate after such a transition. Note that the test conditions go from $-40°C$ to $+65°C$ in a period of only 5 minutes.

Some military equipment must be designed to operate in high-humidity environments such as Central American jungles. Thus, the requirement for military equipment can be more severe than commercial equipment because of this. The humidity conditions in the table are relatively benign.

Table 5.1 Nonoperating Conditions

Test Parameter	Test Publication	Severity or Conditions	Requirements	
			PL1	PL2
Low temperature	IEC 68-2-1 (1990) Test Ab	24 hours	-40° C	-60° C
High temperature	IEC 68-2-2 (1974) Test Bb	24 hours	+65° C	+55° C
Thermal shock	IEC 68-2-14 (1984) Test Na	< 5 min	-40° C to +65° C	N/A
Humidity	IEC 68-2-30 (1980) Test Dd	6 cycles	25° – 55° C 92 ± 3% RH	25° – 55° C 92 ± 3% RH
Free-fall packaged	IEC 68-2-32 (1975)	Corners and faces	Combinations	Faces only
Handling unpackaged	IEC 68-2-32 (1975)	Corners and faces	Combinations	Combinations
Vibration	IEC 68-2-6 (1975)	5–200/300 Hz (see Figure 5.5)		
Shock	IEC 68-2-27 (1987)		30g/20 ms	15g/20 ms
Mixed flowing gas	ASMT 8-827-92		14 days	7 days
Fungus	IEC 68-2-10 (1988) Test J		X	N/A
Flammability	System test		ANSE T1319-1995	UL 94-1991
Electrostatic discharge	IEC 801-2 (1991)	15 kV air discharge	X	X

Source: [3], © 1996, IEEE. Reprinted with permission.

The free-fall tests are designed to evaluate whether the equipment can be dropped under normal handling conditions. Such conditions could be as simple as removing the equipment from the shipping carton and installing it in a rack. The distance the equipment must fall depends on its weight.

The amount of shock and vibration a system must tolerate is dictated to a large extent by the operational environment in which that system is used. For example, a radio intended to operate in a tank must be designed to tolerate the mechanical environment of that tank, which can be quite harsh. On the other hand, a desktop computer designed to never leave the desktop (except for repairs) survives in a benign environment and shipping and handling procedures would probably dictate the amount of shock and vibration it must tolerate.

Shock and vibration are two fundamentally different parameters. Shock refers to the relatively infrequent sharp impulse of mechanical energy coupled into a system from the likes of dropping an electronic enclosure onto a concrete floor. Values of such shock can reach 50g or more, where 1g = the force of gravity. Vibration, on the other hand, refers to the normally repetitive mechanical

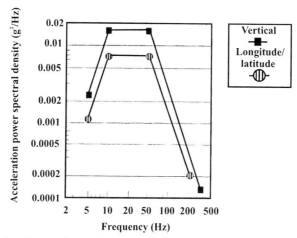

Figure 5.5 Vibration characteristics when a system is nonoperating. (*Source:* [3], © 1996, IEEE. Reprinted with permission.)

oscillation to which a device can be subject. An example of this might be the vibration induced into a system by the rotation of helicopter blades.

Mechanical stability is normally a design requirement whether the EW system is strategically deployed, never to be moved, or tactically designed, intended to be moved frequently. The latter can be over benign traveling conditions or rough, battlefield conditions.

The spectrum of the vibration specified in IEEE Standard 1156.2-1996 is shown in Figure 5.5 for nonoperating equipment. The highest acceleration, and therefore the highest vibration amplitude, is in the frequency range of 10–50 Hz.

Equipment deployed to humid regions frequently encounter conditions that support the growth of fungi. The fungus test will verify that the equipment does not degrade under these conditions. It should be noted that equipment subject to this testing is frequently not usable afterward, for the most part, because of the debilitating effects of the fungus growth.

Flammability is of concern especially for electronic equipment. Such equipment can be expected to operate in hot conditions, and if the components were prone to burn, significant safety issues arise, not to mention whether the equipment will continue to operate. In a military setting, equipment can be exposed to conditions that could promote burning, such as when under hostile fire.

The mixed flowing gas test is a simulation of the effects of being exposed to gases to evaluate metal corrosion. The gaseous test exposes metal components to corrosive gases for a period of time during which thermal cycling is performed. Typical test conditions consist of 25 or more thermal cycles in gaseous mixtures consisting of the low parts per billion of water, nitrogen oxide, and chlorine.

Table 5.2 Operating Conditions

Test Parameter	Test Publication	Severity or Conditions	Requirements PL1	Requirements PL2
Low temperature	IEC 68-2-1 (1990) Test Ad	4 hours	+5°C	+10°C
High temperature	IEC 68-2-2 (1974) Test Bd	4 hours	+50°C	+35°C
Sine vibration	IEC 68-2-6 (1995) Test Fc	5–100 Hz	0.5 mm/0.1g	0.5mm/0.1g
Shock	IEC 68-2-27 (1987)	3 pulses	15g/11 ms	10g/11 ms
Low air pressure	IEC 68-2-13 (1983)	6 hours	10.5–5.5 Pa	10.5–7.85 Pa
Electrostatic discharge	IEC 801-2 (1991)	15 kV/150 pF	X	X
Electromagnetic induction	FCC rules part 15 IEC 801-3 (1984)	Meet	X	X
Immunity	IEC 801-4 (1988)	dc power	X	X
Immunity	IEC 801-4 (1988)	ac power	X	X
Immunity	IEC 801-4 (1988)	I/O ports	X	X
Earthquake	IEC 68-2-57 (1989)		X	N/A

Source: [3], © 1996, IEEE. Reprinted with permission.

Electrostatic discharge testing determines whether the equipment will withstand the typical electrical discharge that occurs when one walks across a carpet and touches something metallic. Electronic equipment designed for human interaction, including operation and/or repair, will encounter such discharge.

Salt fog testing is not included in Table 5.1, but is frequently a requirement of military equipment. It is not so much of a concern to commercial equipment. Military deployments typically involve shipment of electronic equipment, either the system itself, or components of it, say, spare parts, by sea. This environment frequently consists of salt in very humid air.

Similar specifications for electronic equipment under operating conditions are shown in Table 5.2. A few of the conditions are different, there are a few additional tests, and some of the tests have been deleted.

The vibration specification is somewhat different for operating versus nonoperating equipment. The operating spectrum is shown in Figure 5.6. Here the acceleration peak amplitude is about the same as nonoperating, but the spectrum extent of the vibration is extended from 50–500 Hz.

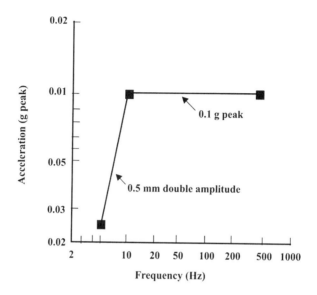

Figure 5.6 Operating vibration characteristics. (*Source:* [3], © 1996, IEEE. Reprinted with permission.)

Low-air pressure tests are included to evaluate the effects of altitude on the operation of the equipment. As the altitude is increased, the air is less dense and there is less air pressure since air pressure is caused by the weight of the air above it and there is less air the higher the altitude. Components can sometimes become inoperable when assembled at one air pressure and operated at a lower (or higher) pressure. Airborne EW equipment is not always located in pressurized space on the aircraft. It therefore can be subjected to very low pressure, as some aircraft fly at significant altitudes. EW equipment that is intended to be used in such applications must be specifically designed for low-pressure operation.

EW systems, by their very nature, are susceptible to unintentional wandering signals. Careful design practices must be followed in order to avoid interfering signals from being generated from within the system itself. Such signals are referred to as EMI.

EW systems are particularly susceptible to EMI because of the very sensitive receivers used for ES. These receivers frequently have wide bandwidths, if not instantaneous, then total bandwidths. Even narrowband receivers have filters at their input that can be as wide as half an octave

One of the design parameters for EW systems is the system sensitivity. The more sensitive the system, the better. However, more sensitivity implies more susceptibility to unwanted noisy EMI signals. Shielding is almost always required at the box level, and frequently at the circuit-board level. Copper cabling that connects the boxes together must be shielded. Fiber-optic interconnects hold promise for alleviating some of the EMI problems when interconnecting boxes together as well as within electronic enclosures themselves.

The high-speed clocks that are designed to approximate square waves in computer and other high-speed digital circuits are a prime culprit for generating EMI. The frequency spectrum of a square wave theoretically extends to infinity in both directions around the fundamental frequency. Some of these components can be significantly large as well. Therefore, careful shielding of such clock signals is required. In fact, in the design of digital circuits, if a slower clock can be tolerated, it should be used. High-speed signal and clock cables should be carefully routed along the chassis, if appropriate, and cross other signal lines at right angles to minimize mutual coupling, for example.

EMC refers to the ability to operate in the presence of other systems without them interfering with the operation of the EW system or the EW system interfering with them. Such systems are said to be compatible with one another. EMC testing is referred to as *immunity* in Table 5.2. EMC is substantially more difficult in EW systems that employ EA because of the high power signals normally associated with EA. Tactical communication EW systems are infrequently required to operate in earthquake conditions, so this specification is rarely applied.

5.2.3 Reliability and Availability

The ability of a system to operate when required is key. This is particularly true for systems discussed herein, where battles can be won or lost and lives saved or lost depending on whether a system operates correctly when required. Reliable systems do not emerge by accident. Conscious effort must be expended to increase the reliability of an otherwise unreliable system. This section introduces the notions and basic terminology associated with designing a system to be reliable.

5.2.3.1 Why Reliable Design Is Necessary

All real systems fail at some point. It is possible to design systems so that their tolerance to failure of components and subsystems can be improved, however. Such design always adds redundancy of some type, thus normally increasing such factors as the cost, the system size, and power consumption. Because of this, redundancy for only critical components, that is, those that are necessary to insure that some facet of the operation of the system is guaranteed to work when required, is added. An example of this might be the central server in a computer network. If its operation is critical for operation of the network, then a second server, performing the same function, could be added. In contrast, one of the work stations using the server on the network might be allowed to fail without causing the functioning of the network to fail totally.

As systems get more complex due to added components and functionality, their reliability declines. Reliability in this sense means the probability that the system will continue to function without failure. Failures can be categorized depending on their importance to the successful completion of the mission of the system. The engine in a tank is a critical subsystem that must continue to function for the tank to move, and normally movement is an important part of a tank's mission. The antenna used for command and control communications for the tank can probably fail and the tank remain useful for most of its intended functions. A failure that precludes a system from accomplishing its mission is called a *critical failure*. One that is not critical is called *noncritical*. The aforementioned tank engine may not be a critical failure if the mission of the tank does not call for it to move. A tank sitting at a road junction in an over watch mission does not need to move, so if its engine fails while it is sitting there, it is a noncritical failure for that mission.

Reliability of complex electromechanical systems, without designing in some degree of fault tolerance, is typically very low, requiring exorbitant amounts of repair actions and nonavailability of the system when it is required. *Built-in test equipment* (BITE) helps, but is intended to assist a technician with repair actions as opposed to continue operation once a failure has occurred. Whereas the downtime of the system can be improved, and thus the ratio of operational time to total time is improved, the mean time between failure is not improved. The reliability of systems can be calculated, or at least estimated, based on the reliability of the individual components used in the systems.

One of the attractions of the evolving *microelectrical-mechanical* (MEM) technology is to design extensive redundancy into an otherwise nonredundant system. Micro technologies such as these facilitates redundancy without unacceptably increasing system size.

5.2.3.2 Reliability

The time between system failures is a random variable with statistical properties. Let $R(t)$ denote the reliability function, which is the probability that the system does not fail at time t. It is frequently assumed that $R(t) = e^{-\lambda t}$ where λ is the failure rate. The amount of time between critical failures of a system is called the *mean time between failures* (MTBF). Then MTBF = $1/\lambda$. For complex electronic systems, typified by modern communication EW systems, the MTBF of systems designed without much fault tolerance could be on the order of 10 hours. The MTBF for the space shuttle, which has extensive fault tolerance built in, is on the

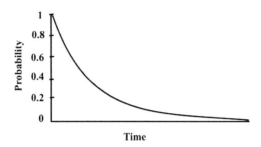

Figure 5.7 A typical reliability function.

order of weeks or months to ensure safe returns. If $\lambda = 1$ failure per 1,000 (1/1,000) hours, then $R = e^{-0.001t}$. This reliability function is shown in Figure 5.7.

Once a system has failed, it would normally undergo some repair action by an appropriate technician. Systems can be designed to be easy to repair and they can be designed to be difficult to repair. An example is automobiles. Modern automobiles are diagnosed by computer, with repairs carried out by mechanics. Having an experienced mechanic carry out the diagnostics would make the cost exorbitant for the repairs. The average time to repair a system is called its *mean time to repair* (MTTR).

5.2.3.3 Tolerating System Failures

There are ways to improve the reliability of systems. It is always true, however, that these ways involve adding components to the system of some variety. Some of the ways of designing these tolerances to faults are introduced in this section.

If the success of accomplishing a mission is dependent on the reliability of two subsystems, then a reliability model of these two subsystems is as shown in Figure 5.8. The system will fail if either of the two subsystems fails. Thus, the reliability of the system is given by

$$R = R_1 R_2 \tag{5.1}$$

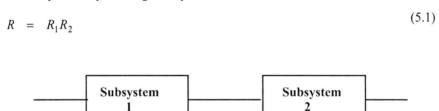

Figure 5.8 Series reliability model.

because for the system to have survived to time t, subsystem 1 and subsystem 2 must have survived. If either of the subsystems surviving causes the mission to be accomplished, then the reliability model of the two subsystems is as shown in Figure 5.9. Noting that $P(\text{failure}) = 1 - R(t)$, then the probability of the system failing is given by $P(\text{failure}) = P_1(\text{failure})P_2(\text{failure})$. Thus, $R(t) = 1 - P(\text{failure}) = 1 - P_1(\text{failure})P_2(\text{failure})$ so

$$R(t) = 1 - [1 - R_1(t)][1 - R_2(t)] \tag{5.2}$$

The reliability model shown in Figure 5.10 will fail only if three of the four subsystems fail. Its reliability can be shown, by derivations similar to that above, to be

$$R(t) = \{1 - [1 - R_1(t)][1 - R_2(t)]\}\{[1 - R_3(t)][1 - R_4(t)]\} \tag{5.3}$$

5.2.3.4 Failure Modes

The failure mode of components sometimes needs to be taken into consideration. The diode shown schematically in Figure 5.11(a) has the ideal performance characteristic shown in Figure 5.11(b). Whenever the voltage on the anode (left) side is greater than the voltage on the cathode (right) side, the diode is a short circuit and exhibits no resistance to current flow. On the other hand, if the voltage on the right is more positive than that on the left, the diode behaves as an open circuit, and no current flows. When two diodes are connected in series as shown in Figure 5.12, and the dominant failure mode of the diode is shorting, then failure of one of the diodes does not cause the circuit to fail. Thus, the probability of failure of the circuit is given by

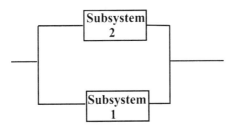

Figure 5.9 Parallel reliability model.

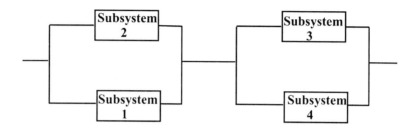

Figure 5.10 Serial/parallel reliability model.

$P = P(\text{Failure of Diode 1 and Diode 2})$ \hfill (5.4)

If it can be assumed that the failure of one of these diodes is independent of the failure of the other (a significant assumption, especially for integrated circuits), then

$P = P(\text{Failure of diode 1})P(\text{Failure of diode 2})$ \hfill (5.5)

and the reliability is given by

$$R = 1 - P$$
$$= 1 - P(\text{Failure of diode 1})P(\text{Failure of diode 2})$$ \hfill (5.6)

On the other hand, if the dominant failure mode of the diode in Figure 5.11 is

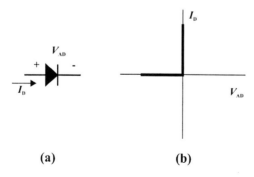

(a) \hspace{4cm} (b)

Figure 5.11 (a) A diode and (b) its (ideal) characteristics.

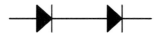

Figure 5.12 When the dominant failure mode of the diodes is shorting, connecting two diodes in series will ensure the circuit will continue to operate.

to open, then the configuration shown in Figure 5.12 does not improve the reliability of the circuit. The circuit shown in Figure 5.13, however, deals with this case, decreasing the overall failure rate and thus increasing the system reliability. If the dominant failure mode of the diodes shown in Figure 5.11 is an open circuit failure, then for the circuit shown in Figure 5.12, the probability of failure of the circuit is the same as the probability of failure of a single diode, and the overall reliability has not been improved by adding the second diode. This is because the probability of failure of the circuit is given by

$$\begin{aligned} P &= P(\text{Failure of Diode 1 or Diode 2}) \\ &= P_1 + P_2 - P_1 P_2 \end{aligned} \quad (5.7)$$

If the dominant failure mode is not known, or is absent, then the circuit shown in Figure 5.14 will handle either when a diode fails by shorting or opening. If any one of the diodes fails, the circuit will continue to operate as designed. For some failure modes, the circuit will continue to operate even if up to three of the diodes fail. For example, if either of the two diodes on the left side shorts and if one of the diodes on the right open-circuits, then the circuit will continue to operate. Clearly this can get to be a fairly expensive process, increasing the component count, in this case, by a factor of four to accomplish the same function.

These examples demonstrate how fault tolerance can be built into electronic circuits if the functioning of a diode is the design component. To show how it can be applied to other types of components, consider the *field effect transistor* (FET)

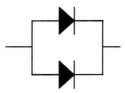

Figure 5.13 Redundant design of a diode arrangement when the dominant failure mode is an open circuit.

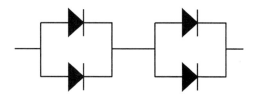

Figure 5.14 Diode arrangement when the dominant failure mode is not known. Open or short failure of any one diode (as well as other selected failures) and the circuit will continue to operate as a diode.

shown in Figure 5.15. The transfer characteristic for this device is shown in Figure 5.15(b). Properly integrated into a circuit, of course, as the voltage between the gate (G) and the source (S) is increased above zero, the channel between the drain (D) and source starts to close and conduct current. The gate settles at some voltage, saturating the device and the drain is essentially at the same voltage as the source.

If the dominant failure mode of the transistor shown in Figure 5.15 is an open channel, then connecting two of the transistors in parallel as shown in Figure 5.16 will provide additional reliability, since if one fails, the circuit will continue to function. In normal circuit operation, when

$$V_{GS} > 0, \text{ then } I_{D1} = I_{D2}, I_D > 0 \text{ and } V_o = 0$$
$$V_{GS} = 0, \text{ then } I_{D1} = I_{D2} = 0 \text{ and } I_D = 0, V_o > 0.$$

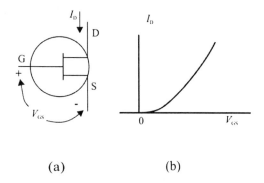

(a) (b)

Figure 5.15 (a) Schematic symbol for an FET and (b) the transfer characteristic of the FET.

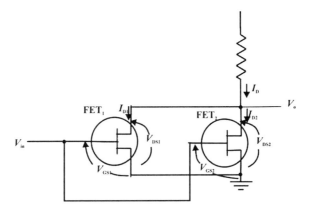

Figure 5.16 If the dominant failure mode of the FETs is open drain to source, then a parallel connection such as this will allow the circuit to continue to operate in the event of a failure.

With a failed FET, say, FET_1, with an open channel then when

$$V_{GS1} > 0,\ I_{D1} = 0,\ I_{D2} > 0,\ I_D > 0 \text{ and } V_o = 0$$
$$V_{GS1} = 0,\ I_{D1} = 0,\ I_{D2} = 0,\ I_D = 0,\ \text{and } V_o > 0.$$

Therefore, the circuit continues to operate as if there were no failure.

Analysis of the reliability of this circuit is the same as above. Likewise, if the dominant mode of failure of the transistor is shorting, the circuit shown in Figure 5.17 will facilitate fault tolerance that can be shown as above. If the FET-dominant failure mode is not known, then the circuit shown in Figure 5.18 can be implemented. In this case any one of the FETs can short or open and the circuit will continue to function correctly.

Extending these ideas to systems is straightforward except that the failure modes are not so easily dealt with. System components normally do not have such simple failure modes as open or short circuits that can be identified as such. Therefore, the failure modes do not normally enter into consideration in the same way.

Shown in Figure 5.19 is a configuration of a subsystem that is critical to the operation of a system. In this system, three (or another odd multiple) of the critical subsystem are performing the same operations in time synchronization. The outputs of the three subsystems are continuously compared, and voting determines the response. If two or more of these outputs are the same, then that is assumed to be the correct subsystem response. If one of the outputs is in disagreement with the other two, then corrective actions are necessary to find out

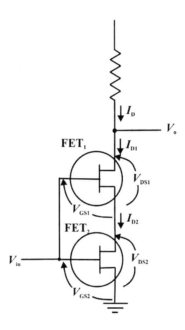

Figure 5.17 Series configuration of FETs will protect against failures if the dominant failure mode is a short drain to source channel.

why this is the case. If it is determined that the subsystem has failed, then it would normally be removed, with the resultant decrease in fault tolerance of the system. If all of the outputs disagree (for the first time), then it can be assumed that more than one of the subsystems has failed.

The reliability of the system shown in Figure 5.19 is given by

$$R(t) = 1 - P(\text{failure}) \qquad (5.8)$$

where failure will occur if all three subsystems fail or two out of the three fail. Therefore,

$$P(\text{failure}) = P(\text{all three fail}) + P(\text{two of the three fail}) \qquad (5.9)$$

But

$$P(m \text{ out of } n \text{ fail}) = \binom{n}{m} p^m (1-p)^{n-m} \qquad (5.10)$$

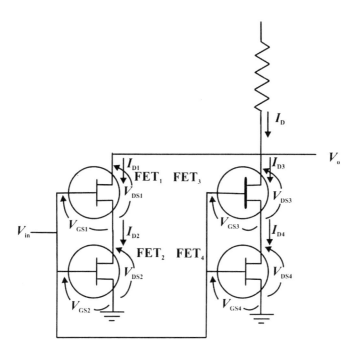

Figure 5.18 A quadruple-redundant FET. Any one of the transistors can open or short and the circuit will continue to function.

where $\binom{n}{m}$ is the binomial coefficient and p is the probability of any one of the subsystems failing, here assumed to be equal. Thus, since $R(t) = e^{-\lambda t}$, $p = 1 - e^{-\lambda t}$, and

$$P(\text{failure}) = \frac{3!}{(3-3)!3!}(1-e^{-\lambda t})^3(1-e^{-\lambda t})^{3-3}$$
$$+ \frac{3!}{(3-2)!2!}(1-e^{-\lambda t})^2(1-e^{-\lambda t})^{3-2} \quad (5.11)$$

Carrying out the algebra yields

$$R(t) = 3e^{-2\lambda t} - 2e^{-3\lambda t} \quad (5.12)$$

170 Introduction to Communication Electronic Warfare Systems

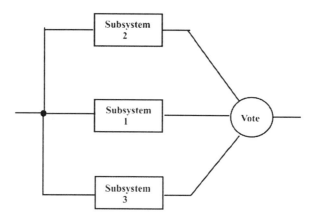

Figure 5.19 A triple-redundant system. Each of the subsystems performs the same function and a vote of their outputs is taken. The largest vote is taken as the output.

This function is shown in Figure 5.20 compared to the reliability of a single subsystem. Note that for short mission times, the triplicated system configuration provides higher reliability, but eventually falls below the single system reliability. This indicates that the fault tolerance is valid only for a limited amount of time.

The reason that the reliability eventually falls below the single subsystem reliability is that there are more components in the triplicated system. That is an unavoidable consequence of a more complicated system.

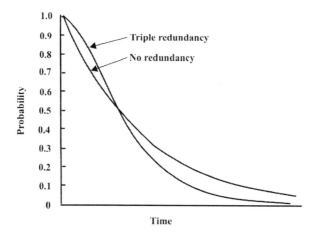

Figure 5.20 Reliability function of the triple-redundant system shown in Figure 5.19. Note that as time goes on, it becomes less reliable than any one of the subsystems alone.

5.2.4 Human-Factors Engineering

Human factors refers to the operation and the maintenance activities of a system. The differences between these two categories are discussed next.

5.2.4.1 Operations

It is always necessary to employ *human factors engineering* (HFE) whenever a human operator is to use an EW system or a maintenance technician is to repair it. One or both of these situations is almost always true (the only exception that is obvious is space applications).

IEEE Standard 1023-1988 [4] outlines six HFE aspects that need to be considered when designing nuclear power stations, which are directly applicable to communication EW systems. These areas are listed as follows:

1. Tasks;
2. Environment;
3. Equipment;
4. Personnel;
5. (Nuclear) Operations;
6. Documentation.

The task aspect includes the allocation of functions in the system to either machines or humans, which, of course, depends on whether the task can be automated or should be automated. It also involves the loading imposed on the operator of the system. The operator can be overloaded with tasks to be performed, or he or she can be underloaded; both conditions lead to underperformance. Also, sometimes machines can be more precise in their actions than humans can, and therefore if the task requires more precision than a human can deliver, it is a task that is a candidate for automation. The feedback to a human on the performance of a task is an important consideration.

The feedback should be timely and significant relative to how well the human performed the task to be useful. In addition, humans tend to be error-prone, and therefore any tasks allocated to a human must be tolerant of this error source. Lastly, the training of humans needs to be considered. It must be thorough and geared toward the level of education the expected system operators will have.

Humans can tolerate only certain variability in their physical environment. Factors such as temperature, humidity, and airflow are important considerations for the environment in which a human is to perform. Communication EW systems normally consist of extensive amounts of electronic equipment that generate heat. This heat source must be taken into account, along with other sources, such as Sun loading, to create an acceptable environment for human operators to occupy.

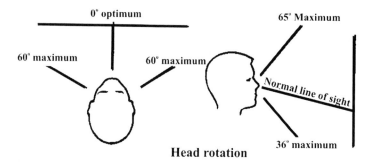

Figure 5.21 Visual ranges for humans. (*Source:* [5], © 1990, Association of Old Crows. Reprinted with permission.)

Lighting must be carefully designed so that screen displays as well as visual indicators on the equipment can be easily read. Lighting should be adequate for the normal operation of the system. If computer displays are involved, the lighting should be on the dim side, allowing the screens to be viewed more clearly. Lights glaring on the screen should be avoided. The reasonable limits on viewing angles are shown in Figure 5.21. These parameters should only be used when the normal limits shown above cannot be attained. These parameters refer to when the head is rotated to visualize displays. Acoustic noise should be minimized so that the operator is not distracted from performance of their normal duties. The size of the area in which the operator is to work needs to be adequate for the performance of the expected tasks.

Layout of equipment must be compatible with the tasks, so that timely and convenient execution of tasks can be performed. Equipment placement for tasks that the operator is to perform frequently should be placed as shown in Figure 5.22, which represents a typical EW system operator console [6]. Specific concerns about equipment placement are illustrated in the figure. Operator positions are normally designed around ninety-fifth percentile humans, be they male or female. Displays must be located for easy viewing, especially if they are to be seen much. The most frequently used controls, such as keyboards, need to be within easy reach and at the right height. The best height for computer keyboards is between 26 and 30 inches, depending on the height of the operator. Therefore, this height must be adjustable. Use of color both for displays as well as warnings and indicators is useful. Green, yellow, and red lights display the obvious meaning at a glance. Finally, safety is always a significant consideration. Exposure to unsafe conditions should be minimized. The capabilities of the personnel expected to operate the system needs to be taken into consideration. To

System Engineering

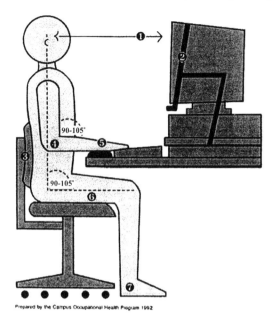

1. Top of screen at eye level for bifocal wearers. Screen distance at arm's length (15-32").
2. Document holder adjustable to screen height.
3. Chair backrest provides firm lower back support. Chair back and seat easily adjustable for height and tilt by user.
4. Keyboard height promotes relaxed arms with forearms parallel to floor.
5. Wrists straight (neutral). Padded, movable wrist rest, same height as keyboard home row, if needed.
6. Thighs parallel to floor. Ample legroom under work surface.
7. Feet rest firmly on floor or foot rest.

Figure 5.22 Workstation layout. (*Source:* [6]. Reprinted with permission.)

a large extent, for military systems, this concern is addressed in a larger context when the operator personnel are trained in their *military occupational specialty* (MOS). There are limits, however, to human capabilities. If lights and indicators are too dim or poorly placed, eyestrain could result, and in the limit, the light may not be seen at all. The amount of information a human operator must remember is a consideration.

The final aspect considered in IEEE Standard 1023-1988 is documentation. The military services have set requirements for operation, maintenance and training documentation. These are normally acquired when the system is first acquired, and updated as the system is updated. Needless to say, however, this documentation must be thorough and accurate. It must reflect all that an operator is expected to do under all conditions.

5.2.4.2 Maintenance

Frequently overlooked in the design of systems is the maintainability design and, in particular, the human factor aspect. How well a system can be maintained is directly related to how well a maintenance technician can repair it. "Well" in this case refers to the ability to determine what is wrong and to repair faulty equipment. A familiar example of this is the placement of oil filters in automobiles. A good design locates these filters where they are readily accessible—after all, it is known from the start that they will be replaced frequently, so why place them at difficult locations? A poor design, from a maintenance point of view, locates the filters at locations that are difficult to get to with the necessary wrench to loosen them.

It is not always possible to design systems so that all the possible maintenance issues are optimized. Some design requirements just do not allow it. Designing systems with maintainability in mind, however, is prudent and should always be considered.

5.2.4.3 Safety

Safety in operation as well as maintenance of a system is of prime importance. Safety issues can arise due to radiation effects, either intentional or unintentional. EA systems, for example, are designed to radiate high levels of energy and, depending on the frequency range, can cause injury to humans. Such a concern is exemplified by the operation of a high power jammer at a frequency range that could potentially injure personnel. The levels of radiation that humans can tolerate for 0.1 hour according to ANSI Standard C95.1-1982 are shown in Figure 5.23 [7].

High voltages are almost always present in EW systems. Contact with these voltages can cause serious injury to both operators as well as maintenance personnel. Wherever high voltages are present, if possible, covers should be used. If not possible, then clear placards must be posted to minimize contact.

Some units are heavy, and require two or more people to handle them. Such cases would be clearly marked as such. When equipment is mounted in racks, heavier units should be placed lower in the rack if possible, precluding the requirement to hold heavy weights high in the air.

If the system allows it, two exits should always be provided. One of these would be the normal entryway, while the other could be an emergency exit in the ceiling, for example.

5.2.5 System Cost

Not much will be said here regarding the cost of communication EW systems, since the details of costs are entirely dependent on the details of the program to develop and deploy the system. A few comments are, however, general in nature.

The total cost of ownership of a system consists of two distinct parts: (1) the initial acquisition costs and (2) the sustainment costs. The first of these addresses what it costs to design and build the systems. It consists of the R&D phase as well as the quantity production. The second phase consists of the costs involved with keeping the systems operational once fielded. There are many cost elements to the

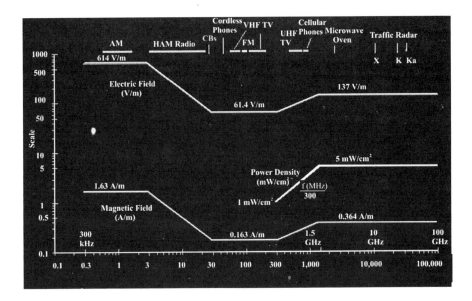

Figure 5.23 Safety limits on the exposure of humans to radiation from ANSI Standard C95.1-1982.

second category, from training operators and maintainers to buying an inventory of spare parts.

In almost all cases, the sustainment costs are orders of magnitude larger than the initial acquisition costs. Therefore, it is important that during the design phase, adequate consideration be given to lowering the sustainment costs. One way of doing this, for example, is to use components or parts that already have been used in other systems so that the infrastructure support is already in place. Using the same part in multiple places reduces the cost of the inventory of spare parts.

5.3 CONCLUDING REMARKS

System engineering as a discipline applies engineering principles to the overall design of systems. A system in this case can be just about anything, but herein it generally refers to a complicated electronic ensemble of equipment. There is a process which, when followed, leads to an orderly life-cycle process of system design, construction, support, and ultimate disposal.

The environment in which a military system is used determines the testing required. In some cases benign environments apply where equipment is installed in office buildings and never moved for example. The other extreme consists of muddy foxhole deployment. The latter requires more stringent design and testing so that the equipment will continue to operate when required.

Designing equipment to be reliable is critical. Complex systems that have many single points of failure are assured to having small MTBFs. Those functions that are necessary for a system to complete its critical mission modes must have redundancy built in. Reliability is designed into a system from the start, not applied late in the design phase.

References

[1] Kayton, M., "A Practitioner's View of System Engineering," *IEEE Transactions on Aerospace and Electronic Systems,* Vol. 33, No. 2, April 1997, pp. 579–586.
[2] IEEE Standard 1220-1998, IEEE Standard for Application and Management of the Systems Engineering Process, 1998.
[3] IEEE Standard 1156.2-1996, "Environmental Specifications for Computer Systems," 1996.
[4] IEEE Standard 1023-1988, "IEEE Guide for the Application of Human Factors Engineering to Systems, Equipment, and Facilities of Nuclear Power Generating Stations," December 12, 1988.
[5] "EW Engineer's Handbook," Association of Old Crows, Alexandria, VA, 1990.
[6] "Ergonomics and VDT Use," Library of Congress Collections Services, VDT Ergonomic Committee, 1991–92.
[7] Chignell, M., "Lecture 12: Standards and Plans," accessed July 2001, http://peach.mie.utoronto.ca/courses/mie240/html/lecture12.html.

Chapter 6

Electronic Support

6.1 INTRODUCTION

For the purposes herein, ES is the new name for what used to be called *electronic support measures* (ESM). ES is comprised of those actions to search for, intercept, identify, and locate intentional and unintentional radiation [1, 2]. There are two fundamental purposes to which the information gleaned from intercepted communication signals is applied. Which purpose is determined by the use made of this information, which in turn is determined by the amount of time taken to extract information. If the signals are analyzed over an extended period of time, then intelligence is generated. If the information is put to immediate use, usually not requiring extensive analysis to put it into context, it is called combat information, not intelligence. Of course, after the signal contents have been analyzed, that information can also be used for combat purposes and combat information can be used to generate intelligence. The key difference here is the time it takes to extract useful information. Combat information is used immediately. Therefore, there is little to no time to perform analysis.

ES generates combat information, whereas SIGINT generates intelligence. Reaction to combat information is immediate, whereas intelligence is used to formulate the long-range picture, and thus can take longer to generate.

6.2 INTERCEPT

The intercept of communication signals is one of the principal functions of an ES system. It is important to note that these communication signals are noncooperative—that is, they generally do not want to be intercepted. This is as opposed to the nodes in a communication system, which generally do want to communicate with each other. Noncooperative intercept is a much more difficult problem, especially with digital communication signals [3–5].

6.2.1 Internals Versus Externals

Sometimes information can be extracted from analysis of the *externals* of a signal. Such externals might consist of such factors as the baud rate, RF bandwidth, and modulation type. Determination of such parameters can yield information about the *electronic order of battle* (EOB) (discussed in Section 6.3) of an adversary, such as the type of military unit one is facing.

The *internals* of a communication signal refer to the information content provided in what was said, or what information was otherwise exchanged (as in modem signals, for example). Meaning can sometimes be extracted from such internals.

6.2.2 Propagation Loss

Radio signals suffer losses as the distance between the transmitter and receiver increases as discussed in Chapter 2 [6]. To the first order, this is due simply to the limited power spreading out in all directions from the transmit antenna. The power density decreases because the surface of the sphere through which the power passes gets larger; thus, the power density must decrease. This is called the attenuation of free space.

Other effects cause further attenuation. Close to the Earth, vegetation and surface irregularities add to the attenuation of free space. Mountains can preclude signal propagation at all, or, in some cases, they can contribute due to a phenomenon called knife-edge diffraction. Generally, the higher the elevation of an antenna, the better it can both transmit and receive radio signals, largely because higher antennas tend to be more free of such limitations.

Because signals get weaker with distance between the transmitter and receiver, ES systems generally must be very sensitive. Friendly communication signals typically are closer to the ES system than the targets of interest. The signals from these transmitters are considered interference to the ES system, and they are typically much stronger. The frequency spectrum is channelized in that different parts of the frequency spectrum are divided according to the needs of that spectrum and the technological ability to provide the channels. In the low VHF between 30 MHz and 88 MHz, the military channels are 25 kHz wide. Transmitters designed to operate in this range must have the majority of their power concentrated in that 25 kHz. This is not all their power. Theoretically it is impossible to have all the energy in a signal within a limited bandwidth while simultaneously having a definite start and stop time for the transmission. Therefore, there is energy from those transmitters that is outside of the 25 kHz channels. In fact, significant energy can extend to several megahertz on both sides of the channel. This energy further exacerbates the interference problem.

Therefore, the ES system must be able to intercept weak signals even though there are very strong signals nearby causing interference and those signals are not necessarily close to the frequency that the ES system is trying to intercept. The parameter that indicates how well an ES system can operate with this near-far problem is its dynamic range. It is difficult to achieve adequate dynamic ranges for ground-based ES systems. The *required* dynamic range might be 120 dB, whereas *achieving* 70 dB can be difficult. Airborne ES systems receive the weak target signals much better than ground systems, so their dynamic range requirements are less, although they are still challenging. In fact, since the airborne systems are higher, they see more of the friendly interfering signals than the ground systems do, so their interference problem can be worse.

6.3 GEOLOCATION

The location of targets is usually considered critical combat information. These locations can be used to track target movement over time as well as to indicate groupings of types of targets into a particular region. Such groupings are often indicative of unit type, intentions, and plans. The amassing of units in a close area, detected and located by EW means, might indicate a specified type of activity [7, 8].

Geolocating communication emitters, known as *position fixing* (PF), is one of the more important functions of ES systems. PF is the process of locating in three-dimensional space the origination of the radiating entity. Such information has several uses. Depending on the use to which the resultant fixes are applied, it determines to a large extent the accuracy requirements. Some of the requirements are much more stringent than others, depending on how precisely the resultant location is required.

By determining the location of emitters in the battlespace, it is frequently possible, normally by combining this information with other information, to determine the disposition of opposing forces—called the EOB. This information is an indication of who, what, when, and where about the opposing force. It also attempts to determine the intentions of the opposing force in terms of their mission. Although knowing the location of entities accurately is useful, it is frequently not required to know their locations as precisely as some other applications. Finding their location for EOB determination on the order of 1 km is generally adequate for this application.

Sometimes weapon platforms have unique communication signals associated with them to exchange command and control information. Locating and identifying these signals facilitate identification of the weapon system. Finding these weapon platforms is useful for focusing the direction of jamming signals as well as, as typified by the SAM case, for threat avoidance. In these applications, precise locations are not necessary either with accuracy requirements on the order of 1–5 km being adequate.

If emitters can be located accurately enough, then that information can form the basis of providing targeting information for attacks by indirect weapons (artillery and missiles mostly). For such dumb weapons, accurate targeting is required—often less than 100m. The location of emitters, however, does not necessarily mean the location of the weapon platform. It is possible to position the transmitting antenna of an emitter quite a distance from the platform. In this case, the resultant geolocation must be accurate enough so that the indirect weapons can hit the target with a reasonable number of dumb shells. Of course, for smart weaponry, such precise locations are not necessary, as the round itself finds the precise location once it gets close enough. Thus, for smart weapons, less precise geolocations are adequate—on the order of 5 km.

In *operations other than war* (OOTW), locating communication emitters might form the basis for tracking individuals or units. Command and control of mobile forces almost always require the use of mobile communications. These communications might take advantage of the commercial infrastructure, such as cellular phone systems. They might be facilitated by *citizen band* (CB) commercial transceivers. They may also be accomplished with transceivers designed specifically for that purpose. Whatever the means, sometimes it is useful to geolocate transmitters even in OOTW.

In the United States the *Federal Communications Commission* (FCC) uses the techniques discussed herein to find illegally operating transmitting stations. To operate an RF transmitter in all but a few frequency bands [called the *instrumentation, scientific, and measurement* (ISM) bands] in the United States, a license is required. This license is issued by the FCC. Operating without a license is illegal, so one of the tasks of the FCC is to geolocate offending transmitting sites.

Another commercial use of geolocation is the cellular phone system and 911 emergency calls. If a 911 call is made from a cellular handset, the signal can frequently be detected at more than one base station. These base stations can triangulate to locate where the emergency call is coming from, so the emergency vehicles will know where to go without relying on the caller to provide that information.

6.4 TRIANGULATION WITH MULTIPLE BEARINGS

One of the most common ways to accomplish emitter geolocation is by *triangulation* [9]. In this, the direction of arrival of an incoming wave front, called the *line of bearing* (LOB) or just *bearing*, also called a *line of position* (LOP) is determined at two or more sites. Where these bearings intersect, which they will always do unless they are parallel lines, is the estimated location of the emitter. A typical example is shown in Figure 6.1, where three sites are obtaining bearings on the radio in the tank shown.

There are always errors present in the measurement of the bearings. Such errors can be systemic errors, errors caused by violation of the assumptions about the wave front or noise induced. The principal assumption is that the wavefront is planar—that is, it arrives at the receiving system as a plane wave. At sufficiently large distances between the transmitter and receiver, this assumption is usually valid relative to the curvature of the wave caused by its spherical propagation. What is frequently not true, however, is that the wave is planar because of unexpected perturbations in the path that the wave takes. Multipath, for example, can cause more than one wave to arrive at the receiving site, probably with different amplitudes, phases, and direction of arrival. These errors cause uncertainty in the calculation of the location of an emitter. Because of this a measure of confidence is usually computed along with the geolocation, or "fix." This measure of confidence is usually a contour describing a region within which

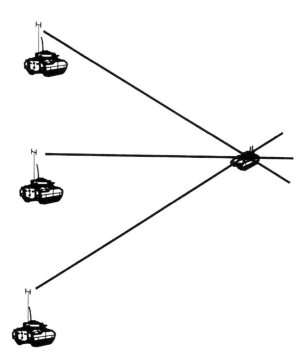

Figure 6.1 Example situation where direction finding is used.

the emitter is located with a prescribed measure of probability. If the region is a circle, then it is called the *circular error probable* (CEP). The CEP is the radius of the circle. If it is an ellipse, then it is called the *elliptical error probable* (EEP). The semimajor and semiminor axes as well as the tilt angle of the major axis describe the EEP. The most common probabilities employed are 50% and 90%.

Typical operational accuracy for RF direction finding equipment depends on the frequency range of interest because of the environmental effects on signal propagation. The HF frequency range, for example, can exhibit considerable reflections from and refraction through the ionosphere and terrain. The *instrumental accuracy* refers to the accuracy of the system itself, without the effects of environmental factors while the *operational accuracy* is a measure of the expected performance once the equipment is put into operational use. There can be considerable differences between the two, and complicating the matter is that the operational accuracy is usually determined by test in field conditions. Unless the field conditions are varied over all of the expected operational environments, there will be biases in the measurements. For modern equipment in the VHF and UHF range the field accuracy is typically 4° to 5° *root mean square* (rms). In the HF range, typical operational accuracy ranges from 10° to 25° rms.

Whereas putting PF assets on airborne platforms normally has significant benefits, putting such assets on UAVs has several additional advantages. The first of these is the required accuracy. Referring to Figure 6.2, if the PF hardware is mounted on an airborne standoff platform that is at a range of, say, 100 km from the target of interest, the situation would be as shown at the top of Figure 6.2. Assuming the (in)accuracy of the PF system is 2°, then $d_1 = 100 \sin 2° = 3.5$ km. On the other hand, if the PF system were on a UAV as shown at the bottom of the

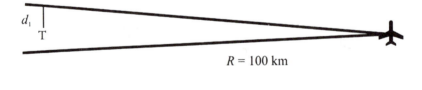

Figure 6.2 Comparison of the miss distance of a 2° accuracy PF system from a standoff platform versus a penetrating UAV.

figure, it can overfly the adversary's air space, getting much closer to the target. When the range to the target is 3 km, then the same PF system at 2° accuracy produces $d_2 = 105$m.

An additional benefit of overflying the target area is the proximity of the UAV to the target and the requirement to obtain multiple LOPs to compute a fix. With more than one system available to obtain simultaneous LOPs on a target, then a single opportunity is adequate to compute a fix. If there is only one asset available, it is necessary for the platform to move some distance to obtain a second (or more) LOP on the target. If the platform is standing off 100 km or so, it will take some time for the system to move an adequate distance to obtain the second LOP. For a good fix, the system would have to move on the order of 100 km before the second LOP could be taken. At 100 km/hr velocity this would take an hour. Chances are good that the target will not be emitting an hour later. On the other hand, if the UAV is close to the target, it would have to move only 3 km for the second LOP. At 100 km/hr this would take only a minute and a half or so.

Alternatively, less accurate PF hardware is required to achieve comparable target location accuracy. If the requirement is for a miss distance of no more than 250m, then a standoff system at a range of 100 km would need PF accuracy at 0.14°, while a penetrating UAV system could achieve this miss distance at a range of 3 km with a PF accuracy of a generous 4.8°.

The second advantage is the system operators can be remotely located on the ground in a relatively safe location. Operations could continue even though the supported force is moving because the UAV can move ahead of them or with them. There is no need for on-the-move operation of the ground segment. If two ground segments were available, they could leapfrog in their operation keeping continuous coverage of the area.

A third advantage is that the sensitivity of the hardware to achieve this accuracy is substantially less. The targets are much closer so for a given amount of emitted power, more power is received by the UAV system than the equivalent standoff configuration. This advantage, however, in some cases will be a disadvantage. Lower sensitivity means smaller instantaneous coverage regions. The UAV may need to move from one area to another to cover the entire region of interest.

One of the more significant shortcomings of overflying targets in a UAV is the depression angle. Standoff systems need to be calibrated in azimuth only, limiting the depression angle requirements to the horizontal plane. A UAV at 3,000-m altitude AGL in Figure 6.2 has a depression angle of about 45°, so substantially more system calibration is required. In addition, the dipoles or monopoles usually used for airborne applications have limited response beneath the aircraft, thus decreasing system sensitivity. The depression angle is shown in Figure 6.3.

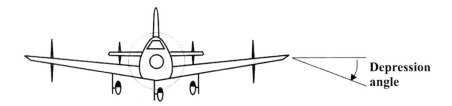

Figure 6.3 Depression angle associated with airborne systems.

Lastly, putting ES assets among the target array minimizes the difficulty of looking through the friendly array of emitters. With U.S. forces using SINCGARS radios, which hop around the low VHF spectrum, considerable friendly RF fratricide occurs. Standoff ES systems must look through this crowded spectrum to find targets. Putting the ES receivers further away from friendly interferers makes this less of a problem.

All of these advantages, among other things, lower the cost of ownership of the system by decreasing the acquisition cost, making a simpler system thereby increasing the reliability and decreasing the amount and number of spare components. Also decreased are the maintenance personnel costs, since fewer will be needed.

6.5 DEPLOYMENT CONSIDERATIONS

Ground-based systems are most useful when extended ranges are not required. Such situations consist of protecting early entry forces by monitoring the region immediately surrounding a landing area. If ground-based systems must be used to cover terrain at extended ranges, then they should be located on the highest terrain possible. These locations must provide for access for such items as fuel, resupply, and rations even under adverse weather conditions. Reverse slopes should be considered with the antenna just reaching over the top of a ridge or hill. Care should be used in such cases, however, because geolocation accuracy can be significantly affected by close proximity to the ground [10–12].

Since antenna height is important in such applications, historically ground-based systems have had to stop and erect antennas in order to get much range. Alternatively, receiving equipment mounted on UAVs, interconnected to ground systems in real time can provide extended coverage while still providing protection for the system operators on the ground. On-the-move operation is even possible in this configuration.

It is possible to remote ES receiving front ends from the back ends, where the operators are located. The interconnects could be terrestrial line-of-sight data links, satellite links, or relayed through a UAV configured for the task. Such configurations not only are safer for the system operators but they facilitate some types of operations that are not possible otherwise. If there is language translation necessary to support an operation, the linguists could all be located at the same location, thereby easing the language mix and skill level issues.

6.6 ELECTRONIC MAPPING

If adequate geolocation accuracy can be achieved and adequate emitter identification can be accomplished, then it is possible to develop and maintain a map of the electronic targets. The difficulty is in obtaining the adequate accuracy. Such a map could indicate the up-to-date disposition of adversary forces as well as possibly show interconnected networks. As will be discussed in Chapter 10, identification comes in many forms, and such identification would add valuable information to the map.

Such a display might be a map or terrain overlay indicating the locations of targets calculated from data obtained from the PF system. If the data is available, in addition to the locations, the type of target could also be indicated. The type could be anything that is available from the ES system, such as frequency range, modulation type, or enemy or friendly. The type of unit could be indicated if associations have been made from several intercepts from the same location of multiple types of emitters. Groupings of targets within particular regions might indicate some sort of specific activity is happening or about to happen.

Groupings of emitters within a region could also be used as a tool for collection management. Geographical filtering could be invoked so that the system resources are only applied to intercepts on targets emitting from that region. Also, other sensors could be deployed to that region to confirm information and otherwise collect information.

6.7 COMMON OPERATIONAL PICTURE

ES can support generation of the *common operational picture* (COP), not only by geolocating targets, but also in some cases, identifying those targets by unit type, or in some cases, by identifying the specific unit. The latter can only occur if specific emitter identification is accurate enough, however. If a target was intercepted previously and the technical capability is present to identify the target as one that was intercepted previously, then the system can automatically indicate that the same target is present. If the target is in a new location, then tracking the target is possible, and this too can contribute to the COP by indicating such movement.

Detection of particular types of emitting targets within a geographical region can sometimes indicate intentions, movements, or other significant battlespace activities. Certain weapon platforms sometimes have unique combinations of emitters that can be intercepted and geolocated indicating that specific weapon platform is present. It also can sometimes indicate that the weapon platform is in a particular state, for example, preparing to fire artillery. This also contributes to the COP generation and maintenance function.

Until reliable automated translation of natural language is available, a person will be required to translate communication intercepts. That is not the situation with computer-to-computer digital signals. Since these signals are processed at both ends by machines anyway, it makes it possible to potentially automatically process them in an ES system.

6.8 OPERATIONAL INTEGRATION WITH OTHER DISCIPLINES

One of the more significant capabilities provided by communication ES systems is their ability to cue other intelligence systems. An EO/IR sensor has a very narrow field of view, at least relative to an ES system. These sensors, however, can be cued by an ES system with a geolocation capability, which reduces the search range required of the EO/IR system by two to three orders of magnitude or more (depending on implementation). An EO/IR system on a UAV has a total scan width of about 10 km or so on the surface of the Earth, while the instantaneous imaging capability is about 0.5 km or so. An ES system on that same UAV has a substantially larger field of view, which can be used to focus the EO/IR payload quickly to targets of interest. TOCs and other formations of interest are frequently guarded by tactical units, which must communicate in order to function. These communications are the targets of UAV ES systems, and they can be used to efficiently cue the EO/IR sensors [13–15].

Radars, such as *synthetic aperture radars* (SARs) or *moving target indicating* (MTI) radars, are sensors with a broad coverage area. They cannot, however, tell much about what an entity is beyond basic indications, such as tracked vehicle versus wheeled, for example. ES systems can sometimes provide that information. Combining this information with that from a radar can indicate that a particular unit is moving or has moved to a new location or that it has been combined with other units. This type of information is useful for event detection, unit tracking, and the like. Like other systems that provide geolocation information, ES systems can provide cues for HUMINT collection by indicating the location of particular units, and, sometimes more importantly, where adversary units *are not*.

6.9 SUPPORT TO TARGETING

Another use for ES, as with other intelligence disciplines, is to support targeting. This is the process of identifying and verifying targets as being high priority—high enough to attack with (usually) indirect-fire weapons, including EA assets. ES sensors are considered wide-area sensors, as opposed to, for example, TV cameras that only cover a small area at a time. As such, ES sensors are most useful for targeting purposes by cueing other, confirming assets such as EO/IR sensors or humans [16].

6.10 CONCLUDING REMARKS

An overview of the operating principles of communication ES systems was presented in this chapter. Chapters 8 to 17 delve into the designs of such systems in much more detail.

ES systems provide two fundamental capabilities to military commanders. The first is the generation of combat information. This information is of immediate battlefield use and requires little or no analysis to be useful. The other type of information is intelligence. The systems to collect these types of information are not necessarily different. It is the utilization of the resultant data that defines the type of information, not the collection means. The amount of data collected, however, may vary depending on the tasks at hand. The ES system design principles described herein apply in both situations.

References

[1] Waltz, E., *Information Warfare Principles and Operations,* Norwood, MA: Artech House, 1998, p. 214.

[2] Schlecher, D. C., *Electronic Warfare in the Information Age,* Norwood, MA: Artech House, 1999, pp. 1–3.

[3] Torrieri, D. J., *Principles of Secure Communication Systems,* 2nd ed., Norwood, MA: Artech House, 1992, pp. 291–366.

[4] Wiley, R. G., *Electronic Intelligence: The Interception of Radar Signals,* Dedham, MA: Artech House, 1985.

[5] Neri, F., *Introduction to Electronic Defense Systems,* Norwood, MA: Artech House, 1991, pp. 30–32.

[6] Stark, W., et al., "Coding and Modulation for Wireless Communications with Application to Small Unit Operations," accessed July 2001, http://www/eecs.umich.edu/systems/TechReportList.html.

[7] Torrieri, D. J., *Principles of Secure Communication Systems,* 2nd ed., Norwood, MA: Artech House, 1992, pp. 364–366.

[8]   Wiley, R. G., *Electronic Intelligence: The Interception of Radar Signals,* Dedham, MA: Artech House, 1985, pp. 107–134.

[9]   Torrieri, D. J., *Principles of Secure Communication Systems,* 2nd ed., Norwood, MA: Artech House, 1992, p. 364.

[10] Torrieri, D. J., "Stastical Theory of Passive Location Systems," *IEEE Transactions on Aerospace Electronic Systems*, Vol. AES-20, No. 183, March 1984.

[11] Adamy, D., *EW 101: A First Course in Electronic Warfare,* Norwood, MA: Artech House, 2001 pp. 145–147.

[12] Neri, F., *Introduction to Electronic Defense Systems,* Norwood, MA: Artech House, 1991, pp. 271–336.

[13] Hall, D. L., *Mathematical Techniques in Multisensor Data Fusion,* Norwood, MA: Artech House, 1992.

[14] Waltz, E., and J. Linas, *Multisensor Data Fusion,* Norwood, MA: Artech House, 1990.

[15] Neri, F., *Introduction to Electronic Defense Systems,* Norwood, MA: Artech House, 1991, pp. 457–472.

[16] Waltz, E., *Information Warfare Principles and Operations,* Norwood, MA: Artech House, 1998, pp. 164–165.

Chapter 7

Electronic Attack

7.1 INTRODUCTION

There are two fundamental types of information denial. One can deny an adversary one's own information and one can deny that adversary his or her own information. The former is called *information protection* and the latter is called *information attack*. There are a variety of ways to execute information attack and protection, such as camouflage, concealment and deception, and EW. The technologies described in this chapter are limited to this last category, attacking an adversary's communication systems to deny information transport. These principles are readily extended to protecting one's own communication from exploitation by an adversary's communication exploitation systems, so this case is covered as well. The other forms of information denial are not covered herein.

7.2 COMMUNICATION JAMMING

The fundamental task to be accomplished with jamming was illustrated with several examples in Chapter 1. Simply put, a communication jammer tries to deny communication over RF links. A jammer attempts to accomplish this by putting unwanted signal energy into the receivers in the communication system. This unwanted energy, if strong enough, will cause the receivers to demodulate the signal from the jammer as opposed to the communication transmitter. Assuming that the jammer signal is not a replica of what was transmitted, communication is denied on the RF link [1–4].

Communication jammers can be used to deny information in two fundamental ways: (1) denying an adversary the ability to talk to another element, thereby limiting command and control, and (2) jamming an ES/SIGINT system. The latter is also known as *communication screening*.

Jamming radio signals has been around almost as long as radio signals themselves. There are several ways one can perform EA against an adversary's RF communication systems. Jammers come in a variety of configurations as well. A jammer that operates from within the friendly held battlespace is called a *standoff* jammer. One that operates within an adversary's held battlespace is called a *stand-in* jammer. A jammer that attacks the target's carrier frequency only is called a *narrowband* jammer, while one that emits a broad range of frequencies simultaneously is called a *barrage* jammer. Due to fratricide considerations, barrage jammers are rarely used in standoff configurations.

Jamming communication nets is always part of a larger military or political maneuver. It is integrated with the battle planning, just as other indirect fire weapons are integrated. The use of jamming must be synchronized with other battlespace activities to maximize its utility and impact on the outcome. The two primary intents of communication jamming are to disrupt an adversary's command and control process—thus, it attacks the connection between the decision-makers and the sensors and/or the implementers of attack. It attacks the neck of an adversary, which carries bidirectional information to and from the decision-makers.

Jamming commercial TV and radio broadcast stations may be included in an EA campaign. TV stations are frequently used to broadcast propaganda to noncombatants in an area, thereby attempting to influence the local populace against a friendly operation. Jamming such signals precludes this propaganda from getting through. Radio Free Europe was jammed for many years by the Soviet Union as are current attempts to communicate with the Cuban populace by the United States.

One of the goals of communications EA is to interrupt communications, or otherwise preclude effective communications, between two or more communication nodes. While digital communications are overtaking the analog forms, the latter is still expected to be around for quite some time, so effective EA against such systems is expected to continue to be required. Also, many of the same principles apply for both forms. The ability to interrupt analog speech signals depends on several factors. Shown in Figure 7.1 are curves for several degrees of interruption along with the rate of that interruption [5]. The intelligibility of the resultant speech is the dependent variable here. The parameter of the curves is the amount (percentage) of the interruption, varying between 25% and 87.5%. The way to interpret this chart is illustrated in Figure 7.2. Suppose there is a 10-second message. The case of the rate = 1 and 25% interruption is shown in Figure 7.2(a). On the other hand, the case for rate = 10 and 25% interruption is shown in Figure 7.2(b). Other rates and interruption fractions are interpreted similarly. It is interesting to note that while the total interruption time shown in Figure 7.2 is the same, the effect on the intelligibility is different by about 20%. Grouping the interference into the same location in time, as it is in

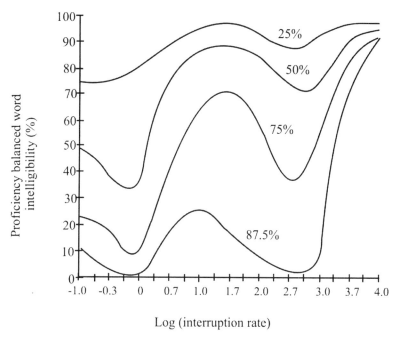

Figure 7.1 Intelligibility of interrupted analog communications as a function of the rate of interruption (expressed as the logarithm) and the degree (percentage) of the interruption per event. (*Source:* [5], © 1992, IEEE. Reprinted with permission.)

Figure 7.2(a), has a greater effect than spreading it out as in Figure 7.2(b). Furthermore, this effect is not the same for all combinations of rate and degree, although the general shape of the characteristics is consistent, so design tradeoffs must be carefully considered to optimize the effects of a jammer.

Experimentation has shown that a percentage of interruption of approximately 30% causes significant degradation of analog voice communications. Therefore, articulations above about 70% are required in order for tactical communications to be effective. It also indicates that a single jammer can simultaneously handle about three analog targets, switching the jamming signal quickly between the three signals. The particular effects are subject to the repetition rate of the interruptions due to the jammer as shown in Figure 7.1. Rates of adequate length (say, 25%) on the order of one interruption per transmission are adequate to reduce the articulation rate sufficiently to preclude acceptable communication.

International treaties may forbid jamming international satellites. Communication satellites are carrying more international communication traffic; the *international telecommunications satellites* (INTELSATs) carry thousands of telephone calls per day all around the world. Jamming one of the transponders on

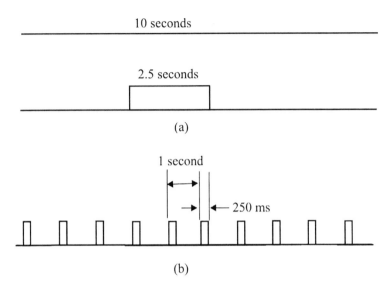

Figure 7.2 Interpretation of the interruption rate and amount terms for Figure 7.1: (a) rate ≠ 1 at 25% interruption and (b) rate ≠ 10 at 25% interruption.

one of these satellites could take out telephone calls from several countries at the same time, some of which would probably be friendly. In fact, that friendly communication may very well be diplomatic or military communications attempting to forward the friendly cause. Thus, extreme care must be taken before attempting to negate such communication paths.

The amount of jamming signal power at the input to the receiver to be jammed relative to the signal power at that input determines the effectiveness of the jamming scenario. The amount of this jamming power depends on several factors:

1. The *effective radiated power* (ERP) of the jammer;
2. The ERP of the communication transmitter;
3. The orientation of the receive antenna relative to that of the jammer antenna;
4. The orientation of the jammer antenna relative to that of the receiver antenna;
5. The intervening terrain.

If the antenna at the receiver has directionality, then it probably would be pointed more or less in the direction of the transmitter to maximize received energy. Unless the jammer were within this beam, it is at a disadvantage because the receive antenna has less gain in the direction of the jammer. The jammer power, therefore, must be increased to overcome this disadvantage. The actual parameter of importance is the J/S ratio, where J is the jammer power and S is the signal power. Both of these power levels are measured at the receiver. This ratio is most often given in decibels:

$$J/S_{dB} = 10\log(J/S) \tag{7.1}$$

7.3 JAMMER DEPLOYMENT

A jammer can be expendable. In that case, it might be delivered to its operating site by a soldier who emplaces it by hand, or it might be delivered by some other means such as deployed from a UAV or shot into place out of an artillery cannon or rocket. Whatever the delivery mechanism, expendable jammers obviously must be economical so that they can be lost. They also must be controllable. Since tactical campaign plans almost always change once the operation is under way, one does not want to emplace active jammers in an area, and then expect one's own soldiers to communicate in that area if they must pass through there with the jammers active [6, 7].

An array of expendable jammers can be formed by any of the emplacement methods mentioned, air delivered by UAVs, hand-emplaced, or artillery-delivered. The array defined by a single artillery shell would be more or less a straight line, whereas an arbitrary pattern can be constructed by the other two methods. The jammers would be placed nominally 500m apart. An area seeded with such jammers could have arbitrary width and depth. Placed within a valley, for example, as an adversary force moves through the valley, they could be precluded from communication by radio among themselves as well as with higher echelon command elements.

Deploying communication jammers onboard penetrating UAVs has several advantages. The first of these is fratricide avoidance. Standoff jammers, being closer to friendly radios than targets, can inject unwanted, interfering energy into friendly communications. By placing the jammer in the adversary's battlespace, nominally all communications can be viewed as targets so the fratricide problem disappears. This does not say, however, that all such communications should be jammed—just that friendly fratricide can be minimized this way.

A second advantage of UAV jammers is their minimal vulnerability. One of the problems with manned jammers is their survivability once they are put into operation. Such high-power transmitters are relatively easy to find both in frequency and geolocation. This makes them particularly vulnerable to enemy attack, from direct fire as well as indirect fire weapons, either of which could involve homing weaponry. Jammers mounted on UAVs are not stationary. In order to eliminate such a system, it would need to be tracked. Furthermore, there are no operators physically colocated with the jammer. These aspects make a UAV-borne jammer substantially less vulnerable and more survivable, to include the operational personnel.

A third advantage of UAV jammers is the relatively smaller amount of effective radiated power required to accomplish the jamming mission. Being substantially closer to the targets than their standoff counterparts, less power from the jammer is required for the same jamming effectiveness. Among other advantages, this makes system design easier, which leads to higher reliability and improved operational utility.

UAV-mounted jammers can be controlled by ground operators substantially distant from the UAV platform. These operators could actually be colocated with the fire support officer in the battalion or brigade *tactical operation center* (TOC) where real-time coordination of EA with other forms of indirect fire can occur. A TOC is where tactical battle management occurs.

Another form of UAV jammer could take advantage of micro-UAVs. Placing dumb, barrage jammers onboard very small UAVs and launching them by hand over a small region could prevent an adversary from communicating for, say, 5 minutes, while a friendly platoon-sized operation transpires in that region.

7.4 LOOK-THROUGH

Look-through is incorporated in jammers to maximize the utility of the available power from them. Once a narrowband target is being jammed, it is normal for that target to change to a prearranged backup frequency that is not being jammed at the moment. The jammer must therefore monitor the target transmission to tell if it is still trying to communicate at the first frequency. A receiver colocated with the jammer cannot monitor at the frequency that is being jammed simultaneously. If it did, it would either be destroyed by the high-power signal at its sensitive front end, or at least it would be desensitized enough so that it could not hear the target. Therefore, for a short period the jammer is turned off so that the monitor receiver can measure the target energy at the frequency. This is called *look-through*. Typically several tens of milliseconds are adequate to determine if the target is still transmitting so the jamming signal need only be disabled for this amount of time.

If look-through were not incorporated, the jammer would waste a good deal of time jamming a target that was no longer trying to transmit at that frequency. This degrades the performance of the jammer in the number of signals it can jam per mission. Techniques incorporated to maximize the utility of the jammer, such as look-through, are called *power management*.

7.5 ANALOG COMMUNICATIONS

The amount of jammer power required to prevent communication depends on the type of modulation used. Some types are more easily jammed than others are. Analog voice, either AM or FM, for example, requires significantly more jamming power than most digital communications—unless the latter contains considerable FEC coding [8, 9].

Modulating a high-power carrier signal with an FM noise signal is the old standby for jamming analog communications. It has been shown several times that this is the best waveform to use against FM targets. Such targets are very plentiful in the VHF band. It has long been known that analog FM communications exhibit a threshold effect wherein if two signals are impinging on a receiver at the same time at the same frequency, the stronger one will dominate the receiver to the more or less complete rejection of the weaker one. The same will happen if one of the signals happens to be a jammer with FM modulation. It would then be expected that in such systems, a jammer broadcasting an FM signal only need be higher in power by some small amount in order to capture the receiver. This indeed does happen. It also occurs in AM systems, however. Shown in Figure 7.3 is the articulation index plotted versus the J/S ratio (noise is considered negligible relative to the jamming signal in this case) [10]. An articulation index of 0.3 or less implies unacceptable performance, between 0.3 and 0.7 is marginal performance, while above 0.7 is an acceptable performance (all from a communicator's point of view). Note that these curves transition from acceptable

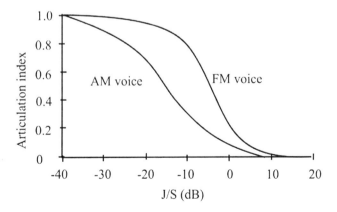

Figure 7.3 Example showing the effects of the J/S ratio on intelligibility in analog communications. (*Source:* [5], © 1992, IEEE. Reprinted with permission.)

articulation to unacceptable fairly sharply, although the capture effect of FM causes it to transition somewhat faster, albeit later. For analog FM, −6-dB J/S or higher is usually considered adequate for disruption of the transmission. At that point, as seen in Figure 7.3, the articulation index is about 50%. The equivalent for AM occurs at about −15 dB, making the latter somewhat less tolerant to jamming than FM.

Other forms of modulated signals are optimal for other forms of modulation, but FM by noise is not a bad choice for most target modulations of interest. Of course, if one knows exactly the waveform that needs to be jammed, one can design an optimal waveform just for that target. This is not usually the case, however.

7.6 DIGITAL COMMUNICATIONS

One of the design criteria for digital communication systems is the BER environment in which they are to operate. Theoretically, if the BER is raised above this level, then communication degradation will occur—usually initially in the form of information transport slowdown and, if the BER gets high enough, complete denial of information exchange [11, 12].

A digital signal need not be jammed continuously in order to generate a high BER. A BER of 0.5 can be achieved against a continuously broadcast digital signal (with adequate signal levels) by only jamming 50% of the time. It can thus be concluded that denying information on digital communication signals should be considerably easier than on analog signals.

In addition to traditional jamming techniques against digital (computer-to-computer) communications, it is also possible to load the transport channel with information to the point where it becomes completely saturated. One would conceivably want to insert bogus data over the channel, thus disallowing the intended information to get through, or at least slow the transport of valid information. Most computers are vulnerable to such data overload, since valid information can be difficult to separate from bogus data, as long as the protocol and format are correct. An invalid report that indicates that there has been a tank battalion spotted at coordinates (x, y) is difficult to detect in a system that is expecting to see reports on tank battalions and their locations.

This is one of the downfalls of the Information Age. Unless and until computers can reason with data, as opposed to just processing it, they will be vulnerable. Indeed, humans are vulnerable to information overload, although to a lesser degree than computers.

Channel coding can be applied to digital signals in order to tolerate higher levels of noise, interference, and jamming. Coding, however, applies to cases where the BER is already rather low (say, less than 10^{-3}). At a BER level of 10^{-1}, there is virtually no protection supplied by most simple coding schemes. The unencoded BER is about the same as the channel BER. It is, however, true that if desired the communicator can apply sufficient coding to beat any jamming scheme, but at that point there is so much coding that there is virtually no information throughput.

7.7 NARROWBAND/PARTIAL-BAND JAMMING

For targets at individual frequencies, a narrowband jammer would be used. The jammer at the frequency used by the targeted communication net transmits a powerful noise signal. The intent is to have the jammer overpower the intended communication signal at the intended receiver. Such a technique is frequently used for standoff jammers to minimize friendly fratricide associated with barrage jammers [13, 14].

This type of jamming should also be useful against DSSS targets as well. DSSS communications systems, as exemplified by the IS-95 cellular radio standard, are particularly sensitive to strong signals close to the receiver: the so-called *near-far problem* inherent in DSSS systems. If an adequately powered narrowband signal can be located close enough to the intended receiver, the jamming margin in the receiver can be overcome, causing unreliable effects to occur in that receiver.

It is possible to utilize the same broadband transmitter and antenna to jam more than one target signal at a time. As mentioned earlier, such techniques carry the appellation power sharing. One exciter is required for each such signal since the exciter determines the frequency to be jammed. The power per signal jammed in this scenario theoretically decreases approximately as the square of the number of signals being jammed with the common power amplifier, however. In practice the power per signal decreases faster than the square of the number of signals because of mismatch issues at the output of the power amplifier. In addition, this technique does not work if the output of the power amplifier is tuned to the jamming frequency unless there is a separate tuned filter at the output of the amplifier for each jamming signal.

It is also possible to time-share a jammer to jam more than one signal essentially simultaneously. Such techniques are called *time sharing*. It is not necessary to preclude 100% of a signal from getting through in order to cause the signal to be unrecognizable. As mentioned, typically, if one-third or so of an analog voice transmission is disrupted, enough is missing to preclude understanding the message and it sounds garbled. Also as discussed above, if 1 bit out of 100 can be disrupted in tactical digital communication systems, communication can be adequately disrupted. Therefore, methods have been devised to rapidly switch an exciter from one frequency to another fast enough so that three analog signals can be disrupted with the same jammer hardware. Against digital signals, perhaps up to 100 targets can be simultaneously precluded from communication. Even with time-sharing, though, look-through techniques are still required to maximize the utility of the jammer. If the output of the power amplifier is tuned to maximize the power transferred to the antenna, it must be rapidly switched as the exciter changes frequency. Retuning high-power filters is difficult at best.

7.8 BARRAGE JAMMING

A barrage jammer is a likely candidate for a UAV-borne jammer that is used against deep, second-echelon forces. This indeed puts the jammer much closer to the adversary than friendly communications, and can essentially completely deny communications within a considerable radius of the jammer. Typically for communication screening a barrage jammer is required. This is when a broad frequency range, perhaps the entire band over which friendly communications will transpire, is transmitted from the screening jammer. The jammer must be much closer to the targeted ES/SIGINT system than the friendly communications, otherwise the jammer will interfere with those friendly communications. One of the ways to implement such a jammer is to generate a relatively narrow (say, 1 MHz) signal comprised of a carrier frequency modulated with noise. This signal then is stepped from one 1-MHz portion of the spectrum to the next, usually, but not necessarily, in succession, dwelling at each step for some period of time, for example, 1 ms. One can cover, say, the lower portion of the VHF frequency band of 30 to 90 MHz in 60 ms [15].

To see how effective such a jammer could be, consider a BFSK signal with incoherent demodulation at the receiver. This is typical for modern-day frequency hopping VHF communication equipment. The demodulator would look something like that shown in Figure 7.4. The two tone frequencies are given by f_1 and f_2, and they occupy a bandwidth of Δf as shown in Figure 7.5. Each channel has a bandwidth in hertz given by B_{ch} and the channel center frequency is f_c. The jamming waveform is stepped at a rate of $R_j = 1/T_j$. The instantaneous bandwidth of the jammer is given by IBW and the total bandwidth covered is given by TBW. The target frequency need not be known exactly and the epoch (timing) information of the target signal is not known. Although the outputs of the detectors in general are correlated due to imperfect filtering, it is assumed for now that these outputs are uncorrelated.

The probability of a symbol error, denoted as P_s, includes the probability of a symbol error when the jamming signal is present, denoted as P_p, as well as the

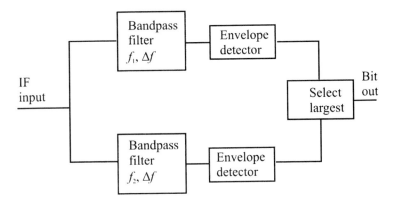

Figure 7.4 Block diagram for a noncoherent detector for FSK.

probability of a symbol error when the jammer is absent, denoted as P_a. Since these events are mutually exclusive

$$P_s = P[\text{jammer present}]P[\text{symbol error}|\text{jammer present}] \qquad (7.2)$$
$$+ P[\text{jammer absent}]P[\text{symbol error}|\text{jammer absent}]$$

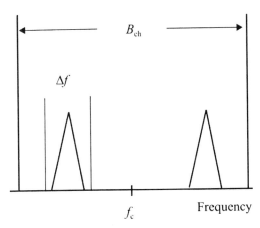

Figure 7.5 Details of the RF channel for the example. The signal is FSK, with each tone occupying a bandwidth of Δf, while the channel bandwidth is B_{ch}.

In general, the probability of a symbol error (for BFSK also a bit error) is given by [13]

$$P_s = \frac{1}{2} e^{-\frac{S}{2N_t + J_1 + J_2}} \qquad (7.3)$$

where S is the power in the signal, N_t is the noise level, and the J_i's are the amount of jammer noise that make it through the bandpass filters. In this case we will assume that $J_1 = J_2 = J$. Thus,

$$P_a = P[\text{symbol error}|\text{jammer absent}] = \frac{1}{2} e^{-\frac{1}{2}\left(\frac{S}{N_t}\right)} \qquad (7.4)$$

$$P_p = P[\text{symbol error}|\text{jammer present}] = \frac{1}{2} e^{-\frac{1}{2}\left(\frac{S}{N_t + J}\right)} \qquad (7.5)$$

The probability of the jammer being present at any particular channel is given by

$$P[\text{jammer present}] = \frac{IBW}{TBW} \qquad (7.6)$$

Since $P[\text{jammer absent}] = 1 - P[\text{jammer present}]$, then

$$P_s = \frac{IBW}{TBW} P_p + \left(1 - \frac{IBW}{TBW}\right) P_a \qquad (7.7)$$

As an example of how well this can work, the results of calculating the probability of producing a symbol error for a variety of different instantaneous bandwidths of the jammer are shown in Figure 7.6 [14]. In this example, the SNR of the intended signal at the receiver is 20 dB. It is well known that the RF environment on a battlefield is very noisy. BERs of 10^{-2} are not uncommon (compare this to telephone quality transmissions where BERs of 10^{-6} are typical). Much of the equipment designed for tactical use is designed for a BER of 10^{-2}. Therefore, achieving BER higher than this can be viewed as successful jamming. As seen from Figure 7.6, this can be achieved with, say an IBW of 5 MHz, at a J/S

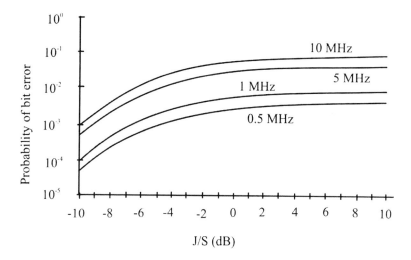

Figure 7.6 Bit error degradation caused by stepped/scanned barrage jamming as a function of the instantaneous bandwidth of the jammer and the J/S ratio. (*Source:* [14], © 1993, IEEE. Reprinted with permission.)

ratio of about −5 dB, relatively easy to achieve with a UAV-borne jammer, for example.

Therefore, as this example shows, disruption of communications can occur for relatively low values of the J/S ratio. If a BER rate of 10^{-2} or better (1 bit in 100 in error or less) is required for reliable communications, then a J/S ratio of −6 to −4 dB, corresponding to instantaneous bandwidths of 5 MHz and 10 MHz, respectively, in this example are all that are necessary to accomplish this amount of degradation.

7.9 JAMMING LPI TARGETS

In the case of jammers that operate against frequency hopping targets, one can either use a follower jammer or a barrage jammer. A follower jammer must detect the frequency of the target, identify that as the target of interest, and then apply narrowband jamming power to that frequency. The problem with barrage jammers is, as mentioned, unintentional fratricide [16–31].

Jamming frequency-hopped targets is dependent on the relative distances between the transmitter and the receiver, the jammer and the receiver and the jammer and the transmitter. If the jammer is too far away from the transmitter relative to the distance between the transmitter and the receiver, the signal will arrive at the jammer after it has already been received at the receiver. The hopping communication system will have already moved to its next frequency, rendering jamming ineffective. The same problem occurs if the distance between the jammer and the receiver relative to the distance between the transmitter and the receiver is too large. Even though the jammer receives the signal in time to ascertain that it is the correct target, the signal emitted from the jammer must travel too far to reach the receiver in time to prevent communication, again rendering jamming ineffective.

The speed at which a frequency-hopped communication network hops is also a consideration for effective EA against it. Since the propagation velocity of the signals is a constant, given a specific configuration of the network and jammer, as the hop rate is increased, a point is reached where the transmitter and receiver hop to the next frequency just as the jamming signal arrives at the receiver at the old frequency, rendering EA ineffective. For that scenario and all faster hop rates the jamming will be ineffective. The closer the jammer is to the receiver and/or transmitter, the faster this rate must be to render the jamming ineffective, but for all scenarios there is such a rate. It is for these reasons that it is important to deploy EA assets as close to the communicating nodes as possible.

7.10 FOLLOWER JAMMER

For LPI targets, particularly those employing frequency hopping, either a barrage jammer or a narrowband jammer can be used, but in the latter case, the ES equipment must be able to follow the transmitter as it changes frequency. This is a matter of identifying the frequency to which the transmitter moved and also identifying that the new signal belongs to the same transmitter, which is no easy task. Such a scheme is called a *follower jammer* [12, 26–30].

Frequency hopping a transmitter is a jamming avoidance ECCM technique. The frequency of the transmitter, as well as the tuned frequency of the receiver, is changed rapidly so that a jammer, fixed on a frequency, has minimum effect on the whole transmission. Follower jamming is a way to combat this ECCM technique, albeit at a fairly high price in EA system complexity (the ES part of the system actually becomes substantially more complex).

The concept of a follower jammer is not new. It has long been standard *communication electronics operating instructions* (CEOI)—the standard operating procedure for tactical communications—practice that, if jammed at one frequency, change to a backup frequency that is not being jammed. This is a form of (very) slow frequency hopping. Tactical jammers have had to track such changes in frequency ever since jammers have been invented.

The receivers used to detect the frequency are the same as those discussed in Chapter 9. Since frequency-hopped radios typically use BFSK as the modulation, a tone jammer can only cover one or the other of the BFSK frequencies. For this reason, modulation is usually added to such jammers so that the whole communication channel is jammed not just one of the tones. Frequently this modulation consists of a noise signal frequency modulated onto the carrier.

7.11 CONCLUDING REMARKS

Successfully denying an adversary the use of the communication spectrum is the purpose of communication EW. It depends on many factors, including the link distances between the transmitter and the receiver relative to the link distance between the jammer and the receiver, the ERP from the jammer relative to the ERP of the transmitter, and the type of communications involved.

Digital communications are easier to jam than analog forms. Creating the necessary BER to deny such communications requires less energy and can be done much more safely than before.

Jamming LPI communications is not difficult, especially if digital modulations are involved. What is difficult with these forms of communication is finding the signal that one wants to jam. With a follower jammer, separating that one signal from all of the RF energy one is likely to encounter in the modern battlespace is difficult. The alternative to surgically jamming a signal such as this is to barrage jam a significant portion of the RF spectrum. This can only be done, however, in situations where friendly communications are not affected.

References

[1] Waltz, E., *Information Warfare Principles and Operations*, Norwood, MA: Artech House, 1998, pp. 165–168.
[2] Schlecher, D. C., *Electronic Warfare in the Information Age*, Norwood, MA: Artech House, 1999.
[3] Adamy, D., *EW 101: A First Course in Electronic Warfare*, Norwood, MA: Artech House, 2001.
[4] Neri, F., *Introduction to Electronic Defense Systems*, Norwood, MA: Artech House, 1991.
[5] Mosinski, J. D., "Electronic Countermeasures," *Proceedings of IEEE MILCOM Conference*, 1992, pp. 191–195.
[6] Torrieri, D. J., *Principles of Secure Communication Systems*, Norwood, MA: Artech House, 1992, pp. 275–288.
[7] Torrieri, D. J., "Fundamental Limitations on Repeater Jamming of Frequency-Hopping Communications," *IEEE Journal on Selected Areas of Communications*, Vol. SAC-7, No. 569, May 1989.
[8] Lathi, B. P., *Communication Systems*, New York: John Wiley & Sons, 1968.
[9] Gagliardi, R. M., *Introduction to Communications Engineering*, 2nd ed., New York: John Wiley & Sons, 1988.
[10] Mosinski, J. D., "Electronic Countermeasures," *Proceedings of IEEE MILCOM Conference*, 1992, p. 193.
[11] Proakis, J. G., *Digital Communications*, New York: McGraw-Hill, 1995.

[12] Simon, M. K., et al., *Spread Spectrum Communications Handbook*, New York: McGraw-Hill, 1994.
[13] Torrerri, D. J., *Principles of Secure Communication Systems*, 2nd ed., Norwood, MA: Artech House, 1992, p. 19.
[14] Poisel, R. A., "Performance Analysis of a Stepped/Scanned Barrage Jammer," *Proceedings of MILCOM 1993*, Boston, MA, 1993.
[15] Peterson, R. L., R. E. Ziemer, and D. E. Borth, *Introduction to Spread Spectrum Communications*, Upper Saddle River, NJ: Prentice Hall, 1995.
[16] Simon, M. K., and A. Polydoros, "Coherent Detection of Frequency-Hopped Quadrature Modulations in the Presence of Jamming – Part I: QPSK and QASK Modulations," *IEEE Transactions on Communications*, Vol. COM-29, No. 11, November 1981, pp. 1644–1660.
[17] Simon, M. K., "Coherent Detection of Frequency-Hopped Quadrature Modulations in the Presence of Jamming – Part II: QPR Class 1 Modulations," *IEEE Transactions on Communications*, Vol. COM-29, No. 11, November 1981, pp. 1661–1668.
[18] Simon, M. K., G. K. Huth, and A. Polydoros, "Differentially Coherent Detection of QASK for Frequency Hopping Systems – Part I: Performance in the Presence of Gaussian Noise Environment," *IEEE Transactions on Communications*, Vol. COM-30, No. 1, January 1982, pp. 158–164.
[19] Simon, M. K., "Differentially Coherent Detection of QASK for Frequency-Hopping Systems – Part II: Performance in the Presence of Jamming," *IEEE Transactions on Communications*, Vol. COM-30, No. 1, January 1982, pp. 165–172.
[20] Lee, J. S., L. E. Miller, and Y. K. Kim, "Probability of Error Analysis of a BFSK Frequency-Hopping System with Diversity Under Partial-Band Jamming Interference – Part II: Performance of Square-Law Nonlinear Combining Soft Decision Receivers," *IEEE Transactions on Communications*, Vol. COM-32, No. 12, December 1984.
[21] Milstein, L. B., and D. L. Schilling, "The Effect of Frequency-Selective Fading on a Noncoherent FH-FSK System Operating with Partial-Band Tone Interference," *IEEE Transactions on Communications*, Vol. COM-30, No. 5, May 1982, pp. 904–912.
[22] Miller, L. E., et al., "Analysis of an Antijam FH Acquisition Scheme," *IEEE Transactions on Communications*, Vol. 40, No. 1, January 1992, pp. 160–170.
[23] Kwon, H. M., L. E. Miller, and J. S. Lee, "Evaluation of a Partial-Band Jammer with Gaussian-Shaped Spectrum Against FH/MFSK," *IEEE Transactions on Communications*, Vol. 38, No. 7, July 1990, pp. 1045–1049.
[24] Viswanathan, R., and S. C. Gupta, "Performance Comparison of Likelihood, Hard-Limited, and Linear Combining Receivers for FH-MFSK Mobile Radio-Base-to-Mobile Transmissions," *IEEE Transactions on Communications*, Vol. COM-11, No. 5, May 1983, pp. 670–903.
[25] Milstein, L. B., and D. L. Schilling, "The Effect of Frequency-Selective Fading on a Noncoherent FH-FSK System Operating with Partial-Band Tone Interference," *IEEE Transactions on Communications*, Vol. COM-30, No. 5, May 1982, pp. 904–912.
[26] Simon, M. K., "The Performance of M-ary FH-DPSK in the Presence of Partial-Band Multitone Jamming," *IEEE Transactions on Communication*, Vol. COM-30, No. 5, May 1982, pp. 953–958.
[27] Torrieri, D. J., *Principles of Secure Communication Systems*, Norwood, MA: Artech House, 1992, pp. 245–257.
[28] Torrieri, D. J., "Fundamental Limitations on Repeater Jamming of Frequency-Hopping Communications," *IEEE Journal on Selected Areas of Communications*, Vol. SAC-7, No. 569, May 1989.
[29] Putnam, C. A., S. S. Rappaport, and D. L. Schilling, "Tracking of Frequency-Hopped Spread Spectrum Signals in Adverse Environments," *IEEE Transactions Communications*, Vol. COM-31, No. 955, August 1983.

[30] Hassan, A. A., W. E. Stark, and J. E. Hershey, "Frequency-Hopped Spread Spectrum in the Presence of a Follower Partial-Band Jammer," *IEEE Transactions on Communications,* Vol. 41, No. 7, July 1993, pp. 1125–1131.

[31] Hassan, A. A., J. E. Hershey, and J. E. Schroeder, "On a Follower Tone-Jammer Countermeasure Technique," *IEEE Transactions on Communications,* Vol. 43, No. 2/3/4, February/March/April 1995, pp. 754–756.

Chapter 8

Antennas

8.1 INTRODUCTION

Antennas are used in communication EW systems to convert electrical signals into propagating EM waves, and vice versa, for converting propagating EM waves into electrical signals. Frequently, antennas for communication EW systems must be broadband, since the frequencies of interest can be located over broad sections of the frequency spectrum. Limited real estate on these systems normally precludes using several antennas, each designed for a relatively narrow frequency band. Thus, antennas that are effective over a wide frequency range are used.

Antenna design historically has been more of an art than a science. Understanding the complexity of especially the near-field components of EM is difficult. The far field is almost always assumed to be a plane wave—an assumption that is reasonable since EM waves propagate in a spherical fashion unless disturbed by an intervening structure.

Antennas are normally treated as reciprocal. That is, they exhibit the same characteristics whether they are used for transmitting or receiving. This is generally true if the antenna is a passive device. It may not be true, however, if the antenna contains active elements such as amplifiers that are unidirectional.

Three of the more important characteristics of antennas are their frequency response, directionality, and impedance characteristics. The first determines the bandwidth over which an antenna will radiate energy effectively. The second describes how an antenna concentrates energy in particular directions while minimizing the propagation of energy in others. The common definitions associated with the important antenna parameters are shown in Figure 8.1.

The impedance characteristics of antennas that are practical for communication EW use present significant design issues. Antennas used for communication purposes can be tuned with appropriate circuitry at the output of the power amplifier so that the output impedance of the amplifier conjugate matches that of the antenna. In communication ES applications, it is frequently

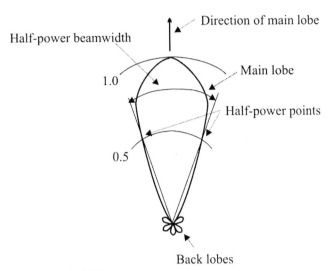

Figure 8.1 Antenna parameter definitions.

required that the antenna have a wide instantaneous bandwidth, precluding the ability to tune the amplifier's output. Without matching the amplifier with the cable and antenna, significant *voltage standing wave ratios* (VSWRs) ensue, which causes energy to propagate back to the amplifier from the antenna, and reduces the amount of energy that is radiated.

Impedance characteristics for small antennas, typical for communication EW applications, frequently have a relatively small resistive component plus a relatively large reactive component. The reactive component can be either inductive or capacitive depending on the frequency and the type of antenna. The output impedance of an amplifier, on the other hand, is typically resistive at 50 ohms, if untuned. This impedance matches that of typical cables used to interconnect the amplifier with the antenna.

The impedance characteristics are important because maximum power is transferred from a source to a load, including antennas, when the impedance of the load is a conjugate match to that of the source. That is, if the output impedance of the source is $Z_s = R_s + jX_s$, then maximum power transfer occurs when $Z_l = R_s - jX_s$. Matching the impedance at one or two frequencies or matching to within reasonable tolerances across a small band of frequencies is relatively easy to achieve. The difficulty of matching the impedances like this arises when trying to match the impedances across a wide bandwidth.

The ability of an antenna to concentrate its energy in a particular direction is referred to as its *directivity*. This directivity is reflected in the gain pattern of an antenna, denoted as G, which is a function of the azimuth and elevation angles of interest relative to some physical parameter of the antenna—usually the direction of maximum gain, or the *boresight*.

Antennas, normally being passive devices, do not increase the amount of power in a signal. They do, however, concentrate the power in certain directions, both for transmitting and receiving signals. There are active antennas, but the active components are not normally utilized to increase power levels—the PA just prior to the antenna can best accomplish that function. Active components are designed into antennas for other purposes, such as impedance matching the transmission line to the impedance of the antenna.

Some of the more common types of antennas, although not all in any sense, are discussed in this chapter. Perhaps the most popular types of antennas used for communication EW systems in the lower frequency ranges are the dipole, monopole, and log periodic, the former two because of their simplicity and 360° azimuth coverage, and the latter because of its relative simplicity, directional gain available, and broad frequency range. As the frequencies of interest increase, other types of antennas are economically feasible (economic in the physical real estate sense).

It should be noted that although the radiation patterns of antennas are approximately known, there are always holes and bumps in the patterns of any real antenna that must be measured. A signal may be received from a particular azimuth very well, yet rotating the antenna slightly could cause the signal to disappear totally.

8.2 ISOTROPIC ANTENNA

An isotropic antenna emits energy equally in all directions [1]. Its radiation pattern is thus a sphere. No physical antenna is truly isotropic, but this antenna configuration forms a convenient standard against which other antennas can be compared.

The power density, in watts per square meter, at a distance R from an antenna in free space is given by

$$P_d = \frac{P_T G_T}{4\pi R^2} \tag{8.1}$$

where P_T is the transmitter power supplied to the antenna. This power is spread out equally by the isotropic antenna onto the surface of a sphere, the area of which is given by $4\pi R^2$. Thus, the power in any selected square meter in free space is decreasing at a rate proportional to $1/R^2$.

The *effective area* of the antenna, denoted as A_{eff}, determines the amount of power an antenna extracts from a passing wave. Ignoring losses, the power out of the antenna is given by

$$P_R = P_d A_{\text{eff}} \tag{8.2}$$

Note that the effective area is not the same as the physical area of the antenna—it typically is between 0.4 and 0.7 × the physical area of an antenna.

The solid angle subtended at the origin of the coordinate system by an area A on the surface of a sphere is given by A/R^2. The *solid beam angle* of an antenna, denoted by Ω, is the solid angle subtended at the base of a cone if all of the energy emitted by an antenna passed through the area A and across A there is uniform power density. The effective area of an antenna is related to Ω by

$$\lambda^2 = A_{\text{eff}} \Omega \tag{8.3}$$

Thus, for a given wavelength, the effective area and the solid beam angle of an antenna are inversely related.

8.3 ANTENNA GAIN

Most antennas have preferred directions for radiating or receiving energy. This preference is referred to as *directivity* and is denoted by D. The directivity at coordinates (x_0, y_0, z_0) or (R, θ, ϕ), where these coordinates are defined in Figure 8.2, of an antenna is the ratio of the power density at that coordinate relative to the power density at the same coordinates due to an isotropic antenna radiating the same total power [2–4].

The *gain* of an antenna indicates this directivity, but it also usually incorporates the efficiency of the antenna. All antennas have some degree of loss associated with resistive losses and losses of structures in the nearby vicinity of the antenna (called its *near field*). G denotes the gain. If the direction is not specified, it is usually assumed to be the direction of maximum gain.

The *efficiency* of an antenna, which is an indication of the losses associated with the antennas, is given by the ratio of the gain of the antenna to its directivity. Efficiency is denoted by ε. Thus

$$\varepsilon = \frac{G}{D} \tag{8.4}$$

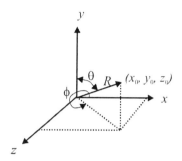

Figure 8.2 Polar and Cartesian coordinate definitions.

Other types of reduction in transmitted or received energy are due to polarization mismatch between the antenna and the EM wave as well as impedance mismatches between the antenna and the feed cable, amplifier feeding the antenna or the receiver for which the antenna provides signals. The cables that connect the antenna to the system electronics also exhibit loss as discussed in Section 2.6.

The gain of an antenna is related to its effective area by

$$G = \frac{4\pi A_{\text{eff}}}{\lambda^2} \tag{8.5}$$

The standard technique for specifying the gain of an antenna is to compare the gain to an isotropic antenna, which is measured in *decibels relative to isotropic* (dBi).

An antenna is said to be *omnidirectional* in a plane if it has no directionality in that plane. That is, if there is no preferred propagation direction, and energy is radiated in all directions in that plane equally. Omnidirectional antennas are frequently used in communication EW systems due to the nature of the communication EW mission—the direction to target emitters is not always known and so must be assumed to be anywhere around the system.

8.4 WIRE ANTENNAS

A wire antenna consists of a length of wire strung between two (or more) support structures. If such an antenna is fed at the center, a dipole is formed as described next. If this antenna is fed at one end, a monopole is formed. More complex structures are common as well [5–8].

8.4.1 Dipole

The dipole is one of the simplest antenna types. It consists of two elements that are aligned in the same direction [9]. Figure 8.3 shows the shape and radiation patterns. For the dipole, as well as the other antennas discussed here, the radiation pattern depends on the physical size of the antenna. For the dipole, in the far field the most important parameter that determines its radiation pattern is its length. These effects are illustrated in Figure 8.3 for two lengths, $L = \lambda/2$ and $L = \lambda$. Since the radiation pattern depends on the length of the antenna, measured in wavelengths, then other factors such as the gain, efficiency, and directivity of these antennas are frequency-dependent. This is true for all known types of antennas, although the characteristic variations with frequency are not the same for all types of antennas as will be shown.

The polarization is the same as the axis of the antenna—if it is vertically oriented relative to the Earth, then the polarization of the signal will be orthogonal

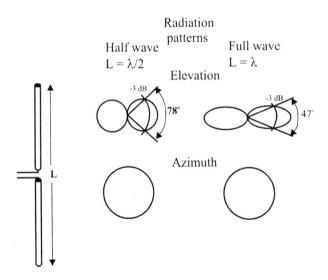

Figure 8.3 Dipole antenna structure and its radiation patterns. The particular radiation pattern depends on the length of the antenna measured in wavelengths. This assumes that the thickness of the antenna is negligible.

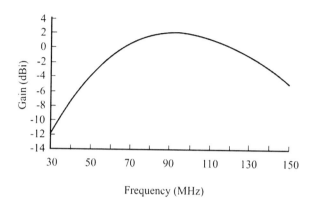

Figure 8.4 Typical gain variation with frequency of a half-wave dipole cut to resonate at $L = \lambda/2$ at 90 MHz.

to the Earth's surface as well. For $L = \lambda/2$ the elevation 3-dB beamwidth is 78°, while obviously the azimuth beamwidth is 360°. The 3-dB point is where the radiation pattern is one-half its value at boresight. A half-wave dipole has about a 2-dB gain over an isotropic antenna, and can be used economically at virtually any frequency where the radiation patterns are adequate. If additional directionality can be allowed or is required, other antenna types provide better characteristics. The effective area of a half-wave dipole is $A_{eff} = 1.64\lambda^2/4\pi$.

Sometimes broadbanding techniques are used with ES antennas. One such technique places resistive loading along the length of the antenna. Such techniques lower the Q of the antenna, since the Q is given by the ratio of the reactance to the resistance. This method, however, also lowers the gain of the antenna at all frequencies; thus, bandwidth is traded for gain. In some cases this tradeoff may be favorable.

Typical gain characteristics of a half-wave dipole antenna, cut for $L = \lambda/2$ at 90 MHz ($\lambda = 3.3$m) are shown in Figure 8.4. At 90 MHz, the antenna has a 2-dBi gain while at the low end it exhibits a 12-dBi loss. At 150 MHz, a loss of 5 dBi is typical. A 10-W transmitter using a tuned dipole antenna at a range of 40 km from a UAV ES system at 3,000m AGL produces a signal with power P_R out of a receive antenna according to the expression

$$P_R = \frac{P_T G_T G_R}{R^4}(h_T h_R)^2 \tag{8.6}$$

This target will produce a signal level out of this antenna of −64 dBm at 90 MHz. At 30 MHz it will be 14 dB less, or −78 dBm. At 150 MHz, it will be 7 dB less or

−71 dBm. For most receiving systems, these signal levels are very adequate for intercept purposes.

In a ground application where h_R = 2m [an *on-the-move* (OTM) application], at 10 km this antenna will produce −104 dBm at 90 MHz, −118 dBm at 30 MHz, and −111 dBm at 150 MHz. These levels are approaching the system noise level and, when the system is used in noisy external environments, probably would not be adequate for intercept.

8.4.2 Monopole

A monopole antenna consists of a single element mounted on a ground plane [10]. It also is a very simple antenna and is a popular configuration for tactical radios in the lower VHF frequency ranges. The characteristics of the monopole are shown in Figure 8.5. The azimuth radiation pattern is also 360°, but the ground plane truncates the elevation pattern. The elevation 3-dB beamwidth is approximately 45° and the maximum gain of these antennas is 0 dBi. Thus, the effective area of a monopole is approximately $A_{\text{eff}} = \lambda^2/4\pi$.

To a large extent the quality of the ground plane determines the antenna pattern close to the horizon. The effect is shown in Figure 8.5. The better the ground plane, the lower to the horizon the antenna pattern falls.

The antennas used in handheld phones of cell systems and PCS are of the

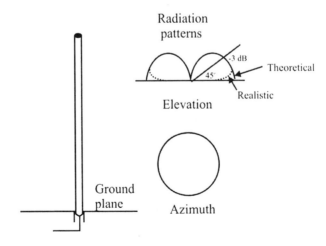

Figure 8.5 Monopole antenna configuration and radiation pattern. This pattern is for a half-wavelength.

monopole variety. In those cases there is a very poor ground plane both in terms of consistency and conductivity. The antenna patterns of such antennas is best modeled as randomly varying.

A variation of the monopole that exhibits a different radiation pattern is the bent monopole. This type of antenna is popular for airborne EW systems where the effective length of the antenna can remain relatively long but it does not protrude from the airframe as far as a linear monopole would.

8.4.3 Loop

A loop exhibits similar radiation patterns as the dipole but does not have the gain [11]. In fact it exhibits a 2-dB loss relative to an isotropic antenna. Loops need not be circular, but can take on arbitrary shapes. Two configurations of loop antennas are shown in Figure 8.6. For the horizontal loop the shape of the E-field forms circles in the horizontal plane, that is, in the same plane as the antenna. The magnetic fields lie in vertical planes for this configuration. The effective area of loop antennas is approximately $A_{eff} = 0.63\lambda^2/4\pi$.

8.4.4 Biconical/Discone

Biconical antennas derive their name from the dual-cone structure of their configuration. Two cones, one inverted on top of the other as shown in Figure 8.7, form this antenna's configuration [12]. The antenna feed point is at the center,

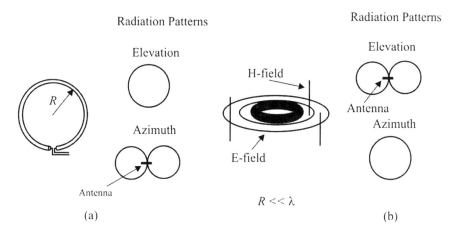

Figure 8.6 Loop antenna patterns: (a) vertical configuration and (b) horizontal configuration.

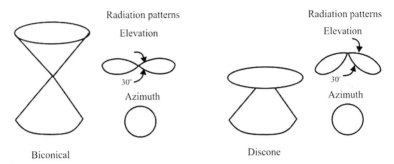

Figure 8.7 Biconical and discone antenna configurations and the radiation patterns.

where the two cones meet. When one of the cones is reduced to a plane, then the discone antenna results as shown.

These antennas have an omnidirectional pattern in the azimuth plane but are more focused (flatter) in the vertical plane than a dipole or monopole. The maximum gain of the biconical antenna is approximately 3–4 dBi, while that of the discone is somewhat less. The biconical effective area is given by $A_{\text{eff}} = 2\lambda^2\pi$. Typically the bandwidth of the biconical is somewhat broader than that of the discone and the discone pattern is more focused downward as shown. The beamwidth in the vertical plane is approximately the same. Of course, the discone antenna can have the opposite orientation by putting the plane on the other side of the cone, forcing the pattern lobes in the upward direction.

8.4.5 Yagi

A Yagi antenna, shown in Figure 8.8, exhibits a relatively narrow frequency bandwidth, typically 5% of its design frequency [13]. Their beamwidth in the elevation plane is approximately 80°–90°, while that in the azimuth plane is about 50°–60° or so. Their gain is between 5 and 15 dBi yielding an effective area of between $3\lambda^2/4\pi$ and $13\lambda^2/4\pi$.

A Yagi antenna consists of one element that is driven. The remainder of the elements are passive and are driven "parasitically." A parasitically driven element absorbs energy from the EM wave that is passing by, and reradiates this energy. The spacing of the elements is such that the phases of the driven element and that of the reradiated waves add properly off the end of the array.

8.4.6 Log Periodic

The log-periodic antenna is the classic TV antenna that has been used for TV reception since the 1950s (see Figure 8.9). The log periodic has similar radiation

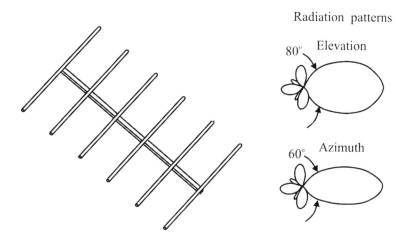

Figure 8.8 Yagi antenna.

patterns to the Yagi. The fundamental difference between the two is the operational bandwidth. The log periodic consists of several dipoles of varying length, separated by a distance that is related to the logarithm of the frequency of the various dipoles (and therefore their length). Log-periodic antennas can be made to cover very broad frequency ranges, even up to 10:1. Its beamwidth in the elevation plane is approximately 80°, while that in the azimuth plane is about 60°. They exhibit maximum gains of approximately 6 dBi. The effective area of a log-periodic antenna is therefore approximately $A_{\text{eff}} = 4\lambda^2/4\pi$ [14].

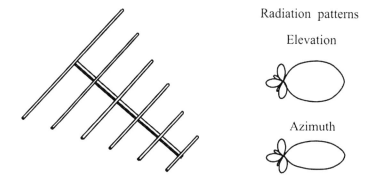

Figure 8.9 Log-periodic antenna. This type of antenna is an example of a frequency-independent antenna.

218   Introduction to Communication Electronic Warfare Systems

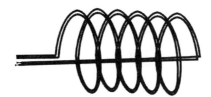

Figure 8.10 Helix antenna.

8.4.7 Helix Antennas

A helix antenna is constructed by configuring several turns of wire into a coil as shown in Figure 8.10 [15]. The normal-mode helix and axial-mode helix antennas are discussed here. Other configurations are possible as well, such as conical shapes. Each such configuration has its own unique characteristics.

8.4.7.1 Normal-Mode Helix

In a normal-mode helix, most of the radiation occurs broadside to the axis of the helix. This requires the diameter of the helix to be much smaller than the

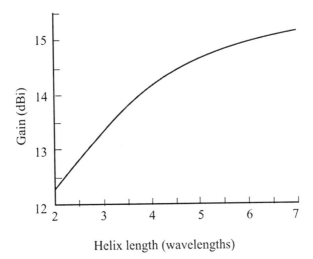

Figure 8.11 Gain variation of an axial-mode helix as a function of the length of the antenna in wavelengths. (*Source:* [16], © 1995, ARRL. Reprinted with permission.)

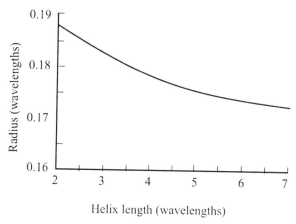

Figure 8.12 Radius of the axial-mode helix at which the gain shown in Figure 8.11 is attained. (*Source:* [16], © 1995, ARRL. Reprinted with permission.)

wavelength. This configuration has similar characteristics to the dipole and monopole.

8.4.7.2 Axial-Mode Helix

In an axial-mode helix, the radiation is predominately off the end of the antenna, or along the antenna's axis. The diameter of the antenna must be approximately the same as the wavelength for this to occur. The shape of the antenna is virtually the same as shown in Figure 8.10. This type of antenna is used to create circular polarization in the transmitted wave. This can also be accomplished with crossed dipoles, which are fed in quadrature (90° out-of-phase). Circular and elliptical polarizations have favorable characteristics in some circumstances. Elliptical polarization is produced by utilizing the proper phasing (other than quadrature) between the feed, or by elliptical construction of the antenna.

The gain of the axial-mode helix for lengths of two through seven wavelengths is shown in Figure 8.11, while the helix radius at which this gain is achieved is shown in Figure 8.12 [16]. The gain is relatively insensitive to variations of the other parameters such as wire diameter or the presence of a short stub at the start of the helix. The reference indicates that a 3.5-dB penalty is incurred if the antenna is operated above radials as opposed to an infinite ground plane. Radials are several long (relative to a wavelength) wires spread out radially from the antenna. Also, a half-wavelength square ground plane is as good as an infinite ground plane. These values of gain yield effective areas between $16\lambda^2/4\pi$ and $32\lambda^2/4\pi$.

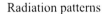

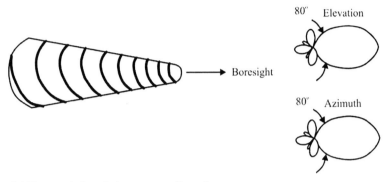

Figure 8.13 Log-periodic spiral antenna configuration.

8.4.8 Spiral Antennas

A spiral antenna, in particular a log-periodic spiral antenna, is shown in Figure 8.13. As in the wire log-periodic antenna, the spacing between windings is varied as the logarithm of the distance from the end of the antenna. Spiral antennas have wide bandwidths, typically covering 3–4 octaves. The regular (non-log-periodic) spiral antennas are typified by gains of about –6 to +1 dBi where the log periodic shown in Figure 8.13 has a higher gain, on the order of 0 to +6 dBi. These antennas exhibit a circular polarized antenna pattern and can be configured as either right-hand circular or left hand circular. As shown in Figure 8.3, the beamwidth in either the elevation or azimuth planes is about 80°.

8.5 ACTIVE ANTENNAS

Shorter antennas can match some of the characteristics of larger (longer) antennas by using an active device (amplifier) with the antenna. Such antennas are called *active antennas*. For short wires, say, less than 10m or so, the impedance of the wire decreases with increasing length. As the length increases, the amount of EM wave energy collected by the antenna increases. For short-wire antennas, then, not much signal energy is collected. The amplifier at the output of the short antenna amplifies what signal is collected, more or less to the same levels as the output of much longer wire antennas. Figure 8.14 shows an example of a short active

Antennas 221

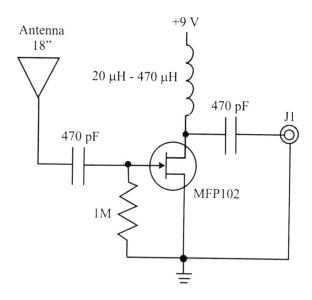

Figure 8.14 An active antenna. The output driver in this circuit converts to a low impedance, matching typical cables. (*Source:* [17], © 1989, Gernsback Publications, Inc. Reprinted with permission.)

antenna [17] for the HF and lower-frequency ranges. The value selected for the inductor depends on the frequency range of the amplifier in Figure 8.14.

The disadvantage of active antennas is that interferers become larger in amplitude as well. Also, the amplifier adds noise to what otherwise might be a low-noise signal. In addition, like all active devices and many devices that are not active, an amplifier is not truly linear over all its operating range. Thus, *intermodulation* (IM) products can be generated in it, particularly by close, strong, unwanted signals. These IM products must be filtered to prevent significant corruption.

For use in arrays, active antennas must be amplitude- and phase-matched in their characteristics. If not, errors will be generated that are difficult to manage, especially over time and temperature.

The principal advantage of active amplifiers is the size of the antenna element. They can be very short relative to their passive counterparts. This is particularly important in the lower frequency ranges (HF and less) because antennas are large in this region. The antenna shown in Figure 8.14 is only 18 inches long in the HF range. In the higher frequency ranges, antennas are relatively small anyway so active antennas are not used much there. Since the first amplification stage in an amplifier chain is the predominant contributor to the noise figure, it is important that the devices used in an active antenna be of the low-noise variety.

8.6 APERTURE ANTENNAS

An aperture antenna presents a surface or two-dimensional structure to a propagating wave, as opposed to wire antennas discussed above. The latter essentially presents a one-dimensional structure. An example of an aperture antenna is a horn illustrated in Figure 8.15. This type of antenna is frequently used at higher-frequencies. They are typically driven by waveguides as opposed to cables that are common at lower frequencies. Signals travel down waveguides by reflecting off the sides of the waveguide. At the end of the waveguide, it is opened up wider and wider, thus forming a horn aperture antenna. Other forms of aperture antennas can be made, such as circular and elliptical. These antennas are frequently used as the feed antennas for reflector antennas such as the parabolic dish. Reflector antennas have reflecting surfaces much larger than these feed antennas, thus producing substantial gain. The effective area for an optimum horn antenna is $A_{eff} = 0.81A$, where A is the physical area of the aperture [7, 18–20].

8.6.1 Parabolic Dish

A parabolic dish antenna is an example of a reflector antenna. This antenna reflects the signal off a surface to form the beam. This type of antenna is popular at the higher-frequency ranges because of its excellent gain and directional

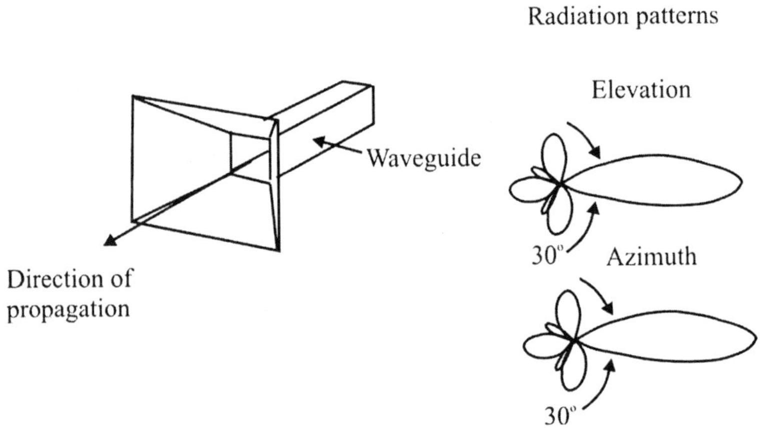

Figure 8.15 Schematic diagram of a horn antenna. The radiation patters are similar in azimuth and elevation.

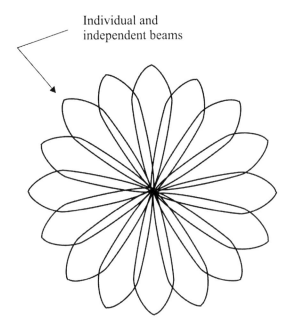

Figure 8.18 Antenna patterns of a switched beam smart antenna array.

patterns to facilitate maximum gain approximately in the direction of the mobile user. Such patterns at a base station might look like that shown in Figure 8.18 [26]. In this example there are 18 individual beams that can be applied.

Adaptive antennas, on the other hand, are a bit more complex. In this type of smart antenna system an RF map of the local RF environment is generated, steering nulls in the specific direction of detected interferers while directing gain in the direction of the intended mobile user.

In both types of antenna systems, the user capacity is increased, which is the motivation behind incorporating them in the first place. Adaptive antennas are more flexible and, in most cases, will increase capacity better than switched beam systems. The latter utilize fixed antenna beams and therefore are less flexible.

8.7 GENETICALLY DESIGNED ANTENNAS

A new antenna design technique that has been devised lately relies on genetic algorithms for determining the structure. In that technique, structure seeds are given to the algorithm and antenna elements are added, one at a time and at random, to the successor structure, the first of which is the seed. A measure of performance is required so that the structures can be evaluated at each stage. If the performance of the new structure is better than the last stage, then the design is kept for the next pass. If not, it is discarded. The normal antenna parameters such as frequency coverage and bandwidth are specified. Some of the performance parameters used to evaluate each antenna configuration are gain and directionality.

These algorithms are called genetic because they mimic the way biological entities evolve. Bad traits typically die off after time while good traits tend to last. The resultant structures can have very interesting shapes, some of which are counterintuitive.

8.8 MORE ON ANTENNA GAIN

An antenna has gain if it has a preferred direction in which to radiate energy over other directions. An isotropic antenna has no preferred direction and radiates equally in all directions. For an antenna with gain, the region in the preferred direction that the signal energy occupies is called the *main lobe*. (See Figure 8.1.) It is defined by the angle formed by the half-power points as shown. The back lobes are minor lobes in the side and back directions, in other words, in directions other than the preferred direction. They represent energy that is wasted.

The Yagi is a popular directional antenna for many applications. This antenna along with its radiation patterns is shown in Figure 8.8. Such an antenna as this typically has a gain of 12 dBi or more, but is typified by a narrow bandwidth. The radiation pattern in the horizontal plane has a beamwidth of approximately 60° and in the vertical plane approximately 90° (see Figure 8.19). A gain of 12 dB represents a gain of $10^{12/10} = 15.8$ in linear terms. There are $6 \times 4 = 24$ sectors defined by 60° in azimuth and 90° in elevation in a sphere. Thus for an isotropic antenna if the radiated power is P_T W, there are $P_T/24$ W radiated in each sector. In the Yagi antenna, in the preferred direction there is 15.8 times more energy radiated than the isotropic antenna. Therefore, the power radiated in the preferred direction is

$$P_{\text{preferred}} = 15.8 \frac{P_T}{24} \qquad (8.7)$$
$$= 0.66 P_T$$

Thus two-thirds of the radiated energy is leaving the antenna through this one sector. Assuming for the moment that the back lobes are not lobes at all but are represented by a single segment of a smooth sphere, one-third of the energy is radiated through this segment. This region is comprised of the remaining twenty-

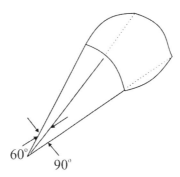

Figure 8.19 Rectangular approximation to a sector in the sphere around an antenna.

three $60° \times 90°$ sectors. In any one of these sectors, then, the amount of power radiated is

$$P_{back} = \frac{1}{23}\frac{1}{3}P_T$$
$$= \frac{P_T}{69} \quad (8.8)$$
$$= 0.014 P_T$$

As an example, suppose $P_T = 10W$. Then with this Yagi, $P_{preferred} = 6.6W$ and $P_{back} = 0.14W$.

Another popular antenna for point-to-point communication links is the horn-fed parabolic dish. A rendering of one of these is shown in Figure 8.16. These antennas have a very narrow beamwidth and are used at higher frequencies. Typical characteristics are 2° beamwidth in elevation, 2° beamwidth in azimuth, and 40-dB gain in the preferred direction at 4 GHz. Using the same approximate analysis as above, there are 32,400 $2° \times 2°$ sectors in a sphere. Thus, an isotropic antenna radiates $P_T/32,400W$ in any of these sectors. The power radiated in the preferred direction of the directional antenna is thus

$$P_{preferred} = 10^{40/10}\left(\frac{P_T}{32,400}\right) \quad (8.9)$$
$$= 0.309 P_T$$

and the power in the other sectors is given by

$$P_{\text{back}} = \frac{1}{32{,}399} 0.691 P_T \qquad (8.10)$$

If $P_T = 10$W, then $P_{\text{back}} = 2 \times 10^{-4}$W.

This quick analysis points out why it is difficult to intercept the sidelobes of directional signals. With the Yagi antenna there is 47 times more power in the preferred direction than in a sidelobe direction. For the parabolic dish there is over 15,000 times more power radiated in the preferred direction than in a sidelobe.

8.9 CONCLUDING REMARKS

Antennas come in a wide variety of forms and configurations, each with its own characteristics. Only a few of the types of antennas available were included here to illustrate those that are typically used with communication EW systems.

Antennas in communication EW systems are used to convert radiated energy as manifest in propagating EM waves into electrical signals for further signal processing, and vice versa. They are also used to convert electrical signals into forms suitable for propagation as EM waves. Antennas are generally complex systems to design and analyze and frequently simplifying assumptions are necessary to facilitate any kind of analysis on all but the simplest of configurations. Antennas store energy temporarily until it can be radiated—thus, they exhibit characteristics similar to capacitors and inductors; the particular design dictates which. All real antennas exhibit a loss as well.

The type of antenna used in any given application will depend on several factors, such as the frequency range of operation, the real estate available for the antenna, and the required gain. The desired spatial coverage is also an important consideration. Omnidirectional antennas emit radiation equally in all directions in one or more planes.

References

[1]   Stutzman, W. L., and G. A. Thiele, *Antenna Theory and Design*, New York: John Wiley & Sons, Inc., 1981, p. 33.
[2]   Stutzman, W. L., and G. A. Thiele, *Antenna Theory and Design*, New York: John Wiley & Sons, Inc., 1981, pp. 32–40.
[3]   *Reference Data for Radio Engineers*, 6th ed., Indianapolis, IN: Howard W. Sams & Co, Inc., 1975, p. 27–33.

[4] Pender, H., and K. McIlwain, *Electrical Engineers' Handbook Electric Communication and Electronics*, 4th ed., New York: John Wiley & Sons, 1963, pp. 6-71–6-72.
[5] *Reference Data for Radio Engineers*, 6th ed., Indianapolis, IN: Howard W. Sams & Co, Inc., 1975, p. 27–25.
[6] Pender, H., and K. McIlwain, *Electrical Engineers' Handbook Electric Communication and Electronics*, 4th ed., New York: John Wiley & Sons, 1963, pp. 6-65–6-71.
[7] Gagliardi, R. M., *Introduction to Communication Engineering*, New York: John Wiley & Sons, 1988, pp. 105–117.
[8] Fujimoto, K., et al., *Small Antennas*, Letchworth, Hertfordshire, U.K.: Research Studies Press, Ltd, 1988.
[9] *Reference Data for Radio Engineers*, 6th ed., Indianapolis, IN: Howard W. Sams & Co, Inc., 1975, p. 27-7.
[10] Stutzman, W. L., and G. A. Thiele, *Antenna Theory and Design*, New York: John Wiley & Sons, 1981, pp. 92–94.
[11] Stutzman, W. L., and G. A. Thiele, *Antenna Theory and Design*, New York: John Wiley & Sons, 1981, pp. 95–104.
[12] Stutzman, W. L., and G. A. Thiele, *Antenna Theory and Design*, New York: John Wiley & Sons, 1981, pp. 270–278.
[13] Stutzman, W. L., and G. A. Thiele, *Antenna Theory and Design*, New York: John Wiley & Sons, 1981, pp. 220–229.
[14] Stutzman, W. L., and G. A. Thiele, *Antenna Theory and Design*, New York: John Wiley & Sons, 1981, pp. 287–303.
[15] Stutzman, W. L., and G. A. Thiele, *Antenna Theory and Design*, New York: John Wiley & Sons, 1981, pp. 261–270.
[16] Emerson, D., "The Gain of the Axial-Mode Helix Antenna," accessed March 2001, http://ourworld.compuserve.com/homepages/demerson/helix.htm.
[17] *Popular Electronics*, July 1989.
[18] Stutzman, W. L., and G. A. Thiele, *Antenna Theory and Design*, New York: John Wiley & Sons, 1981.
[19] *Reference Data for Radio Engineers*, 6th ed., Indianapolis, IN: Howard W. Sams & Co, Inc., 1975, pp. 27-40–27-43.
[20] Pender, H., and K. McIlwain, *Electrical Engineers' Handbook Electric Communication and Electronics*, 4th ed., New York: John Wiley & Sons, 1963, pp. 6-73–6-80.
[21] Stutzman, W. L., and G. A. Thiele, *Antenna Theory and Design*, New York: John Wiley & Sons, 1981, pp. 422–440.
[22] Torrieri, D. J., *Principles of Secure Communication Systems*, Norwood, MA: Artech House, 1992, pp. 367–462.
[23] *Reference Data for Radio Engineers*, 6th ed., Indianapolis, IN: Howard W. Sams & Co, Inc., 1975, pp. 27-28–27-43.
[24] Stutzman, W. L., and G. A. Thiele, *Antenna Theory and Design*, New York: John Wiley & Sons, 1981, pp. 160–167.
[25] Rappaport, T. S. (ed.), *Smart Antennas: Adaptive Arrays, Algorithms, & Wireless Position Location*, Piscataway, NJ: IEEE Press, 1998.
[26] Goldburg, M., and M. Lynd, "Tutorial On Smart Antennas," wireless design on-line, accessed July 2001, http://news.wirelessdesignonline.com/design-features/19980127-92.html.

Chapter 9

Receivers

9.1 INTRODUCTION

Receivers used for the intercept of communication signals come in a variety of forms, depending on the specific functions to be performed and the signal type to be intercepted. For narrowband signals, the most popular receiver type over the last several years has been the superhetrodyne. For LPI signals, where the instantaneous frequency of operation of the emitter is unknown and must be determined or where the bandwidth of the signal is intentionally made larger than that necessary to transmit the signal, wideband receivers are necessary—the types of these discussed here are the compressive receiver, the Bragg cell receiver, and the digital receiver. Note that for frequency-hopping LPI signals, once the instantaneous transmission frequency is determined, system configurations are frequently such that handoff to a fast-tuning superhetrodyne receiver is made so that the wideband system assets are not tied up at that one frequency.

The fundamental function that a receiver performs is to convert a signal from the one present at the antenna to a signal in more usable forms, usually demodulated. This often involves frequency conversion and demodulation. Frequency conversion is frequently necessary because it is often easier and cheaper to produce components in one frequency range rather than others. Standardization of components reduces the cost of those components. Demodulation extracts the information carried on the signal. Typical examples are to recover the AM or FM audio signals on a signal, or to reconstruct a bit stream of a digital signal.

An overview of the types of receivers typically used in communication EW systems is presented in this chapter. There are several good books that go into much more detail than presented here. The reader is referred to [1].

9.2 RECEIVERS

Illustrated in Figure 9.1 is perhaps the simplest receiver. This receiver connects directly to an antenna. The diode and capacitor combination, with the speaker serving as a resistive load, perform AM (envelope) detection of any and all signals at the antenna. If there is just one such signal around and it is sufficiently strong, then the detected/demodulated signal will be heard from the speaker.

Primarily because of its simplicity this receiver has several drawbacks. The first was mentioned above—all signals are detected, allowing no selectivity possibilities among signals. The second is that it is not very sensitive. Signals must be strong in order to be detected. The last major drawback for the use of such a receiver in communication EW systems is that it detects only AM signals.

Selectivity can be included by adding a filter as shown in Figure 9.2. (The box with the squiggly lines in it is the symbol for a bandpass filter.) The bandpass filter can be set at whatever frequency is desired with a bandwidth that is matched to whatever signal is at that frequency. A variation of this receiver is to make the bandpass filter tunable, thereby facilitating detection of signals at different frequencies. The problem here is if the frequency spectrum of interest is at all large, over which this receiver must be tuned, such tunable bandpass filters are difficult to make.

While addressing the selectivity problem, this receiver is still relatively insensitive and detects only AM signals. To make the receiver more sensitive, RF amplifiers can be added as shown in Figure 9.3. These fixed-tuned amplifiers are in fact a practical configuration for a receiver that is used to intercept targets at a known frequency that are AM-modulated.

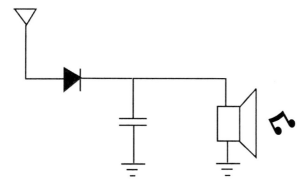

Figure 9.1 Block diagram of a simple AM receiver.

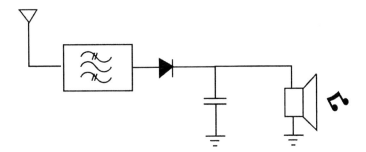

Figure 9.2 Block diagram of a simple AM receiver with bandpass filtering for increasing the selectivity.

To facilitate easier construction of components for receivers, the *superhetrodyne receiver* was invented; it is a modification of the receiver shown in Figure 9.3. The block diagram of this type of receiver is shown in Figure 9.4, where additional types of detectors have been included. The circle with the × in it is called a *mixer* and it multiplies the incoming signal from the antenna with a *local oscillator*. The object is to have the output of the mixer at a constant frequency for subsequent filtering and amplification; this is called the *intermediate frequency* (IF). Fixed-frequency filters and amplifiers are much easier to build than tunable ones. The local oscillator is the only component in such a receiver that is tuned.

Frequently filtering preceding the mixer is included. This is called *preselection filtering* and reduces the amount of interference that is input to the receiver—more on this in Section 9.2.1. Figure 9.4 is oversimplified in that there is typically more than one IF frequency used, and there are several stages of amplification and filtering.

This type of receiver is the most prolific receiver used in communication EW systems. It provides excellent selectivity and sensitivity while being flexible in

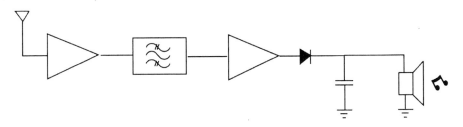

Figure 9.3 Block diagram of a selective AM receiver with amplifiers to increase the sensitivity.

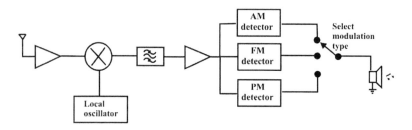

Figure 9.4 Simplified block diagram of a superheterodyne receiver with multiple demodulators/detectors.

applications. The use of common frequency plans gives this receiver flexibility in use at a relatively inexpensive price.

9.2.1 RF Amplification and Filtering

When signals first enter a receiver, they are amplified and a stage of filtering called preselection filtering is applied. It is the first such stage that plays the largest role in establishing the noise figure for the receiver, so it is important that this stage have low noise.

9.2.1.1 Preselection

The filters prior to the mixer are called *preselection filters*. They are used to eliminate or reduce signals from producing unwanted mixing products within the receiver. They are either suboctave filters that are lowpass filters that cover a bandwidth that is less than twice the center frequency, or they are tuned filters. The tuned filters have a bandwidth that is about 5 to 10% of the tuned frequency. Both types of filters eliminate any signals that are twice the tuned frequency or higher.

9.2.1.2 Noise Figure and Noise Factor

Noise is always present in communication systems and the sensors that target them. The discussion of noise, probability of detection (P_d), probability of false alarm (P_{fa}), and *receiver operating characteristic* (ROC) curves presented in Chapter 10 applies here. That discussion assumes that the signals to be received are pulses. The results are directly applicable to the receipt of digital signals and are generally assumed as close approximations for other types of signals as well.

Every electronic device generates internal noise due to the excitation of electrons when the temperature is raised above absolute zero. This added noise decreases the quality of the signal and makes subsequent signal processing, such as demodulation, more difficult. The amount of noise that a receiver adds to a signal is indicated by the *noise figure*, denoted by F. When expressed in decibels relative to kTB, it is known as the *noise factor,* denoted by N. That is, $N = 10 \log F$ dB$_{kTB}$. kTB is a term that represents ambient conditions for a set noise bandwidth where

1. $k = 1.38 \times 10^{-23}$ Joules/K;
2. $T =$ Ambient temperature in K;
3. $B =$ Noise bandwidth in the receiver stage under consideration, in hertz.

Specifically the noise figure represents the amount of noise that must be present at the input to a noiseless receiver stage, relative to kTB, in order to produce the amount of noise observed at the output of that stage.

When n noisy, linear stages are cascaded, which is usually the case, then the resultant noise figure is given by

$$F = F_1 + \frac{F_2 - 1}{G_1} + \frac{F_3 - 1}{G_1 G_2} + \cdots + \frac{F_n - 1}{G_1 G_2 \ldots G_{n-1}} \qquad (9.1)$$

where the gains of the individual stages are given by G_i and their noise figures are given by F_i. In a typical communication EW receiver, the linear stages to which this noise figure calculation applies consist of the RF amplification stages as well as the IF amplifier stages—everything before the demodulation of the signals. Therefore, to minimize the noise figure of the system, it is important to use high gain stages early in the RF path.

When a receiver is embedded into a system, then the system noise figure is defined in an equivalent fashion, but cable losses, signal distribution losses, preamplification gains, and the like must be taken into consideration. The components in an RF receiver chain prior to the receiver are shown in Figure 9.5. They include antennas, cables, and perhaps some amplification at the antenna. Losses prior to the first amplification stage are simply added to the noise figure of that stage. With all parameters expressed in decibels, the system noise figure can be computed by using the following equation

$$N(\text{in dB}) = 10 \log \left[\begin{array}{l} 10^{N_p/10} + 10^{L_1/10} + 10^{(L_2 - G_p)/10} \\ + 10^{(N_R - G_p)/10} - 10^{-G_p/10} \end{array} \right] \qquad (9.2)$$

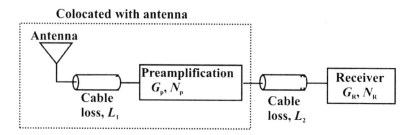

Figure 9.5 Components in an RF chain in an EW system prior to the receiver.

N_R is the receiver noise figure, N_p is the noise factor of the preamplifier, and L_2 is the loss between the preamplifier and the receiver. This equation assumes that the gain of the receiver is sufficiently high that all stages following the receiver contribute negligibly to the system noise figure.

9.2.1.3 Noise-Power Ratio

The *noise-power ratio* is a way to measure the overall performance of a receiver. It is a test procedure that measures the amount of noise introduced by the nonlinearities of a receiver. It is the same test that is used to measure the noise contributions in amplifiers, which is described in Section 13.3.10.

9.2.2 Sensitivity

The sensitivity of a receiver is defined as the smallest signal that a receiver needs at its input to adequately process the signal. It is frequently expressed in decibels relative to 1 mW measured at the RF input terminals of the receiver.

The thermal noise level, caused by excitation of electrons when raised above a temperature of absolute zero, is the minimum noise level achievable without explicit signal processing to remove it (some of which are described in Section 10.5). It is given by the above expression for kTB.

The SNR required to demodulate a signal is signal-dependent. Herein it will be denoted by SNR_{req}. Some signals require more than others. The level present however must be higher than $kTB + N + SNR_{req}$ (all in decibels) in order for the receiver to demodulate the signal.

9.2.2.1 Minimum Detectable Signal

Another definition of sensitivity that is sometimes used is the *minimum detectable signal* (MDS). The implication here is that all that the receiver needs to do is

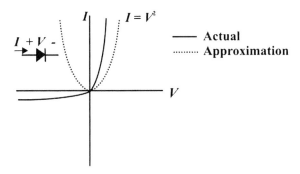

Figure 9.6 The characteristics of a diode and an approximation.

detect a signal, as opposed to the more general case given in Section 9.2.2 where the required SNR for demodulation is specified.

9.2.3 Dynamic Range

The *dynamic range* of a receiver is an indication of how well it will receive weak signals in the presence of stronger ones at its input. Strong signals can cause spurious signals to be generated within the receiver, which can override weak signals, among other deleterious effects. There are several ways to specify the dynamic range and some of them are described here.

9.2.3.1 Intermodulation Distortion

When two (or more) signals are input to a nonlinear device, mixing products are generated that may not be wanted. An example of a nonlinear device is the diode that is shown in Figure 9.6. Suppose an approximation to this device is to have a transfer function given by the square of the input, and further suppose that two simple signals are input to the device. Thus the input is

$$\sin(2\pi f_1 t+\theta_1)+\sin(2\pi f_2 t+\theta_2) \qquad (9.3)$$

The output is given by output = input^2. Thus,

$$\begin{aligned}\text{output} &= [\sin(2\pi f_1 t+\theta_1)+\sin(2\pi f_2 t+\theta_2)]^2 \\ &= \sin^2(2\pi f_1 t+\theta_1)+2\sin(2\pi f_1 t+\theta_1)\sin(2\pi f_2 t+\theta_2)+\sin^2(2\pi f_2 t+\theta_2)\end{aligned} \qquad (9.4)$$

The middle term can be expanded into

$$\cos[2\pi(f_1 - f_2)t+(\theta_1-\theta_2)] - \cos[(2\pi(f_1+f_2)t+(\theta_1+\theta_2)] \tag{9.5}$$

The first and third terms can be expanded as

$$\sin^2(2\pi f_i t+\theta_i) = \frac{1}{2}[\cos(2\pi f_i t-\theta_i - 2\pi f_i t+\theta_i)-\cos(2\pi f_i t+\theta_i +2\pi f_i t+\theta_i)$$
$$= \frac{1}{2}[1-\cos(2\pi f_i t+2\theta_i)] \quad i=1,2 \tag{9.6}$$

These terms thus produce a signal at the output that is at twice the input frequency, with twice the phase, as well as a term at $f = 0$. Thus, signals at $f_1 - f_2$ and $f_1 + f_2$ have been generated at the output, in addition to the expected signals at twice the input frequency. These signals are referred to as *intermodulation distortion*.

Whereas this simple analysis used a square of the input signal as the nonlinearity, in general any nonlinearity creates similar effects. The specific degree of the signals thus generated is different for other types of nonlinearities.

9.2.3.2 Intercept Points

The *intercept points* of a receiver are the input signal levels at which the first and second and the first and third transfer characteristics of the receiver intersect. Figure 9.7 shows the transfer function characteristics of a receiver. The *fundamental* is the expected, normal transfer characteristic with a slope given by the gain of the receiver. The second-order transfer characteristic describes how the receiver reacts to a signal if the output is the square of the input, likewise for the third order where it is the cube of the input. The higher these points, the larger the dynamic range of the receiver, since larger input signals are required to put the receiver into its saturation region. Generally it is desired to make the dynamic range as high as possible in communication EW receivers. Although not unique to the receiver itself, as other components must also have a wide dynamic range, such a range is required because close signals can saturate the receiver, overpowering the distant, weaker target signals of interest.

It is not always possible to make these points arbitrarily high, however. One of the many factors involved concerns power consumption. The input stages to a receiver are typically low impedance, say, 50 ohms. This is to match the impedance of interconnecting cables. However, to make the intercept points high,

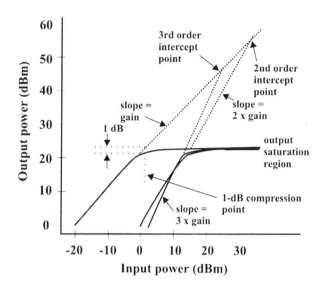

Figure 9.7 Transfer characteristics of receivers and amplifiers.

higher voltages are necessary. The net result of higher voltages across relatively low impedances is higher power consumption as $P = V^2/Z$. In those cases where the receiver is battery-powered, there are definite limits on the amount of power consumption allowed.

9.2.3.3 1-dB Compression Point

The *1-dB compression point*, also shown in Figure 9.7, is the input level at which the fundamental response differs from linearity by 1 dB. It is an indication of the onset of nonlinear performance, or the end of the linear range of operation.

9.2.3.4 Two-Tone Dynamic Range

The *two-tone dynamic range* is the difference, in decibels, between the magnitudes of two equal amplitude tones input to the receiver to the magnitude of the first spurious product. This notion is illustrated in Figure 9.8. These spurious signals are the result of taking the receiver into its nonlinear region of operation as described above.

240   Introduction to Communication Electronic Warfare Systems

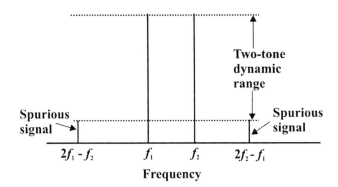

Figure 9.8 Definition of two-tone dynamic range. It is a measure of the linear range of a receiver.

9.2.4 Mixing/Frequency Conversion

The most common way to convert the frequency of a signal is to multiply two signals together in a mixer, as shown in Figure 9.9. The input signal is multiplied by the signal from a *local oscillator*, whose frequency has been set to produce the desired frequency of the output signal. Tuning a receiver this way is accomplished by changing the frequency of the local oscillator, which in effect causes signals at a different RF frequency to be presented to the stages following the mixer. The output frequency is held constant to facilitate amplification at a constant

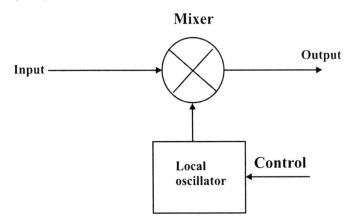

Figure 9.9 A local oscillator and mixer are used to convert frequencies. The control is used to set the frequency of the local oscillator thus changing the frequency to which the receiver is tuned.

frequency, thus allowing narrow bandwidth amplifiers to be used. The output frequency is referred to as the IF.

When two signals, represented mathematically as two sine waves, are multiplied together, the output signal contains components at the sum and difference frequencies of the input. That is,

$$\sin(2\pi f_1 t+\theta_1)\sin(2\pi f_2 t+\theta_2) = \frac{1}{2}\{\cos[(f_1-f_2)t+(\theta_1-\theta_2)] \\ -\cos[(f_1+f_2)t+(\theta_1+\theta_2)]\} \quad (9.7)$$

Typically one or the other of these output components is filtered out, and the remaining signal is used for further processing. This is the foundation of the superhetrodyne receiver described above.

9.2.5 Intermediate Frequency Filtering and Amplification

Standard IFs are used so that common, easy-to-manufacture parts can be used for filtering and amplification. A typical IF for HF receivers is 455 kHz, whereas for VHF and above, 10.7 MHz and 21.4 MHz are typical. In some applications, 70 MHz is used.

9.2.5.1 Automatic Gain Control

Automatic gain control (AGC) is frequently employed to keep the signal levels within the receiver small enough so that the amplifier stages remain in their linear range. The AGC controls the gain of the RF and IF portions of the receiver.

9.2.5.2 Selectivity

Perfect, also called *brick-wall*, filters are idealized devices. In reality there are no such devices. If there were, however, then the spillover from one frequency channel to the next would be avoided and there would be no interference from this source. Since real filters are not brick-wall, some spillover from adjacent channels occurs. This spillover can occur over several frequency channels. If there is energy in one of these channels, some of it will show up on the other channels as shown in Figure 9.10.

The *selectivity* of a receiver is the degree to which signals in adjacent frequency cells are rejected. It is determined principally by the filters used in the IF chain. How well a receiver rejects this adjacent channel energy is its selectivity.

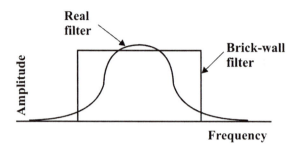

Figure 9.10 IF filter characteristics. The brick-wall filter is an idealization of a real filter response.

9.2.6 Detection/Demodulation

The detection and demodulation function extracts the information content from the carrier. For the demodulation process to work correctly, the demodulator must complement the modulator at the transmitter. How well this works depends to a large extent on the SNR at the receiver.

As previously indicated the SNR requirements for demodulation or other signal processing can be, and often are, different from those at IF. Therefore the relationship of the post-detection, or demodulated, SNR to the predetection, or IF, SNR, also called the *carrier-to-noise ratio* (CNR) is important. The relationship between the postdetection SNR and the CNR depends on the type of modulation employed.

9.2.6.1 Amplitude Modulation

Demodulation of AM is relatively easy. A simple, noncoherent demodulator is shown in Figure 9.11, although more sophisticated forms are available. For other forms of AM, such as SSB, more sophisticated demodulators are required.

In AM, the information is carried in the variations in the amplitude of the carrier. The demodulator employed with AM is an envelope detector. If P_m is the power of the modulating signal at the transmitter, and CNR_{IF} is the SNR coming from the IF section of the receiver, then the postdetection SNR is given by [2]

$$SNR_s = 2\left(\frac{P_m}{P_m+1}\right)CNR_{IF} \qquad (9.8)$$

In this equation P_m is normalized so that $P_m \leq 1$, so the maximum postdetection SNR that can be achieved is equal to CNR_{IF}.

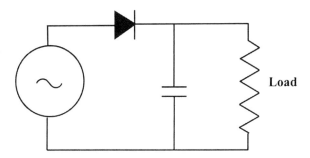

Figure 9.11 Simple noncoherent AM DSB demodulator.

For coherent demodulation, where the carrier phase is used to synchronize the demodulation process with the carrier waveform,

$$\text{SNR}_d = 2 \, \text{CNR}_{IF} \tag{9.9}$$

Therefore, coherent demodulation of AM signals produces a 3-dB improvement in sensitivity (the output signal is 3 dB, or two times better than with incoherent demodulation).

9.2.6.2 Frequency and Phase Modulations

A simple way to demodulate FM is to impress the modulated carrier across an inductive load. The voltage produced across such a load is proportional to frequency, so measuring the output voltage appropriately will generate the modulating signal. This circuit is called a (frequency) discriminator. Since the impedance of an (ideal) inductor with inductance L is given by $Z = j2\pi fL$, the impedance then linearly increases with frequency so the voltage across it will also. Therefore, the circuit shown in Figure 9.12 can be used to demodulate an FM carrier.

Phase demodulation has characteristics similar to FM, in that there is a capture effect and the output SNR can be much larger than the CNR_{IF}. In wideband FM the frequency deviation can be made as large as required to attain the desired performance. The equivalent is not present for PM, however. The maximum phase deviation must be limited to $\pm\pi$ radians for unambiguous demodulation.

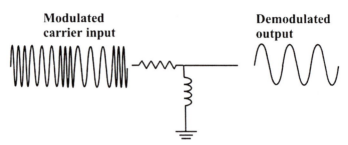

Figure 9.12 Demodulation of an FM signal with a discriminator.

In FM and PM, there is an improvement over AM that manifests itself as a threshold effect. If the modulation index is denoted as β, that is,

$$\beta = \frac{\text{maximum frequency deviation of RF carrier}}{\text{maximum frequency deviation of modulating signal}} \tag{9.10}$$

then the postdetection SNR is given by [3]

$$\text{SNR}_d = 6\beta^2(\beta+1)\text{CNR}_{IF} \tag{9.11}$$

Thus, the output signal to noise ratio in this case can be much larger than the CNR_{IF}. This analysis applies, however, if the amplitude of the signal is much larger than the maximum amplitude of the noise present. For example, for voice communications in the VHF frequency range, intelligible voice can be band-limited to about 2,000 Hz. In tactical FM radios, the channel allocation in the VHF range has typically been 25 kHz. If one-half of this is the assumed maximum frequency deviation and the transmitter is designed to exactly meet these specifications, then $\beta = 12{,}500/2{,}000 = 6.25$. Then the postdetection SNR is given by SNR_d (in decibels) = $10 \log[6 \times 6.25^2 \times (6.25 + 1)]$ + CNR_{IF} (in decibels) = 32.3 + CNR_{IF}. The postdetection SNR is 32 dB higher than the CNR_{IF}. Therefore, a CNR_{IF} of 10 dB would produce a postdetection SNR of about 42 dB, adequate for understanding audio signals.

FM is an example of Shannon's basic theory that bandwidth can be traded for SNR. PM has similar characteristics except there is a limit to the modulation index of PM signals. If wide bandwidths can be tolerated (an assumption that is rapidly decreasing in reasonableness), then lower SNR levels can be tolerated. Lower SNRs mean that lower transmitter powers can be used, or, in a military setting, higher levels of noise are tolerable.

Receivers 245

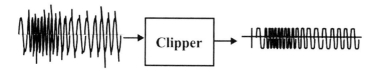

Figure 9.13 Since FM is a constant-amplitude (modulus) modulation, amplitude noise can be removed by clipping. The information content is in the zero crossings.

The information in an FM or PM signal is carried in the deviation of the frequency or phase from a nominal value. Thus, the information is retained even if there are amplitude variations, which could be desired or undesired. If it is undesired, then eliminating the amplitude variations in the demodulator can decrease demodulated noise. This can be done with a *clipper* as shown in Figure 9.13.

A *phase-locked loop* (PLL) is a device that responds to the changes in phase of an input signal. A block diagram of a PLL is shown in Figure 9.14. At steady state, the output of the VCO is at exactly the same frequency and is in phase with the input signal. Recall that the mixer, the same as the mixer previously analyzed, produces output frequencies that are the difference and the sum of the input frequencies of the two signals. The output of the lowpass loop filter is zero, because the sum frequency signal is filtered out. Recalling that frequencies are the time derivative of the change in phase, as the input signal phase changes, the output of the loop filter either increases or decreases, depending on the direction of the change of the phase. This signal is slowly changing relative to the carrier frequency and the natural frequency of the VCO. It is proportional to the frequency modulation, and it causes the VCO to change frequency, attempting to follow the phase changes in the input signal. The output, then, is the frequency- or phase-demodulated signal.

Reverse-biased varacter diodes are frequently used as the frequency-controlling element in the VCO. The capacitance of these devices changes with

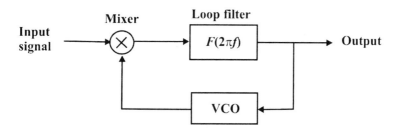

Figure 9.14 Block diagram of a PLL.

the bias voltage, making them strong candidates for this function.

For FM discriminators (Figure 9.12) the threshold for which the above applies is about 12 dB [4]. For PLL demodulation of FM signals (Figure 9.14) the threshold is about 4 dB.

The signal processing prior to detection in most cases converts the signal in frequency and amplitude, but otherwise does not change it (approximately). Therefore, once the CNR_{IF} is known, it is possible to specify the amount of IF amplification necessary in order to detect the minimum detectable signal.

9.2.6.3 Required SNR

The required *postdetection* SNR for digital signals is frequently expressed as E_b/N_0, or energy-per-bit to noise-level ratio, expressed in decibels. To convert this to power levels, it is necessary to divide this ratio (in linear form, not in decibels) by the data rate. The postdetection SNR can be substantially greater than the RF SNR_{req} defined here.

For digital communications, the BER is one of the fundamental design parameters. Coding and other techniques can be incorporated to ensure that a specified BER is accomplished. The BER versus postdetection SNR, as specified by E_b/N_0, for some of the more common digital modulations is shown in Figure 9.15 [5]. This chart indicates that in order to achieve the BER (P_e in the chart) then the SNR must be at least that shown in the abscissa. Using the E_b/N_0 ratio facilitates better comparisons of the various modulation schemes, since it is independent of the data rate.

Control of the gain in the demodulator is sometimes necessary to avoid driving the amplifiers into saturation. Such gain control is referred to *automatic level control* (ALC).

9.3 TYPES OF RECEIVERS

There are various types of receivers that are used for several different purposes in EW systems. Discussed next are salient characteristics of the more prevalent kinds of receivers used. This discussion is not all encompassing, but is intended to point out the fundamental characteristics of receivers used in communication ES applications.

9.3.1 Narrowband Receivers

Narrowband receivers are used primarily for processing signals when the frequency and other parameters of the signal are known. "Narrow" in this case can actually be quite wide—for example, when processing DSSS signals. The

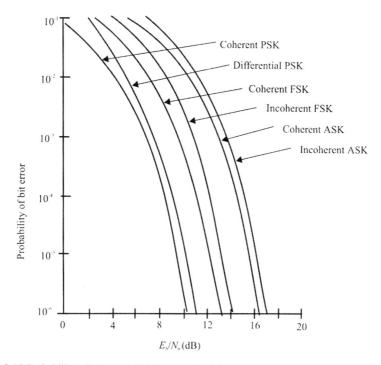

Figure 9.15 Probability of bit error (P_e) for several digital modulation types versus the SNR, as expressed by the E_b/N_0. (*Source:* [5], © 1986, John Wiley & Sons. Reprinted with permission.)

distinguishing characteristic in this sense is in the use of the receiver in the EW system—intercept or searching.

9.3.1.1 Superhetrodyne Receiver

As described above, superhetrodyne receivers are probably the most common receivers for EW use around. They exhibit good selectivity and sensitivity in most frequency ranges.

9.3.1.2 Zero IF

A zero-IF receiver is constructed when the signal in a superhetrodyne receiver is moved all the way to a zero IF. When phase-locked to the input signal, zero-IF receivers are also known as *homodyne* receivers. Amplification and filtering are then possible at low frequencies. The filtering following the mixer is comprised of lowpass filters.

Noise processing is particularly difficult in zero-IF receivers both because of noise sources inherent in semiconductors, which is exacerbated at low frequencies, and the wide bandwidth needed in the front end, which increases the noise generated by resistive components [6]. Notwithstanding the problems with noise, these receiver architectures have several attractive features, especially for monolithic implementations.

Another significant problem with zero-IF receivers is *local oscillator* (LO) leakage. If part of the locally generated signal used for moving the signal to zero IF leaks into the signal path in front of the mixer, that signal energy will be transformed to zero IF along with the signal.

$$\begin{aligned}\cos(2\pi f_{LO}t)\cos(2\pi f_{LO}) &= \frac{1}{2}\left[\begin{array}{c}\cos(2\pi f_{LO}t + 2\pi f_{LO}t) \\ + \cos(2\pi f_{LO}t - 2\pi f_{LO}t)\end{array}\right] \\ &= \frac{1}{2}\left[\cos(4\pi f_{LO}t) + 1\right]\end{aligned} \quad (9.12)$$

Therefore, a dc term is generated (in addition to a double-frequency term that can be filtered) in the mixing process. Frequently this dc term is much larger than the signal of interest at the input to the mixer, effectively saturating the mixer. Therefore isolation of the local oscillator signal from the other input to the mixer is critically important in such designs.

9.3.1.3 Fixed-Tuned Receivers

As described above, a *fixed-tuned receiver* (FTR) is a priori tuned to some frequency and stays there. They are for EW applications where the frequency of the target is known and does not change.

9.3.1.4 Scanning Superheterodyne Receivers

This type of receiver is used for searching the spectrum, looking for energy in frequency channels. Thus, frequencies of signals of interest as well as an estimate of their amplitudes can be determined. Scanning receivers of this type have long been used to implement spectrum analyzers, a common piece of RF electronic test equipment. A block diagram of such a receiver is shown in Figure 9.16. Torrieri [7] provides a detailed analysis of this receiver.

The swept local oscillator causes signals within a frequency band to be mixed within the mixer. The preselector filters must also be tuned along with the local oscillator. A narrowband signal at the input will be mixed whenever the local

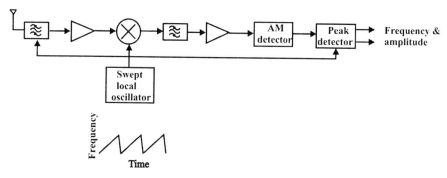

Figure 9.16 Simplified block diagram of a scanning superheterodyne receiver.

oscillator tunes to the IF offset from the signal (recall that the mixer output is a constant frequency). At that point in time the AM detector will detect the peak amplitude of the signal and the peak detector will measure the amplitude. By measuring where the peak occurs in time, that time can be compared with where the scanning local oscillator is at that time so a frequency estimate can be computed.

As derived in Torrieri, if μ represents the scanning rate of the receiver in hertz per second and B represents the 3-dB bandwidth of the bandpass filter after the mixer, the normalized peak value α (normalized relative to the amplitude of the input signal) is given by

$$\alpha = \left(1 + 0.195 \frac{\mu^2}{B^4}\right)^{-1/4} \tag{9.13}$$

and the frequency resolution is given by

$$\Delta = B\left(1 + 0.195 \frac{\mu^2}{B^4}\right)^{1/2} \tag{9.14}$$

These expressions are plotted in Figures 9.17 and 9.18 when the bandpass bandwidth is 25 kHz. Thus the normalized amplitude peak value decreases as the scan rate increases, while the resolution increases in bandwidth (decreases in selectivity). In a dense RF environment a resolution bandwidth of less than 50 kHz or so is desirable in order to minimize adjacent channel interference. Thus in the military VHF range where the signals typically have a bandwidth of 25 kHz,

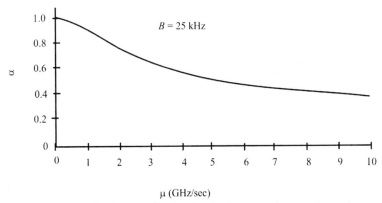

Figure 9.17 Example plot for the normalized peak value for a scanning superheterodyne receiver when $B = 25$ kHz.

the scanning rage must be kept at about 3 GHz per second or less in order to maintain the required resolution.

9.3.2 Wideband Receivers

Wideband receivers are normally used for searching the RF spectrum for energy. Whereas a wideband receiver can stare at a portion of the spectrum indefinitely, in most cases further processing of the signals received is required. This further processing normally deals with only a portion of the signals, dividing the time

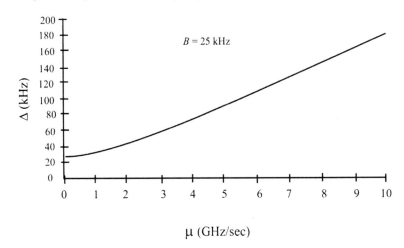

Figure 9.18 Scanning superheterodyne receiver frequency resolution when $B = 25$ kHz.

scale into segments. What happens to the time signals when this is done is illustrated in Figure 9.19. In this case an input signal in the form of a cosine wave at frequency f_0 is illustrated; however, Fourier analysis shows that any deterministic periodic signal can be described in terms of sine and cosine functions, so the conclusions are general in nature. Of course, stochastic signals are not deterministic so a different approach is required in that case. Shown at the top of Figure 9.19(b) is the frequency representation of the cosine wave.

The cosine signal is effectively multiplied by a time gate that defines the segment of the signal that is to be processed. For example, an A/D converter may change this analog signal into digital words over only this time segment for further digital processing. The frequency representation of the time gate is a sinc function shown in the middle of Figure 9.19(b). Since multiplication in the time domain is equivalent to convolution in the frequency domain, the two frequency representations at the top and middle of Figure 9.19(b) are convolved (denoted by *), resulting in the frequency representation of the gated cosine waveform shown at the bottom of Figure 9.19(b). Two sinc regions emerge, located at $\pm f_0$.

This illustration is ideal, assuming that the time gate includes an integer number of complete periods of the incoming signal. This would not normally be the case for a searching wideband receiver. The more normal case is illustrated in Figure 9.20, where the input waveform has been expanded for clarity. Only portions of the first and last period are included within the time gate that causes

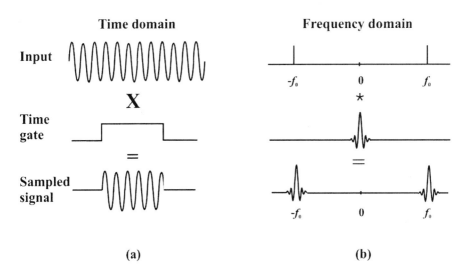

Figure 9.19 The effects on a signal that is sampled in the time domain are to add sidelobes in the frequency domain: (a) time domain and (b) frequency domain.

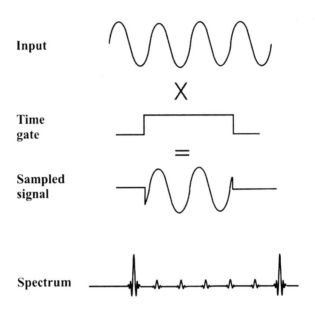

Figure 9.20 Effects of misalignment of time sampling.

portions of the spectrum of the signal to appear at other places in the frequency spectrum [8].

9.3.2.1 Channelized

Channelized receivers are essentially a bank of fixed-tuned (normally narrowband) receivers. The RF spectrum is divided into channels, the bandwidth of which depends on the signal types of interest. For VHF communication channels, the bandwidth is typically 25 kHz, although that is getting narrower.

Implementation of channelized receivers can be in a variety of forms. The above TRF receivers can be configured into banks, where each one of the banks covers one channel. Alternately a digital receiver can be used where one or more Fourier transform points cover each channel.

When the channelized receiver is implemented by taking the Fourier transform, unless the frequency point is precisely at the frequency of the input signal, splitting of the energy in the signal to several points occurs. This is illustrated in Figure 9.21.

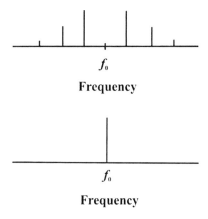

Figure 9.21 Effects of not sampling the input signal at the proper frequency.

9.3.2.2 Bragg Cell

This type of receiver takes advantage of the interaction of acoustic signals and lasers on the inside of a Bragg (acousto-optic) cell. RF signals are converted into acoustic waves by transducers on the surface of the cell. These waves propagate orthogonal to a laser interior to such a cell causing the laser direction to be changed proportionally to the wavelength of the acoustic wave. The laser is deflected an angle given by

$$\theta = \frac{\lambda_{laser}}{\lambda_{acoustic}} \tag{9.15}$$

Thus, the deflection angle is quite small, and in order to build reasonably sized detector arrays, this light must traverse a considerable distance (on the order of 1m) before being detected. At the output of such a receiver is an array of photodetectors, which convert the impinging laser signal into electrical signals. The amplitude of these electrical signals are proportional to the number of photons detected, which, in turn, is proportional to the amount of light collected by that detector. When combined with a lens, such a receiver computes the Fourier transform of the RF signal in essentially real time. A block diagram of this type of receiver is shown in Figure 9.22.

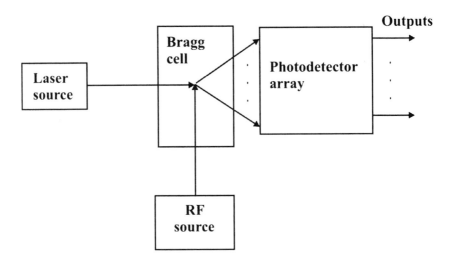

Figure 9.22 Block diagram of a Bragg cell receiver.

9.3.2.3 Compressive Receivers

A compressive receiver essentially computes the Fourier transform of an incoming signal continuously. The result is the frequency spectrum, which is comprised of a plot of energy distribution versus frequency of whatever is input to the receiver. This information is used to ascertain where signals are present and if so, at what frequency they are located [9–11].

In a compressive receiver the received signals are combined with a rapidly changing locally generated signal within the receiver. The entire spectrum in use by military tactical communication equipment can be scanned in 100 µs or less. The processing is such that a signal can be detected even though it is itself changing frequency as in, for example, the frequency hopping case. The compressive receiver allows for the determination of active frequencies. All signals within the range of the receiver are detected, subject, of course, to hearability constraints that are determined by system parameters. It is necessary to sort all of these signals with processing techniques that follow the compressive receiver [7, 11, 12].

There are essentially two ways to construct a compressive receiver. One version, *convolve-multiply-convolve* (C-M-C), does a convolution of the incoming signal first, and then multiplies by a linear chirping local oscillator (linear chirp means that the frequency is increasing or decreasing linearly with time). Lastly, another convolution is performed. The other version is called an M-C-M, and the

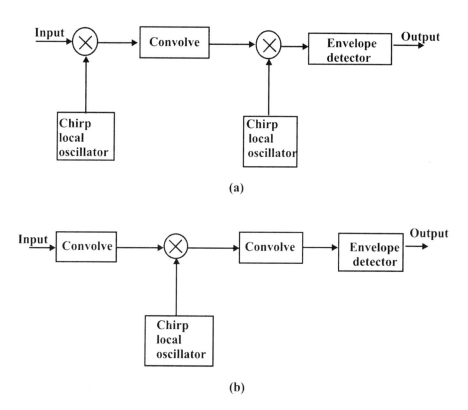

Figure 9.23 Two architectures of compressive receivers: (a) M-C-M configuration and (b) C-M-C configuration. (*Source:* [10], © 1980, IEEE. Reprinted with permission.)

architecture is obvious from these letters. Block diagrams of these architectures are shown in Figure 9.23.

Frequently in a compressive receiver it is not necessary to calculate the actual Fourier transform of the signal, and the last convolution or multiplication is not implemented. The output of an envelope detector still contains the necessary amplitude versus frequency information. This configuration is shown in Figure 9.24.

The operation of a compressive receiver can be explained as follows. Two signals at the input to the mixer at two different frequencies are shown in Figure 9.25. As the local oscillator sweeps from a low frequency to a high frequency over the range Δf, the frequencies of these two signals at the output of the mixer follow the curves shown in Figure 9.26. The dispersive delay has the impulse response

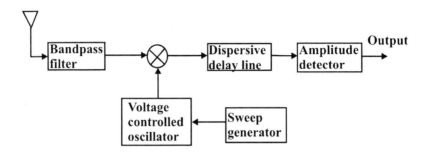

Figure 9.24 Simplified block diagram of the compressive receiver.

shown in Figure 9.27 and the slope of the frequency versus time response, shown in Figure 9.26, is matched to that of the sweeping local oscillator. That is,

$$\frac{B}{T_D} = \left|\frac{\Delta f}{T_{\text{sweep}}}\right| \qquad (9.16)$$

Thus, the lower frequency signal is delayed more than the higher frequency signal, and it emerges from the filter later in time than the high frequency signal. The output of the detector would look something like that shown in Figure 9.28. The 3-dB width of the main lobes of the signals emerging from the detector is approximately $1/B$. The time between these pulses of energy is given by their frequency difference divided by the slope of the response of the dispersive delay line. Thus,

$$T = \frac{f_2 - f_1}{B/T_D} \qquad (9.17)$$

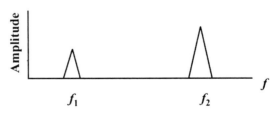

Figure 9.25 Signals at the input of a compressive receiver.

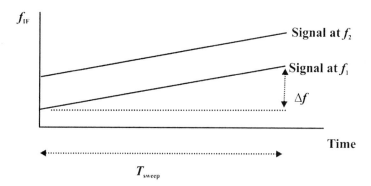

Figure 9.26 Time delays through a dispersive device such as an SAW.

If the resolution is defined as the closest two signals can be and still be resolved as two pulses of energy, then clearly it is approximately $1/B$, the width of the main lobes and, more importantly, one-half the null-to-null separation of these pulses. Shown in Figure 9.29 are two such pulses separated by exactly $1/B$. The second pulse occurs $1/B$ seconds later than the first pulse. If they were any closer, then the second pulse would occur within the 3-dB width of the first pulse and they could not be resolved, at least not very well.

The total number of these resolution cells is given by

$$N = \frac{T_D}{1/B} \qquad (9.18)$$

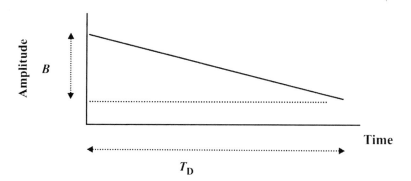

Figure 9.27 Impulse response of an SAW.

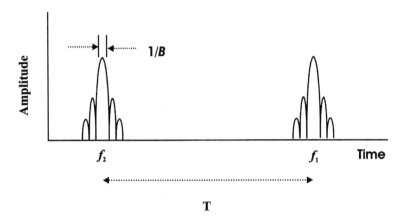

Figure 9.28 Time response of a compressive receiver. Two signals separated in frequency will emerge from the receiver at different times.

or

$$N = T_D B \qquad (9.19)$$

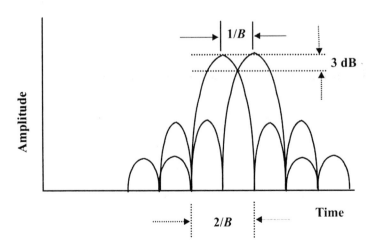

Figure 9.29 Definition of the frequency resolution of a compressive receiver.

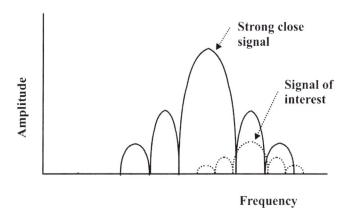

Figure 9.30 Masking of target signals by the sidelobes.

Another important parameter is the frequency resolution (versus time resolution just discussed). If f_1 and f_2 are at the opposite ends of the delay line response, then $f_2 - f_1 = B$. The frequency resolution is therefore $f_r = B/N = 1/T_D$, the reciprocal of the response time of the dispersive delay line.

One of the issues with compressive receivers is their first sidelobe dynamic range, which is given by the relative amplitude of the main lobe to the amplitude of the first (assumed to be the highest) sidelobe, one on each side. Without further processing it is theoretically 13 dB. The frequency response of a compressive receiver, indeed, any processing system that samples time waveforms, is shown in Figure 9.30. The high close-in sidelobe levels are visible. The first sidelobe, for example, is down from the peak of the main lobe by only about 13 dB. For an intercept receiver, if there is a signal present producing the response of the main lobe, any signal that is in either adjacent channel, as shown, defined by the first sidelobe, would be masked by this response and would not be detected. What this means practically is that, since signal energy falls off with range proportionally to $1/R^4$ close to the ground and $1/R^2$ in the air, close to the ground there is a 12-dB loss every time the distance is doubled (an octave) (40 dB per decade) and in the air the loss is 6 dB per octave (20 dB per decade).

Thus, if there is a friendly transmitter close to this receiver that radiated about the same amount of power as a target transmitter, then if the friendly transmitter is, say, 500m away, any target transmitter that is further than about 1 km away from the intercept receiver will be masked. In the case of an airborne platform, if the friendly transmitter is on the ground 5,000 feet below the aircraft, then any target that is further away than 10,000 feet (slant range assuming the target is on the

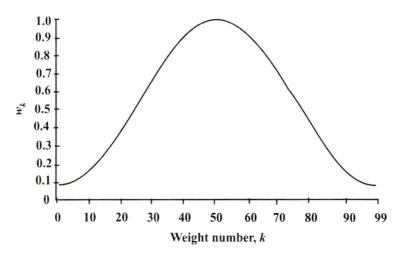

Figure 9.31 Weighting coefficients for a Hamming window.

ground) will be masked. These considerations make compressive receivers of little utility in many practical applications.

This sidelobe characteristic is not unique to compressive receivers, however. It is a characteristic of any system that samples rather than takes data continuously. Digital systems and Bragg cell receivers also have the characteristic. The advantage of digital receivers, however, is that with further signal processing the first sidelobes can be theoretically made arbitrarily small.

Weighting the frequency response can reduce the sidelobes. This is accomplished in an analog compressive receiver at the SAW manufacturing stage by weighting the interdigital transducers of the SAW itself. In the case of a digital implementation of a compressive receiver, or a direct digital receiver, the windowing would be applied during the calculations. In any case, the weighting is normally applied in the time domain, and the sidelobes are reduced by an amount that depends on the characteristics of the weighting function. Such reductions do not come freely, however. The greater the reduction in the close-in sidelobe

Table 9.1 Sidelobe and Main Beamwidth Characteristics of Some Common Weighting Functions

Window Type	Peak Sidelobe Amplitude (dB)	Width of Main Lobe Relative to Rectangular
Rectangular	−13	1
Bartlett	−25	2
Hanning	−31	2
Hamming	−41	2
Blackman	−57	3

Source: [13], © 1975, Pearson Education, Inc. Reprinted with permission.

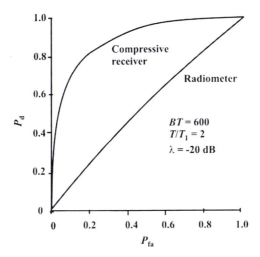

Figure 9.32 ROC for a compressive receiver against a tone target signal. (*Source:* [11], © 1991, IEEE. Reprinted with permission.)

amplitude, the wider the main lobe becomes smeared. Table 9.1 shows the amount of sidelobe reduction and main lobe widening that occurs for some common forms of weighting functions [13]. If the main lobe becomes smeared too much, then the benefits of weighting disappear because both the signals get combined in the same (main) lobe, so discerning the target signal from the interfering signal is not possible.

$$w_k = 0.54 - 0.46 \cos\left(\frac{2k\pi}{N-1}\right) \quad k=0,1,2,...,N-1 \quad (9.20)$$

The weights corresponding to Hamming window, for example, are shown in Figure 9.31 for $N = 100$. The window coefficients are given by [14] where N is the number of sample points, or in the case of a compressive receiver, the number of interdigital transducers on the SAW. The other weighting functions are similar in shape. Typical performance specifications for modern compressive receivers that operate against communication signals would be (1) 40-dB highest sidelobe dynamic range, (2) 1-MHz/µs sweep speed, (3) input (total) bandwidth 500 MHz, (4) 25-kHz signal resolution, and (5) 70-dB instantaneous spur free dynamic range (noise limited).

The theoretical signal detection performance of compressive receivers is illustrated in Figures 9.32 and 9.33 compared to a radiometer. In Figure 9.32 [11], the signal is a tone with a signal to noise ratio of −15 dB. For a narrowband tone in, say, a 25-kHz channel, and a compressive receiver with a total bandwidth of $2B$

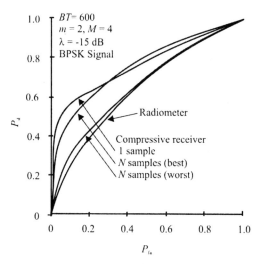

Figure 9.33 ROC for a compressive receiver against a direct spread spectrum BPSK signal. (*Source:* [11], © 1991, IEEE. Reprinted with permission.)

= 500 MHz, this represents a signal level of about 25 dB above the thermal noise floor within its bandwidth. Also in this chart, $BT = 600$, corresponding in this case to $T = 2.4$ μs. This duration provides a signal resolution of about 400 kHz.

In Figure 9.33 the signal is a direct sequence BPSK signal [12]. Again in this case the SNR is −15 dB. If T_b represents the bit time, then $T = MT_b$, and $T_{sweep} = T_1 = mT_b$.

9.3.2.4 Digital Receivers

A digital receiver converts the incoming signals into digital form as soon in the receiving chain as possible. This allows for subsequent processing to be performed in the digital domain, which has desirable characteristics, such as repeatability and stability with temperature and time. One of the most common digital processing techniques is to perform an FFT on the incoming signals. This calculates the frequency spectrum, from which it can be determined if there is a signal at particular frequencies, or alternately, to determine at what frequencies signals are located. Once this is determined, other narrowband receivers can be tuned to those frequencies for such purposes as subsequent analysis and listening. A simplified block diagram of a digital receiver implemented with the FFT is shown in Figure 9.34.

Digital receivers are becoming more popular than most other wideband receiver types primarily because the shrinking size and increased functionality of

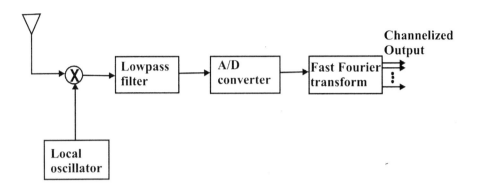

Figure 9.34 Simple block diagram of a digital receiver.

microelectronics. Fast A/D converters and reasonably sized FFT processors can be obtained on single chips. Furthermore, this trend will continue.

9.4 CONCLUDING REMARKS

There are many types of receivers and the one selected for a particular application depends, for the most part, on that application. For general-purpose narrowband applications the superheterodyne has become the most favored because of its versatility, sensitivity, and selectivity. For applications that do not need the tuning versatility, the fixed tuned receiver is the one of choice due to its low cost.

Driven to a large extent by the explosion in the personal portable phone market, that is, cell phone systems and PCS, research and development of the zero IF receiver is emerging. This particular architecture has many favorable attributes for portable phone applications. These applications have much in common with many EW system requirements—size, for example. Hundreds or more of these receivers could be put into a signal chassis, for example, providing broadband ES coverage with narrowband receivers in a small package.

For communication EW wideband applications the most prolific receiver type is the digital receiver, again because of its versatility. Analog receivers such as the Bragg cell or the compressive receiver have some advantages, such as smaller size and lower power consumption. Their biggest disadvantage is their limited dynamic range caused by component availability.

Dynamic range is an important parameter for communication EW system applications because of the wide variety of deployments such systems encounter. In standoff applications or any other operational scenario where extensive range is required, close-in interference caused by friendly or neutral transmitters can easily overwhelm a receiver if it has inadequate dynamic range.

References

[1] Erst, S. J., *Receiving Systems Design*, Dedham, MA: Artech House, 1985.
[2] Gagliardi, R. M., *Introduction to Communication Engineering*, 2nd ed., New York: John Wiley & Sons, 1988, p. 200.
[3] Gagliardi, R. M., *Introduction to Communication Engineering*, 2nd ed., New York: John Wiley & Sons, 1988, p. 209.
[4] Adamy, D., "Sensitivity of FM and Digital Receivers," *Journal of Electronic Defense*, April 1996, p. 64.
[5] Das, J., S. K. Mullick, and P. K. Chatterlee, *Principles of Digital Communication*, New York: John Wiley & Sons, 1986, p. 290.
[6] Razavi, B., *RF Microelectronics*, Upper Saddle River, NJ: Prentice Hall, 1998, p. 129.
[7] Torrerri, D. J., *Principles of Secure Communication Systems*, 2nd ed., Norwood, MA: Artech House, 1992, pp. 323–328.
[8] Tsui, J., *Digital Techniques for Wideband Receivers*, Norwood, MA: Artech House, 1995.
[9] Luther, R. A., and W. J. Tanis, "Advanced Compressive Receiver Techniques," *Journal of Electronic Defense*, July 1990, pp. 59–66.
[10] Jack, M. A., P. M. Grant, and J. H. Collins, "The Theory, Design, and Applications of Surface Wave Fourier-Transform Processors," *Proceedings of the IEEE*, Vol. 68, No. 4, April 1980, pp. 450–468.
[11] Li, K. H., and L. B. Milstein, "On the Use of a Compressive Receiver for Signal Detection," *IEEE Transactions on Communications*, Vol. 39, No. 4, April 1991.
[12] Dillard, R. A., and G. M. Dillard, *Detectability of Spread Spectrum Signals*, Norwood, MA: Artech House, 1989.
[13] Oppenheim, A. V., and R. W. Shafer, *Digital Signal Processing*, Englewood Cliffs, NJ: Prentice Hall, 1975, p. 250.
[14] Feldman, M., and J. Henaff, *Surface Acoustic Waves for Signal Processing*, Norwood, MA: Artech House, 1986, p. 152.

Chapter 10

Signal Processing

10.1 INTRODUCTION

The processing of signals within a communication EW system comes in a variety of forms. Received signals must be processed to extract information from them. Signals to be transmitted from a communication EW system must be processed to be put in the correct format. Thus, understanding the basics of signal processing is important to understand the design and operation of such systems.

Normally signals convey information, thus the interest in signal processing for IW. The light that impinges on our eyes that allows us to see objects are signals. The varying air pressure that carries audio tones detected by our ears are signals. Some characteristic of some medium is changed to facilitate information transfer with signals. The air just mentioned is an example of this. Signal processing is performed to change a signal from one form to another, to impose some characteristic onto a signal, or to facilitate the measurement of some feature of a signal. Features obvious in one domain may be hidden from others, and thus it is necessary to convert the signal to recover such.

When a signal is analog in nature, generally speaking it can be processed in analog form or it can be converted into digital form, in certain cases. An example of an analog signal is the way voice is sent from a telephone—at least prior to the 1990s. Broadcast TV signals in the United States today are analog, although there is a rapid movement to send digital TV signals to the home via cable and direct broadcast satellite.

Some signals originate in digital form, under which circumstances they would have to be converted to analog form in order to process them with analog technologies. Digital signals are normally generated in that form to take advantage of digital signal processing, so conversion to analog is rarely done. Some examples of digital signals being generated that way are random sequences of numbers and digital clocks that display characters rather than sweeping hands.

10.2 ORTHOGONAL FUNCTIONS

The set of functions $\Phi = \{\phi_1, \phi_2, \ldots, \phi_N\}$ defined over an interval $x_1 \leq x \leq x_2$, is *orthogonal* (*unitary* if the functions are complex) if the following is satisfied

$$\int_{x_1}^{x_2} \phi_i(x) \phi_j^*(x) dx = k_{ij} \delta_{ij} \tag{10.1}$$

where δ_{ij} is the Dirac delta function defined as

$$\delta_{ij} = \begin{cases} 1 & i = j \\ 0 & \text{otherwise} \end{cases} \tag{10.2}$$

k_{ij} is some constant, and * is complex conjugate. If $k_{ij} = 1$, then Φ is said to be *orthonormal*. Because the results can be easily scaled to or from the unit interval on the real line, it is assumed without loss of generality that $x_1 = 0$ and $x_2 = 1$.

A function $f(x)$ can be approximated by a linear combination of these functions as

$$\hat{f}(x) = \sum_{i=1}^{N} c_i \phi_i(x) \tag{10.3}$$

where

$$c_i = \int_0^1 f(x) \phi_i^*(x) dx \tag{10.4}$$

Φ is thus a set of *basis functions* for $f(x)$. This linear combination representation for $f(x)$ is said to be a *projection* of $f(x)$ onto Φ. Φ is said to be *complete* if any (piecewise) continuous function $f(x)$ can be represented this way and that the *mean square error* (MSE) given by

$$\text{MSE} = \int_0^1 \left| f(x) - \hat{f}(x) \right|^2 dx \tag{10.5}$$

converges to zero for some N.

When $f(x)$ is defined only at discrete points, then the above integrals are replaced by summations. Orthonormality is defined when

$$\frac{1}{N}\sum_{j=1}^{N}\phi_i(x_j)\phi_k(x_j) = \delta_{ik} \tag{10.6}$$

and the approximation function $\hat{f}(x_j)$ is given by

$$\hat{f}(x_j) = \sum_{i=1}^{N} c_i \phi_i(x_j) \tag{10.7}$$

where

$$c_i = \frac{1}{N}\sum f(x_j)\phi_i^*(x_j) \tag{10.8}$$

Parseval's relationship says that the energy in the x-domain must be equal to the energy in the transformed domain. That is,

$$\sum_{i=1}^{N}|c_i|^2 = \frac{1}{N}\sum_{i=1}^{N}|f(x_i)|^2 \tag{10.9}$$

For example, the x-domain could be the time domain and the transformed domain could be the frequency domain.

10.3 TRANSFORMS

A transform is a mathematical manipulation of a signal to convert it from one domain to another. There are many ways to change the representation of a signal from one form to another. Such transformations are performed for a variety of reasons. One of these is to ascertain the frequency content of the signal. Another is to compute more efficient ways to transmit the signal from one place to another.

Many transforms have been discovered that are useful for processing signals. Only a few of the many transforms will be discussed here.

10.3.1 Trigonometric Transforms

Several transform relationships are based on the trigonometric functions, in particular the cosine and sine functions. Some of these are presented in this section.

10.3.1.1 Fourier Transform

One of the primary functions that most RF EW systems perform in signal processing is detecting the presence of signals while simultaneously determining the frequency of these signals. This is frequently accomplished by calculating the *Fourier transform* of the signals. One of the things for which this transform is useful is determining the energy versus frequency content of a spectrum, and therefore can give an indication of the center frequency as well as a signal's bandwidth.

Perhaps the most prolific transform in signal processing is the Fourier transform, and in particular, the *fast Fourier transform* (FFT). Cooley and Tukey discovered the FFT in 1965 at Bell Laboratories as a fast method of computing the *short-time Fourier transform* (STFT) of a signal on a digital computer. Compressive receivers discussed in Section 9.3.2.3 are essentially Fourier transform calculators, as is the acousto-optical processor based on a Bragg cell. The digital receiver normally computes the Fourier transform to determine the signal spectrum. The FFT is typically calculated with a digital signal processor, usually implemented on one or more semiconductor chips.

Knowing this frequency content is sometimes useful for a variety of reasons. One simple way to visualize this is to consider a signal that has the full bandwidth of the U.S. FM radio band. That is, all of the signals from 88 to 108 MHz are present. Of course, listening to this signal would be nonsensical since no intelligibility could be ascertained. Taking the Fourier transform of such a signal, however, would reveal all of the FM radio channels that are present. Putting an appropriate filter around one of these channels would allow only that signal to get through and therefore it could be clearly heard.

The basis functions for the Fourier transform are $\sin(x)$ and $\cos(x)$. The Fourier transform is a complete transform. Orthogonality dictates that the inner product of the basis functions must equal zero unless the two functions are the same basis function. Let T denote an integral number of periods of the sine and cosine functions. In this case

Signal Processing

$$\int_0^T \sin x \cos x\, dx = \frac{1}{2}\int_0^T \sin 2x\, dx + \frac{1}{2}\int_0^T \sin(0)\, dx \qquad (10.10)$$

The first integral on the right side equals zero because any sine function integrated over an integral number of periods is zero. The second integral is zero since sin(0) = 0. Thus, the sine and cosine functions are orthogonal.

The Fourier transform of a time signal $s(t)$ is given by

$$S(f) = \int_{-\infty}^{\infty} s(t) e^{-j2\pi ft}\, dt \qquad (10.11)$$

where f is the frequency in hertz, and t is time in seconds. Even though this integral goes from minus infinity to plus infinity, most signals of practical interest have a limited time extent. Some mathematical approximations to real signals do include the unlimited time, however. Limiting the time extent can cause some unexpected results since it is mathematically impossible to simultaneously limit a signal in time and frequency extent. Thus, a signal defined over a time interval (t_1, t_2) has infinite frequency content. Likewise, a signal with frequency extent (f_1, f_2) must exist for all time.

The magnitude of the Fourier transform of an unmodulated cosine wave of frequency f_0 is two impulse functions as shown in Figure 10.1 (phase is ignored for now). The negative frequency might appear odd to some. There is no physical manifestation of negative frequency that is currently known —it is necessary for

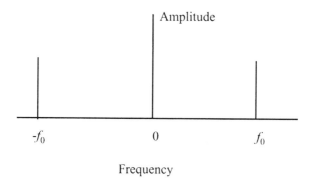

Figure 10.1 Fourier transform of the cosine function.

the mathematics to work out. The Fourier transform is particularly useful for processing smooth, continuous functions. It does not, however, work well with discontinuous functions or nonstationary data. This can be illustrated with trying to approximate a square pulse with sinusoidal functions. The result is known as *Gibb's phenomenon,* which describes oscillations at the points of discontinuity as shown in Figure 10.2. No matter how many sinusoidal functions are included in the transform, the overshooting oscillations remain. Their amplitude is about 18% of the amplitude of the pulse.

Since practical implementations of the calculation of the Fourier transform must by necessity cover only a finite amount of time, the STFT was devised. This transform essentially puts a time window around the signal, which has a duration that is shorter than the duration of the whole signal. This calculates the Fourier transform over just this time interval. The effect is to multiply the signal by the unit pulse time window, which has amplitude of 1 and an extent from 0 to T or $-T/2$ to $T/2$ (the same results ensue). Since multiplication in the time domain equates to convolution in the frequency domain, and the spectrum of the unit pulse has a sin x/x shape, the resultant magnitude of the spectrum of the STFT of a sine wave is as shown in Figure 10.3. The spectrum extent (bandwidth) of each impulse is larger than the theoretical single line. The windowing has essentially spread the signal over more frequencies. These frequencies were not part of the original signal, therefore artifacts were introduced into the process.

One of the properties of the STFT is that it is not possible to simultaneously select an arbitrary time window and arbitrary frequency extent. These two parameters are related to each other. The smaller the time resolution, the larger the frequency resolution, and vice versa, the larger the time resolution, the lower the frequency resolution. This is expressed by the uncertainty inequality, which states that in Fourier analysis

$$\Delta f \Delta t \geq \frac{1}{4\pi} \tag{10.12}$$

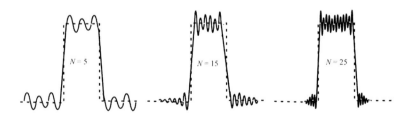

Figure 10.2 Gibb's phenomenon is manifest when trying to approximate a discontinuity with Fourier coefficients.

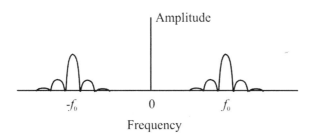

Figure 10.3 Effects of time windowing the cosine function.

where Δf and Δt are defined by their second moments divided by their energies of the signal being analyzed.

It is frequently assumed that signals are *stationary* and *ergodic*.[1] When a signal is stationary, the Fourier transform statistics do not vary with time since the signal does not. In practice this is rarely true; however, the assumption that the signal is not changing with time is common. Since this is an STFT, the assumption of stationarity is only necessary over the interval for which data samples of the signal are taken. The ergodic property allows the statistics of a particular realization of a signal from the ensemble to be calculated and used to represent the entire set.

To help deal with nonstationary signals, more complex transformations than the STFT have been devised, such as the Wigner-Ville distribution (the term distribution refers to the distribution of energy with frequency and is irrelevant here). These transforms determine the frequency content versus time. Unfortunately such transforms are nonlinear, and what is called *cross-product terms* are generated if two or more signals are analyzed simultaneously [1]. The continuous Wigner-Ville transform for signal $s(t)$ is given by

$$W(t,f) = \int_{-\infty}^{\infty} s(t+\tau/2)s^*(t-\tau/2)e^{-j2\pi f t}d\tau \qquad (10.13)$$

Thus, the Wigner-Ville distribution is the Fourier transform of the autocorrelation function of the signal. Its discrete form for a time series $x(n)$ is given by

---

[1] A sequence is *stationary* if its statistical properties are constant with time. A sequence is *ergodic* if its stastical properties calculated over time are the same as those calculated across the ensemble of sequences.

$$W_d(n,f) = 2\sum_{k=-\infty}^{\infty} h_N^2(k)x(n+k)x^*(n-k)e^{-j4\pi fk} \qquad (10.14)$$

where $h_N(k)$ is a data window for smoothing the frequency response. One of the advantages of this transform is it is real with no imaginary component as in the STFT.

To help cope with the cross product problem, the Choi-Williams transform was developed. It is similar to the Wigner-Ville except the cross-product terms are suppressed (but not eliminated). The Choi-Williams transform is given by

$$C(t,f) = \int_{-\infty}^{\infty} e^{-j2\pi ft} \int_{-\infty}^{\infty} \sqrt{\frac{\sigma}{4\pi\tau^2}} e^{-\frac{\sigma(\mu-t)^2}{4\tau^2}} s(\mu+\tau/2)s^*(\mu-\tau/2)d\mu d\tau \qquad (10.15)$$

In this expression σ is a variable that controls the amplitude cross-product terms as well as the frequency resolution. As $\sigma \to \infty$, this transform becomes the Wigner-Ville distribution.

10.3.1.2 Hartley Transform

The Hartley transform also computes a representation of a signal that displays its energy versus frequency contents. In fact, the Fourier and Hartley transforms are mathematically related to each other. An advantage of the Hartley transform is that the way it can be calculated sometimes is more efficient than the Fourier transform. The Hartley transform works with real numbers, whereas the Fourier transform applies to complex numbers. Another advantage of the Hartley transform is its ease of implementation with real hardware.

The *Hartley transform pair* is given by

$$H(f) = \int_{-\infty}^{\infty} x(t) \operatorname{cas} 2\pi ft \, dt \qquad (10.16)$$

and

$$x(t) = \int_{-\infty}^{\infty} H(f) \operatorname{cas} 2\pi ft \, df \qquad (10.17)$$

where cas x = cos x + sin x (<u>c</u>osine <u>a</u>nd <u>s</u>ine). These are similar to the Fourier transforms but are not the same. For example, $H(f)$ is real where, generally, the Fourier transform is complex [2].

When the input sequence is given by x_k, the *discrete Hartley transform* is given by the relations

$$H_{N,f}(x) = \frac{1}{N} \sum_{k=0}^{N-1} x_k \operatorname{cas}\left(\frac{2\pi}{N} fk\right) \tag{10.18}$$

and

$$H_{N,k}^{-1}(v) = \sum_{f=0}^{N-1} v_f \operatorname{cas}\left(\frac{2\pi}{N} fk\right) = x_k \tag{10.19}$$

where, as above, $\operatorname{cas}(x) = \cos(x) + \sin(x)$. The Hartley transform can be calculated with a recurrence relationship which facilitates the fast Hartley transform. This relationship is

$$H_k(v) = H_{k-1,\text{odd}}(v) \cos\left(\frac{2\pi}{N} v\right) + H_{k-1,\text{even}} \sin\left(\frac{2\pi}{N} v\right) \tag{10.20}$$

The discrete Hartley transform is quite fast with relatively low computational requirements, especially if the $\operatorname{cas}(x)$ terms are precomputed and stored in a table.

10.3.1.3 Discrete Cosine Transform

The discrete cosine transform is a popular transform in modern image compression algorithms such as JPEG and MPEG. The reason for this is the relatively light computational load involved. The one-dimensional transform relationships are given by

$$X_n = \frac{C(n)}{2} \sum_{k=0}^{N-1} x_k \cos\left[\frac{(2k+1)n\pi}{2N}\right] \tag{10.21}$$

and

$$x_k = \sum_{n=0}^{N-1} \frac{C(n)}{2} X_n \cos\left[\frac{(2k+1)n\pi}{2N}\right] \tag{10.22}$$

where

$$C(n) = \begin{cases} \dfrac{1}{\sqrt{2}} & n = 0 \\ 1 & \text{otherwise} \end{cases} \qquad (10.23)$$

For transformation and exchange of images, the two-dimensional discrete cosine transform is used. The transform coefficients are determined and those are transmitted versus the raw image samples themselves. This is a more efficient way of transmitting the images. The two-dimensional transform and its inverse are given by

$$X_{k,m} = \frac{2}{N} C(k)C(m) \sum_{i=0}^{N-1} \sum_{j=0}^{N-1} x_{i,j} \cos\left[\frac{(2i+1)k\pi}{2N}\right] \cos\left[\frac{(2j+1)m\pi}{2N}\right] \qquad (10.24)$$

and

$$x_{i,j} = \frac{2}{N} \sum_{k=0}^{N-1} \sum_{m=0}^{N-1} C(k)C(m) X_{k,m} \cos\left[\frac{(2i+1)k\pi}{2N}\right] \cos\left[\frac{(2j+1)m\pi}{2N}\right] \qquad (10.25)$$

For image processing, these two-dimensional transforms are taken over an $N \times N$ block of pixels denoted by $x_{i,j}$.

There is a fast version of the discrete cosine transform available. Furthermore, this transform is separable in that one-dimensional transforms can be obtained for the rows of the image and then down the columns of the image. Doing so reduces the number of computations involved.

10.3.2 Haar Transform

The Haar transform is useful for processing functions with discontinuities and functions that are nonstationary. It is a complete transform and is one of the fastest available because the only numbers involved are 0, 1, −1, $\sqrt{2}$ and $-\sqrt{2}$. As an example, the 8 × 8 Harr transform matrix is given by

$$\mathbf{H}_8 = \frac{1}{\sqrt{8}} \begin{bmatrix} 1 & 1 & 1 & 1 & 1 & 1 & 1 & 1 \\ 1 & 1 & 1 & 1 & -1 & -1 & -1 & -1 \\ \sqrt{2} & \sqrt{2} & -\sqrt{2} & -\sqrt{2} & 0 & 0 & 0 & 0 \\ 0 & 0 & 0 & 0 & \sqrt{2} & \sqrt{2} & -\sqrt{2} & -\sqrt{2} \\ 2 & -2 & 0 & 0 & 0 & 0 & 0 & 0 \\ 0 & 0 & 2 & -2 & 0 & 0 & 0 & 0 \\ 0 & 0 & 0 & 0 & 2 & -2 & 0 & 0 \\ 0 & 0 & 0 & 0 & 0 & 0 & 2 & -2 \end{bmatrix} \quad (10.26)$$

When the Haar matrix multiplies a signal vector, the signal vector is sampled from low to high frequencies. That is, if

$$\mathbf{f}(x) = \mathbf{H}_8 \mathbf{g}(x) \quad (10.27)$$

then

$$\begin{bmatrix} f_1 \\ f_2 \\ f_3 \\ f_4 \\ f_5 \\ f_6 \\ f_7 \\ f_8 \end{bmatrix} = \frac{1}{\sqrt{8}} \begin{bmatrix} 1 & 1 & 1 & 1 & 1 & 1 & 1 & 1 \\ 1 & 1 & 1 & 1 & -1 & -1 & -1 & -1 \\ \sqrt{2} & \sqrt{2} & -\sqrt{2} & -\sqrt{2} & 0 & 0 & 0 & 0 \\ 0 & 0 & 0 & 0 & \sqrt{2} & \sqrt{2} & -\sqrt{2} & -\sqrt{2} \\ 2 & -2 & 0 & 0 & 0 & 0 & 0 & 0 \\ 0 & 0 & 2 & -2 & 0 & 0 & 0 & 0 \\ 0 & 0 & 0 & 0 & 2 & -2 & 0 & 0 \\ 0 & 0 & 0 & 0 & 0 & 0 & 2 & -2 \end{bmatrix} \begin{bmatrix} g_1 \\ g_2 \\ g_3 \\ g_4 \\ g_5 \\ g_6 \\ g_7 \\ g_8 \end{bmatrix} \quad (10.28)$$

$$\begin{bmatrix} f_1 \\ f_2 \\ f_3 \\ f_4 \\ f_5 \\ f_6 \\ f_7 \\ f_8 \end{bmatrix} = \frac{1}{\sqrt{8}} \begin{bmatrix} g_1 + g_2 + g_3 + g_4 + g_5 + g_6 + g_7 + g_8 \\ g_1 + g_2 + g_3 + g_4 - g_5 - g_6 - g_7 - g_8 \\ \sqrt{2}g_1 + \sqrt{2}g_2 - \sqrt{2}g_3 - \sqrt{2}g_4 \\ \sqrt{2}g_5 + \sqrt{2}g_6 - \sqrt{2}g_7 - \sqrt{2}g_8 \\ 2g_1 - 2g_2 \\ 2g_3 - 2g_4 \\ 2g_5 - 2g_6 \\ 2g_7 - 2g_8 \end{bmatrix} \quad (10.29)$$

Thus, f_1 is a measure of the average, or mean, value, f_2 is a measure of the differences of the means of the first four samples compared to the last four, f_3 is a measure of the differences of the mean values of the first two samples compared with the second two, and so on. The higher the index on f_i, the higher the "frequency" of the measurement of the data. If the input samples, g_i, correspond to a set of pixels from an image, the 8 × 8 Haar transformation produces information about the makeup of the set of pixels. The average value would indicate the brightness of the set, while the last four would facilitate edge detection, for example.

10.3.3 Wavelet Transforms

Wavelet transforms are linear transforms as is the STFT. The transforms deal with the problem of selecting time and frequency windows by using short time windows for high frequencies and long time windows for lower frequencies. The Fourier transform assumes that the signal has been present in its current form forever in the past (stationary), and when it is applied to signals that do not conform to this assumption, then errors occur. Wavelets avoid this situation by not making such an assumption. Therefore, wavelets are useful for nonstationary signal analysis. In particular, they are useful for analysis of signals with discontinuities (or almost discontinuities) in them. A discontinuity is an instantaneous change in a signal that is abrupt and not smooth. Discontinuities do not really exist in real situations, but there are cases where almost discontinuities occur—close enough that the engineering analysis of the signals must treat them as discontinuous. Therefore, the ability to transform such signals is important [3].

Variable-sized time windows are used in wavelet analysis. Short windows are used at high frequencies and long windows at low frequencies. This is like maintaining a constant Q in filters, where the fractional bandwidth is maintained relative to the center frequency. The result of wavelet transforming is time-scale representation versus the time-frequency of Fourier analysis. In the case of Fourier analysis to obtain a realistic representation of the signal in the frequency domain, typically many terms of the transform are required—that is, many of the sine and cosine terms need to be retained. In wavelet analysis typically many fewer terms need to be maintained.

Another advantage of the wavelet transform over the Fourier transform is that the time and frequency resolution of the latter cannot be independently set, whereas in the former they can be. In Fourier analysis, if one wants fine frequency resolution, then time resolution must be sacrificed. If one wants fine time resolution, then frequency resolution must be given up.

Once a window has been selected for the STFT, the time and frequency resolutions remain fixed. For continuous wavelet analysis, it is the ratio $\Delta f/f$ that remains constant, or the frequency resolution is proportional to the frequency. Time resolution becomes arbitrarily good at high frequencies and frequency resolution becomes arbitrarily good at low frequencies.

Shown in Figure 10.4 [4] is the wavelet composition for the sawtooth signal shown at the top [4]. Notice how the basis functions are associated with different portions of the wave. At the discontinuity, the short wavelets are large, while during the longer duration ramps the long wavelets have a larger amplitude.

Noise reduction (filtering) is possible using wavelets as shown in Figure 10.5 [4]. This is useful in many areas in communications where signals are corrupted by noise either intentionally or unintentionally. This noise reduction is accomplished by zeroing the basis function coefficients when they are below a certain threshold value. The signal is then reobtained by inverting the transform. Wavelets have found application in characterizing acoustic signals for computer synthesis of music [5], coding for communications over digital channels [6, 7], computer vision [8], and graphics [9] and elsewhere.

Compression of data is one of the more useful and interesting properties of the wavelet transforms. Most of the energy in image data is located in very few wavelet coefficients. In a typical application, only 3–5% of the wavelet coefficients are necessary to accurately recreate the original data sequence. The remainder of the coefficients are set to zero by some mechanism (e.g., thresholds and percentage of coefficients).

If the wavelets that make up the basis are suitably chosen, representation of functions by these coefficients can be very efficient. That is to say, very few of the coefficients are nonzero. Thus, storing the coefficients or transmitting the coefficients rather than the original function can be executed with fewer symbols than the original function.

There are an infinite number of families of wavelets such as the Haar wavelet discussed below. Selection of the right wavelet family to match the problem at

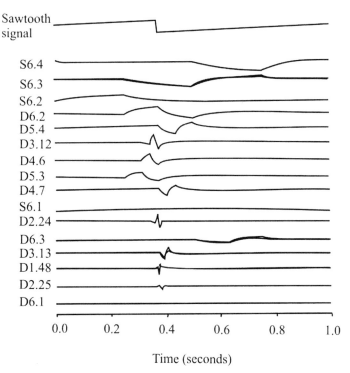

Figure 10.4 Example dilation and scaling of a mother wavelet function. (*Source:* [4], © 1996, IEEE. Reprinted with permission.)

hand is one of the challenges of application of the technology. As opposed to the Fourier transform that is defined for an infinite *x*-scale, wavelets have only *compact support*. That is to say that they are nonzero for only a segment of the *x*-axis. The advantage of compact support is the ability of wavelets to represent functions that have characteristics that are localized on the *x*-axis. Also, when functions are correctly matched to the wavelet basis, most of the energy in the function is localized to a few coefficients. Noise, on the other hand, is normally distributed evenly everywhere. Thus, to denoise a function, it is only necessary to retain the appropriate few coefficients and zero all the rest. Considerable SNR improvements can ensue.

For the Fourier transform, the basis functions are sine and cosine functions. Wavelets also have basis functions. If $h(t)$ is the prototype wavelet function, also called the *mother wavelet*, the other wavelet basis functions are given by

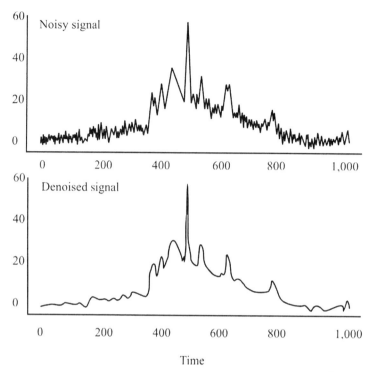

Figure 10.5 Smoothing effects of wavelet transforming a time waveform, removing the coefficients that aree below the threshold, and reassembling the waveform. (*Source:* [4], © 1996, IEEE. Reprinted with permission.)

$$h_a(t) = \frac{1}{a^{1/2}} h\left(\frac{t}{a}\right) \qquad (10.30)$$

where a is a scale factor. Therefore, once the prototype basis function $h(t)$ is specified, the remainder of the basis functions are obtained by expanding and contracting in time and amplitudes as well as by time shifts of the prototype. Thus, the *continuous wavelet transform* (CWT) is given by

$$\text{CWT}_x(\tau,a) = \frac{1}{|a|^{1/2}} \int x(t) h^*\left(\frac{t-\tau}{a}\right) dt \qquad (10.31)$$

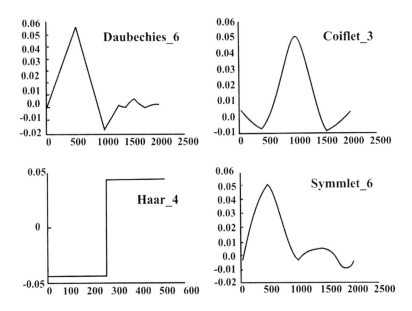

Figure 10.6 Four mother wavelet functions. The other wavelets are scaled and stretched versions of the mother wavelet. (*Source:* [10], © 1995, IEEE. Reprinted with permission.)

Shown in Figure 10.6 are four example mother wavelet functions [10]. The _N_ notations specify the order of the wavelet. Note that just as other basis function sets need not be orthogonal, wavelet families also need not be orthogonal.

The *discrete wavelet transform* (DWT) was discovered by Daubechies [11]. If the mother wavelet is given by $\varphi(x)$, then the DWT of a discrete function $f(i)$ using the wavelet basis functions is given by

$$\text{DWT}(j,k) = 2^{-\frac{j}{2}} \sum_{i=0}^{N-1} f(i)\phi(2^{-j}i - k) \tag{10.32}$$

where the basis functions $\varphi(x)$ are recursively given by

$$\varphi(x) = \sum_{i=0}^{M-1} c_i \phi(2x - k) \tag{10.33}$$

where M is the number of nonzero coefficients, also called the *order* of the wavelet. The inverse transform is given by

$$f(i) = \sum_{j=0}^{N-1} \sum_{k=0}^{M-1} \text{DWT}(j,k) 2^{-\frac{j}{2}} \phi\left(2^{-j} i - k\right) \tag{10.34}$$

These functions form an orthonormal basis. Like the CWT, they are amplitude weighted and shifted versions of the mother wavelet, where the weighting and shifting are powers of two. In this expression j and k are integer coefficients that scale and dilate the mother function. The parameter j determines the width of the wavelet while k controls its location on the x-axis.

A scaling function $w(x)$ is defined based on the mother wavelet as

$$w(x) = \sum_{k=0}^{N-1} (-1)^k c_{1-k} \phi(2x - k) \tag{10.35}$$

The c_k are called wavelet coefficients. These coefficients satisfy constraints such as

$$\sum_{k=0}^{N-1} c_k = 2 \qquad \sum_{k=0}^{N-1} c_k c_{k-2j} = 2\delta_j \tag{10.36}$$

where δ is the delta function and j is the location index. The latter of these two equations is referred to as *normalization*. The set of coefficients $\{c_0, c_1, \ldots, c_{N-1}\}$ are often thought of as a filter that is applied to the raw data. In doing so, intermediate results ensue that represent smoothed data and detailed data.

Any function $g(x)$ can be written as a series expansion, just as in the Fourier transform above, by

$$g(x) = c_0 + \sum_{j=0}^{\infty} \sum_{k=0}^{2^j - 1} c_{j,k} \phi_{j,k}(x) \tag{10.37}$$

for some appropriate coefficients $c_{j,k}$. The wavelet coefficients for some common wavelet families are given in Table 10.1.

The aforementioned Haar transform is a wavelet transform, although wavelets were not invented when Haar discovered his transform. The Haar transform is based on Haar functions shown in Figure 10.7 where the first eight functions are shown. Harr functions are defined over the interval (0, 1). Clearly these functions

Table 10.1 Wavelet Coefficients for Some Wavelet Families

Wavelet	c_0	c_1	c_2	c_3	c_4	c_5
Haar	1	1				
Daubechies_4	$\dfrac{1+\sqrt{3}}{4}$	$\dfrac{3+\sqrt{3}}{4}$	$\dfrac{3-\sqrt{3}}{4}$	$\dfrac{1-\sqrt{3}}{4}$		
Daubechies_6	0.332671	0.806801	0.459877	−0.135011	−0.085441	0.035226

satisfy the above requirement that they all are shifted and/or amplitude varied version of the same function.

10.3.3.1 Haar Wavelet

The function shown in Figure 10.8 is a wavelet function from the Haar family of wavelet functions. The family is comprised of this function and all functions made

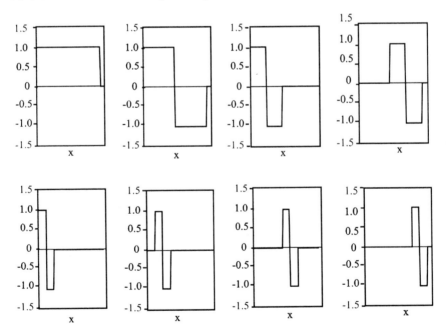

Figure 10.7 Haar functions forming the basis of the Haar wavelet transform.

Signal Processing

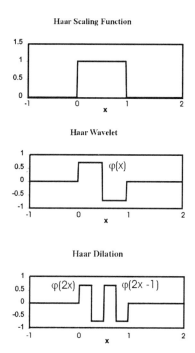

Figure 10.8 The Haar wavelet functions.

by shifting this function along the x-axis in integer amounts, stretching/dilating the wavelet out by factors of 2^k, and shifting in multiples of 2^m units, where k and m are integers.

The dilation shown in Figure 10.8 can be expressed as

$$d(x) = \varphi(2x) + \varphi(2x-1) \qquad (10.38)$$

for example. Any dilation can be expressed in this form from the mother wavelet. As is immediately obvious from Figure 10.8, $\varphi(2x)$ has twice the frequency as $\varphi(x)$ and $\varphi(2x-1)$ is simply an x-shifted version (by one place) of $\varphi(2x)$.

The Haar basis functions are defined as follows. The Haar mother wavelet is given by

$$\varphi(x) = \begin{cases} 1 & 0 \leq x < \frac{1}{2} \\ -1 & \frac{1}{2} \leq x < 1 \\ 0 & \text{otherwise} \end{cases} \quad (10.39)$$

and

$$\phi_{j,k}(x) = \phi(2^j x - k) \quad (10.40)$$

The functions shown in Figure 10.7 are then

$$\begin{aligned} \phi_{0,0} &= \phi(x) \\ \phi_{1,0} &= \phi(2x) \\ \phi_{1,1} &= \phi(2x-1) \\ \phi_{2,0} &= \phi(4x) \\ \phi_{2,1} &= \phi(4x-1) \\ \phi_{2,2} &= \phi(4x-2) \\ \phi_{2,3} &= \phi(4x-3) \end{aligned} \quad (10.41)$$

These functions are orthogonal over (0, 1) because

$$\int_0^1 \phi(x)\phi_{j,k}(x)dx = 0$$

$$\int_0^1 \phi_{j,k}(x)\phi_{n,m}(x)dx = 0 \quad (j,k) \neq (n,m) \quad (10.42)$$

which can be easily verified by examination of Figure 10.7.

The Haar transform is extensively used in image processing, where an image is converted via the Haar transform, and the converted coefficients are transmitted or stored instead of the original image. A fairly good reconstruction of the original image is possible.

The Haar transform is particularly applicable to a problem in simultaneous data smoothing and sharpening. Consider the simple case of $N = 4$, and a set of linear equations given by

$$y(0) = \frac{1}{2}[x(0) + x(1)] \qquad y(2) = \frac{1}{2}[x(2) + x(3)]$$
$$y(1) = \frac{1}{2}[x(0) - x(1)] \qquad y(3) = \frac{1}{2}[x(2) - x(3)] \tag{10.43}$$

In matrix form this is

$$\begin{bmatrix} y(0) \\ y(1) \\ y(2) \\ y(3) \end{bmatrix} = \frac{1}{2} \begin{bmatrix} 1 & 1 & 0 & 0 \\ 1 & -1 & 0 & 0 \\ 0 & 0 & 1 & 1 \\ 0 & 0 & 1 & -1 \end{bmatrix} \begin{bmatrix} x(0) \\ x(1) \\ x(2) \\ x(3) \end{bmatrix} \tag{10.44}$$

The sum equations perform averaging of two sequential time samples and the difference equations are of the form of a discrete differentiation. The net effect of this transformation is to smooth the data with the sum equations and to sharpen the data differences with the differentiation. The differentiation will tend to enhance sharp transitions in the input data. The sum expressions implement a lowpass filter, which suppresses high-frequency information in the input data, while the difference expressions form a highpass filter. This filter suppresses low-frequency information while enhancing high-frequency information.

For larger data sets the recursion equations for this problem are given by

$$y(2n) = \frac{1}{2}[x(2n) + x(2n+1)] \qquad \text{lowpass filter}$$
$$y(2n) = \frac{1}{2}[x(2n) - x(2n+1)] \qquad \text{highpass filter} \tag{10.45}$$

10.3.4 Fast Transforms

The transforms discussed above, indeed most transforms common in communication EW signal processing, require on the order of N^2 [denoted as $O(N^2)$] operations to complete where N is the transform size. For large values of N, such computations can become unmanageable, especially when the signal processing must be done in real or near-real time. For most of the unitary transforms when N is a power of 2, however, fast versions exist. They are based on factoring the transform matrix into subproblems, and the results of these subproblem computations can be used in subsequent processing without the need to redo them. The fast Fourier transform, for example, can be computed in $O(N\log N)$ computations. A comparison of these quantities is informative. One such comparison for several values of $N = 2^M$ is presented in Table 10.2. Clearly, even for relatively small values of M, the number of operations in the long (nonfast) transforms quickly becomes prohibitive.

The fast transforms are often represented in the form of flow diagrams. As an example, consider the 8 × 8 Haar transform above. Equation (10.26) gives the transform matrix. The flow matrix for this transform is given in Figure 10.9 [12]. The $1/\sqrt{8}$ is ignored here. This fast transform requires $2(N - 1)$ additions and subtractions, the fastest unitary transform available.

10.4 CYCLOSTATIONARY SIGNAL PROCESSING

Most analysis of stochastic (random) signals assumes that those signals have stationary properties—that is, the properties do not change with time. This assumption frequently does not apply. In fact, a generalization of the stationarity property for most signals of practical interest is periodic variation of the stastical properties. Such signals are called *cyclostationary* because their stastical

Table 10.2 Comparison of the Approximate Number of Computations to Compute a Transform

M	$N = 2^M$	N^2	$N \log N$
1	2	4	0.60
2	4	16	2
3	8	64	7
4	16	256	19
5	32	1,024	48
6	64	4,096	116
7	128	16,384	270
8	256	65,536	617
9	512	262,144	1,387
10	1024	1,048,576	3,083

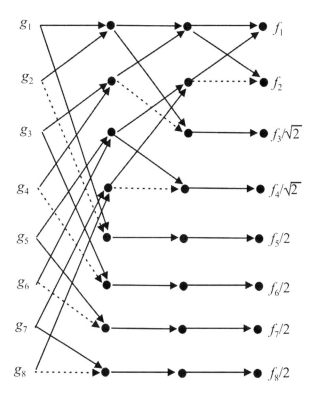

Figure 10.9 Flow diagram for the fast Haar transform. Solid lines are additions while the dashed lines are subtractions. (*Source:* [12], © 1999, R. Bock. Reprinted with permission.)

properties are cyclic—they are periodic with time. Computation of the cyclic properties can yield results than can be used for a variety of purposes. One of these is to classify signals by modulation type. Shown in Figure 10.10 [13] are the cyclostationary power spectral densities of a QPSK signal and an MSK signal. Note that while these characteristics for $\alpha = 0$ are essentially the same, for $\alpha \neq 0$ they are dramatically different. For QPSK there are distinctive features at $\alpha = nf_0$ for integer n, whereas for MSK the features are only present for even values of n, and at $\alpha = \pm nf_c \pm nf_0$ for odd values of n. Here f_c is the carrier and f_0 is the baud rate of the digital signal. These differences can be used to discern the type of signal being analyzed. The power spectral density when $\alpha = 0$ corresponds to the normal noncyclic PSD that is usually analyzed. This spectrum is unique for this

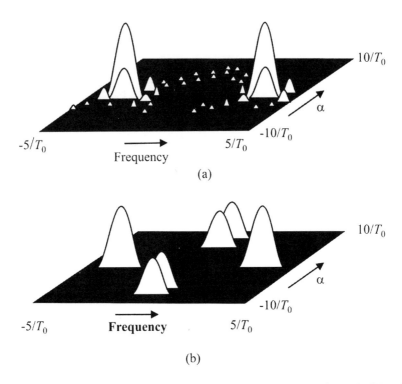

Figure 10.10 Examples of the cyclostationary spectrum of two digital signals: (a) QPSK and (b) MSK. The dramatic difference in form facilitates modulation recognition. The MSK signal in (b) is an SQPSK signal with a raised cosine carrier envelope. (*Source:* [13], © 1988, IEEE. Reprinted with permission.)

type of modulation and if automated means are available to evaluate this spectra, then the modulation can be uniquely identified.

Some of the uses that have been identified include signal detection and classification, parameter estimation, TDOA estimation, spatial filtering, direction finding, frequency shift filtering for signal extraction, and frequency shift prediction [13].

In a tactical military setting, identification of the modulation type is extremely useful for a variety of reasons. First, it provides an indication of the possible military unit type with which the radio is associated and this can provide information of the echelon and other relevant information known as the EOB. Second, in those cases where demodulation assets are assigned automatically, determination of the modulation type is first required in order to properly assign the resources. This latter property can be useful in a nonmilitary setting when using general-purpose modems, for example, and it is necessary to know the modulation type in order to select the demodulator.

Another advantage is that Gaussian noise does not exhibit cyclostationary characteristics. Therefore, the cyclostationary spectrum is noise-free and the intercept range of ES systems is thus extended, which equates to increased sensitivity. This noise-free characteristic can be seen in Figure 10.10.

10.5 HIGHER-ORDER STATISTICS

Calculation of the statistical properties of random processes of which most of us are aware are called *first-order statistics*. This notion can be generalized to higher orders by generalizing the equations used to calculate the statistics. If $x = \{x_i\}$, $i = 1, ..., N$, represents random sequences or samples of a signal, then the moments of that sequence are given by

$$m_n = E\{x^n\} \tag{10.46}$$

where $E\{\ \}$ represents the familiar expected value operator

$$E\{x^n\} = \int_{-\infty}^{\infty} x^n p(x) dx \tag{10.47}$$

and where $p(x)$ is the *probability density function* (pdf) of x^2 [14].

If the signal is ergodic and stationary, then these moments can be estimated by

$$m_n = \frac{1}{N} \sum_{i=0}^{N-1} x_i^n \tag{10.48}$$

If the sequence has zero mean, then the *cumulants* of order n of that series are given by

$$c_n(t_1, t_2, ..., t_n) = E\{x(t)x(t+t_1)x(t+t_2)\cdots x(t+t_n)\} \tag{10.49}$$

---

[2] x^n is shorthand notation for $\{x_1^n, x_2^n, ..., x_N^n\}$.

In particular, if $n = 2$, the *second-order cumulant* of $x(t)$ is

$$c_2(t_1) = E\{x(t)x(t+t_1)\} \qquad (10.50)$$

Thus, the second-order cumulant is the *autocorrelation* function of x. The Fourier transform of $c_2(t_1)$ is the familiar *power spectral density* (PSD) function, which displays the power distribution in x as a function of frequency:

$$\text{PSD}(f_1) = \int_{-\infty}^{\infty} c_2(t_1) e^{-j2\pi f_1 t} dt_1 \qquad (10.51)$$

In general, the Fourier transforms of the higher-order spectra are called *polyspectra*.

The *third-order cumulant* of x, given by

$$c_3(t_1, t_2) = E\{x(t)x(t+t_1)x(t+t_2)\} \qquad (10.52)$$

and its Fourier transform is known as the *bispectrum*, denoted as B_x.

$$B_x(f_1, f_2) = \int_{-\infty}^{\infty}\int_{-\infty}^{\infty} c_3(t_1, t_2) e^{-j2\pi(f_1 t_1 + f_2 t_2)} dt_1 dt_2 \qquad (10.53)$$

Normally a more efficient way to compute B_x is

$$B_x(f_1, f_2) = E\{X(f_1)X(f_2)X^*(f_1+f_2)\} \qquad (10.54)$$

where $X(f)$ is the Fourier transform of $x(t)$ and, as usual, * denotes complex conjugate.

The first four cumulants can conveniently be calculated using

$$\begin{aligned}
c_1 &= m_1 \\
c_2 &= m_2 - m_1^2 \\
c_3 &= m_3 - 3m_1 m_2 + 2m_1^3 \\
c_4 &= m_4 - 3m_2^2 - 4m_1 m_3 + 12 m_2 m_1^2 - 6m_1^4
\end{aligned} \qquad (10.55)$$

The fourth-order spectrum is called the *trispectrum*.

There are several potentially useful properties of cumulants. If $x(t)$ is a random process with a symmetric pdf, which includes the Gaussian random variable case, then all cumulants higher than the second are equal to zero. In addition, the cumulants of the sum of two random variables are equal to the sum of the cumulants of the individual variables.

If Gaussian noise is added to a non-Gaussian communication signal, then the cumulants of this sum will be those of the signal, since the cumulants of the noise component are all zero. Almost Gaussian noise is generated in the front-end electronics of receiving systems. It is almost Gaussian because, in fact, it is band-limited, whereas true Gaussian noise is not. It is this noise source, in addition to noise from external sources, such as the electrical fields set up by close-by generators, that limit the range of ES systems. Thus, calculating the cumulants of such a time series can extend the dynamic range of EW systems.

Since the higher-order statistics of Gaussian noise are zero, if they are computed for signal detection purposes, there should be no noise present and the detrimental effects of noise are therefore mitigated. In practice, of course, noise is not truly Gaussian and there are usually some residual effects present in the higher-order statistics.

Thus, the higher-order statistics associated with a random sequence x can be used for several purposes. They can be used to minimize or eliminate some forms of additive noise and can be used for signal detection in otherwise noisy environments. Some types of signals have features that show up only in higher-order stastical calculations that can be used for signal classification.

10.6 APPLICATIONS

Some applications of the above generic signal processing techniques in communication EW systems are presented in this section. These include signal detection, signal classification, entity recognition and identification, language identification, and emitter identification.

10.6.1 Signal Detection

In many communication EW systems a receiver is used to search the RF spectrum to seek channels where energy is present. This is the signal detection problem, and it is one of the first functions such systems must perform. Signal detection is frequently combined with signal classification discussed subsequently.

Searching the RF spectrum can be further divided into general search and directed search. In *general search* one or more scan ranges are usually specified in the form of f_{start} to f_{stop}. A frequency step is also sometimes specified, which is usually one channel wide (e.g., in the military VHF frequency range the channels are 25 kHz wide). Searching then starts at f_{start} and stops at f_{stop}. Frequently this searching is continuous and repetitive. General search is usually performed when the specific frequencies of the targets are not known or are only partially known. Note that this form of searching, although described here as sequential, could also be simultaneous if a wideband, channelized receiver is used. Channelization in that case would normally also correspond to one channel width according to where in the spectrum the searching is taking place.

For *directed search*, specific discrete frequencies are used for the searching. These frequencies form a list and the receiver tunes to each frequency in turn, looking for the presence of signals of interest. This mode is used when the frequencies are known.

Combinations of general and directed search are also possible. Indeed, this would likely be the normal mode of operation as even though many frequencies might be known ahead of time, some important targets may have changed frequency and those are not known. A typical example of this would be as a result of EA operations against a target net.

Both of these search strategies require the measurement of energy in the channel to which the receiver is tuned. Since noise is always present along with the signals, and since noise can only be described statistically, signal detection is a probabilistic process. The ability of a receiver to detect signals can only be described in statistical terms, even if the signals of interest are deterministic with parameters that are totally known.

Before any processing of a signal can occur, that signal must first be detected. That is to say, it must first be ascertained if there is a signal present. More generally, detection refers to deciding which of two or more hypotheses is correct. Sometimes, however, signal detection is included as one of the outcomes of the signal classification problem, where the signal absent hypothesis is one of those tested. A classification of any other than signal absent then implies there is a signal present, signal detected.

There is another dimension to signal detection in communication EW system design, however. Since the target environments are typically noncooperative, it is often necessary to scan the RF spectrum looking for signals. When a signal is encountered, that signal is detected. Since the RF spectrum is everywhere channelized (although the size of the channels varies), this signal detection is normally accomplished by measuring the energy in these channels.

10.6.1.1 Hypothesis Testing

The presence of a signal $s(t)$ in a channel is characterized by two hypotheses, denoted by H_0 and H_1, depending on whether the signal is present or not:

$$H_0 : \quad r(t) = n(t)$$
$$H_1 : \quad r(t) = s(t) + n(t) \quad (10.56)$$

In these equations, $r(t)$ is the received signal that consists of either $s(t)$ accompanied by noise $n(t)$, or just the noise alone. Hypothesis testing is the statistical process of estimating the validity of some hypothesis based on some set of measurements. Suppose that the measured parameters are manifest in a variable x. That is, the decision variable is $x = f(p_1, p_2, \ldots, p_N)$ where p_i is a measured parameter.

When the decision rule is "decide the hypothesis that has the most likely probability of occurring," it is called the MAP criterion. Thus,

Decide H_0 if $\quad P(H_0|x) > P(H_1|x)$, or $P(H_0|x) / P(H_1|x) > 1$ (10.57)
Decide H_1 if $\quad P(H_0|x) < P(H_1|x)$, or $P(H_0|x) / P(H_1|x) < 1$

Using the definition of conditional probabilities

$$P(H_0|x) = P(H_0) P(x|H_0) / P(x) \text{ and}$$
$$P(H_1|x) = P(H_1) P(x|H_1) / P(x) \quad (10.58)$$

$P(H_0)$ is the a priori probability of H_0 occurring at all, while $P(H_1)$ is the a priori probability of H_1 occurring at all. Decide H_0 if

$$\frac{P(H_0|x)}{P(H_1|x)} = \frac{P(H_0)P(x|H_0)/P(x)}{P(H_1)P(x|H_1)/P(x)} > 1 \quad (10.59)$$

or

$$\frac{P(x|H_0)}{P(x|H_1)} > \frac{P(H_1)}{P(H_0)} \quad (10.60)$$

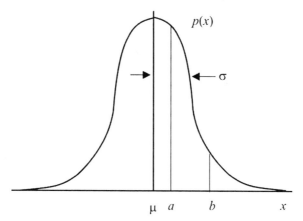

Figure 10.11 The probability of an event is given by the area under the density function between two points.

Recall from the definition of the probability density function (see Appendix A) that a probability is given by the area under the density function. That is, if $p(\xi)$ is the density function, then (referring to Figure 10.11)

$$P(a \leq x \leq b) = \int_a^b p(\xi) d\xi \tag{10.61}$$

As $\Delta x = a - b$ becomes smaller and smaller, then

$$\lim_{\Delta x \to 0} \frac{P(x|H_0)}{P(x|H_1)} = \lim_{\Delta x \to 0} \frac{\int_a^b p(\xi|H_0) d\xi}{\int_a^b p(\xi|H_1) d\xi} = \frac{p(x|H_0)}{p(x|H_1)} \tag{10.62}$$

This ratio is known as the *likelihood ratio*, and is denoted by λ_0. Thus,

$$\lambda_0 = \frac{p(x|H_0)}{p(x|H_1)} \geq \frac{P(H_1)}{P(H_0)} \quad \text{for hypothesis } H_0 \tag{10.63}$$

$$\lambda_1 = \frac{p(x|H_1)}{p(x|H_0)} > \frac{P(H_0)}{P(H_1)} \quad \text{for hypothesis } H_1 \tag{10.64}$$

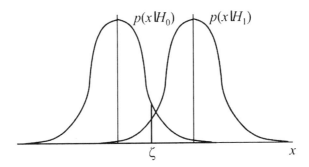

Figure 10.12 Probability density functions for H_0 and H_1.

Although this analysis compares the ratios to a threshold value of one, it obviously can be trivially extended for any threshold ζ. This parameter then becomes a multiplier for the right side of these inequalities.

Suppose that $p(x|H_0)$ and $p(x|H_1)$ are as shown in Figure 10.12. Let P_m denote the *probability of missed detection*. This is the probability that, given that a signal is present (H_1), it is not detected. As shown in Figure 10.12, this is the area under the $p(x|H_1)$ curve to the left of the threshold point ζ. Thus,

$$P_m = \int_{-\infty}^{\zeta} p(\xi|H_1)d\xi \tag{10.65}$$

Let P_{fa} denote the *probability of false alarm*. This is the probability, given that there is no signal present (H_0), that the decision is made that there is. In Figure 10.12 this is the area under the $p(x|H_0)$ curve to the right of the threshold ζ. Thus,

$$P_{fa} = \int_{\zeta}^{\infty} p(\xi|H_0)d\xi \tag{10.66}$$

Both of these probabilities are a measure of an error being made. Thus, the overall *probability of error*, denoted by P_e, is the sum of these two.

$$P_e = P_m + P_{fa} \tag{10.67}$$

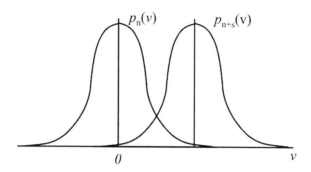

Figure 10.13 Detector output probability density functions with noise only [$p_n(v)$] and signal plus noise [$p_{n+s}(v)$].

The *probability of detection*, denoted here by P_d[3] is equal to one minus the probability of missed detection, namely,

$$P_d = 1 - P_m = 1 - \int_{-\infty}^{\zeta} p(\xi|H_1)d\xi = \int_{\zeta}^{\infty} p(\xi|H_1)d\xi \qquad (10.68)$$

10.6.1.2 Effects of Noise on Detection

The SNR is the dominating parameter to determine signal detectability. Suppose $p_n(v)$ represents the probability density function of the output power of a receiver when there is no signal present—that is, the output consists of noise only. Further suppose $p_{n+s}(v)$ represents the probability density function of this output when there is a signal represent. Figure 10.13 shows the two receiver output density functions under these two assumptions. When there is a signal present, the mean value of the detector output will be larger than when there is no signal present. Here, the probability density function for noise only is assumed to have a zero average (mean) value.

Characteristics of receivers utilizing these concepts are typically presented on what are called ROC curves. Figure 10.14 shows a typical ROC curve and clearly shows how P_d and P_{fa} can be traded off. If one wants to achieve a high P_d, then a high P_{fa} is typically required, and vice versa. If a low P_{fa} is desired, one usually must accept a low P_d. Optimizing both is usually not possible.

---

[3] Not to be confused with the power density function, also denoted by P_d—the usage should make clear which is intended.

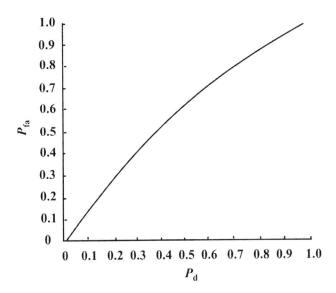

Figure 10.14 Typical ROC curve.

10.6.1.3 Radiometer

A diagram of a *radiometer* is shown in Figure 10.15. Radiometers are frequently used as a standard against which other types of signal detectors are compared, although real implementations of radiometers are possible (or at least close approximations). A radiometer is an energy detector since the energy, E, in signal $s(t)$ is given by

$$E = \int_0^T s^2(t)\,dt \tag{10.69}$$

where it is assumed that the signal lasts only from 0 to T or, equivalently, the analysis of the signal is over the interval $(0, T)$. The radiometer is an optimum detector if it is assumed that the signal is a stationary process with a flat spectral density and the noise is Gaussian [15]. The input to the radiometer is the signal to be detected. It is sent through a bandpass filter, of width wide enough to pass $r(t)$ without significant distortion, yielding $s(t)$. This signal is then squared and integrated from 0 to T. After T seconds, the integrated results $y(t)$ are compared

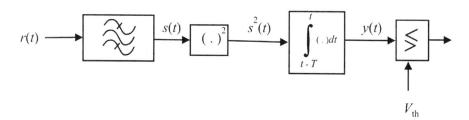

Figure 10.15 Block/flow diagram of a radiometer that measures energy over a time interval (0, T).

with a threshold V_{th}. If $y(t) > V_{th}$, then the signal is declared present. If $y(t) < V_{th}$, then the signal is declared absent.

For realistic communication EW systems, wideband radiometers are idealized devices, having limited practical use. What is more common is to measure the energy in a frequency channel using a narrowband radiometer. Such a configuration is often referred to as a *channelized radiometer*. The performance of such a configuration depends, along with other parameters, on how long the signal is present. This could be a considerable length of time for such signals as special mobile radios, for example, or it could be very short in the case of fast frequency-hopping targets.

The performance of a radiometer can be determined as follows [16]. Assuming there is noise only at the input, then the output statistics have chi-squared characteristics with $2TW$ degrees of freedom where T is the integration time and W is the noise bandwidth. If there is a signal present, then the output statistics are also chi-square but are noncentral. The noncentrality parameter is $2E/N_0$, where E is the energy in the input signal given by

$$E = P_s T \tag{10.70}$$

N_0 is the one-sided noise spectral density in watts/hertz and P_s is the power in the signal.

When TW is large, the output statistics are approximately Gaussian from central limit theorem arguments. These statistics are characterized by a factor d, which is a measure of the output SNR. It is defined as

$$d = Q^{-1}(P_{fa}) - Q^{-1}(P_d) \tag{10.71}$$

where Q^{-1} is the inverse normal cumulative distribution function. For the radiometer the performance is given by

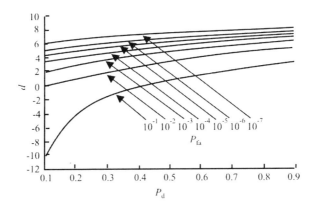

Figure 10.16 A measure of the output SNR of a radiometer. (*Source:* [16].)

$$d^2 = \frac{T}{W}\left(\frac{P_s}{N_0}\right)^2 \tag{10.72}$$

Thus, the SNR required to achieve the specified P_{fa} and P_d is given by

$$\frac{P_s}{N_0} = d\sqrt{\frac{W}{T}} \tag{10.73}$$

The parameter d can be determined numerically. Some values for typical performance parameters of interest in the design of communication EW systems are shown in Figure 10.16.

Typically for communication EW applications, TW is not large. Some numbers for the low VHF frequency range might be $T = 10$ ms and $W = 25$ kHz for channelized radiometers. Thus, $TW = 2.5$. In this case the above analysis generates values for d that are too optimistic [16]. This is optimistic from an intercept point of view—from a communicator's point of view, they are pessimistic. In that case, a factor is provided in Woodring to multiply d to correct the analysis. Charts of this factor, η, for two cases, $P_d = 0.9$ and $P_d = 0.1$ are shown in Figures 10.17 and 10.18. Therefore, for small TW, $d' = \eta d$ is used instead.

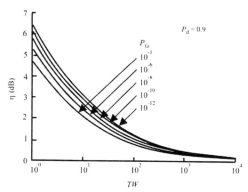

Figure 10.17 Adjustment factor η for small values of TW when $P_d = 0.9$. (*Source:* [16].)

10.6.1.4 Detection of Frequency-Hopping Signals

The wideband radiometer discussed above is an optimum detector for signals if there is little known about the signals. A channelized radiometer is constructed if the input is first filtered. One approach for constructing a detector for frequency-hopping signals is to assemble G bandpass radiometers that operate on the incoming signal, where G is the number of channels used by the frequency-hopping target. Each of these channels has a bandwidth matched to the channelization of the target radio. After N_h hop intervals of duration T_h each, the outputs of the channelized radiometers are compared with a threshold η, and under hypothesis H_0 (signal absent), the output should be less than η, and under H_1

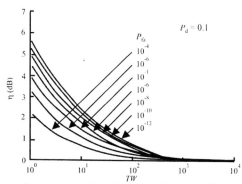

Figure 10.18 Adjustment factor η for d when $P_d = 0.1$. (*Source:* [16].)

(signal present), it should be greater than η. If more is known about the signal of interest, then the radiometer is not an optimum detector for the signal. The optimum detector will, in general, be different for each different type of signal. These optimum detectors can become quite complex. Figure 10.19 shows the configuration for the optimum detector for slow frequency-hopping signals, which employ multiple frequencies within a hop (MFSK) and also employ continuous phase within a hop [17]. In this case, the detector is optimum in the *average likelihood ratio* (ALR) sense. For this figure, E_h is the energy in each hop, $N_0/2$ is the two-sided noise level, $I_0(\ )$ is the modified Bessel function of the 0th kind, and N_d is the number of data symbols in each hop. The frequency corresponding to the ith channel is denoted by f_i while θ_i corresponds to the local oscillator phase offset. The function $\varphi(d_i, t)$ is the baseband data modulation.

Thus, for each of the G channels, N_d noncoherent detectors are implemented. In each of these N_d detectors, there is a noncoherent detector required for each possible data sequence that can occur. Each of these detectors is configured as shown in Figure 10.19. Each consists of two parallel channels of multipliers, integrators and squaring circuits. These two channels are added and then the square root of the result is computed. This is followed by multiplication by the SNR estimate, which is followed by computation of the Bessel function. All of these computations are executed in parallel in each channel as shown. Within each channel the outputs of the data pattern channels are added together, followed by addition of the results from each channel. After the N_h hops the product of the results from each hop is computed and compared with the threshold.

The difference between optimization based on the ALR and the *maximum likelihood ratio* (MLR) is that for the former the average ratio is determined for all possible input sequences whereas for the latter the most likely sequence is determined. MLR detection is suboptimal compared with ALR, because not all of the possible input sequences are considered in the analysis. Performance of the above optimum detector compared with other techniques is illustrated in Figure 10.20 [18].

The above detector is optimum for slow frequency-hopping targets with continuous phase from one symbol to the next within a hop, but noncoherent from hop to hop. The case of fast frequency-hopping targets was analyzed by Beaulieu et al. [19] The most common definition of slow frequency hopping versus fast is based on the number of data bits per hop. If there are more than one such data bit per hop, it is called slow frequency-hopping, while if there is more than one hop per data bit (less than one data bit per hop), it is called fast frequency-hopping. The optimum detector structure for this case is shown in Figure 10.21 [17].

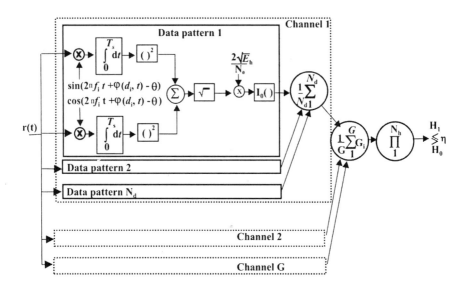

Figure 10.19 Optimum receiver for slow frequency hopping targets. (*Source:* [17], © 1992, IEEE. Reprinted with permission.)

10.6.2 Signal Classification

Another important function that is performed in the receiving subsystem of communication EW systems is *signal classification*. Classifying signals can help in determining the EOB by indicating what type of communication equipment is in use. Knowing the type of equipment an adversary has sometimes will also indicate who that adversary is, or perhaps indicate the measure of lethality or danger it presents.

Signal classification can be further broken down into several types of processes as listed here (not necessarily mutually exclusive or exhaustive):

- Modulation recognition (e.g., AM, FM, or PM);
- Signal type recognition (e.g., analog or digital);
- If digital, then which type (e.g., BPSK or QPSK);
- If modem, then which standard (e.g., V.32 or V.90);
- If multiplexed, then which type (e.g., FDM or PPM);
- If FDM, the type of signal in each channel;

- Specific emitter identification (e.g., to the serial number of the transmitter).

In general, there are two approaches to address the signal classification problem. The first approach is somewhat ad hoc. Signal features are determined largely by reasoning about what might make sense—amplitude histograms for example might make a good discriminate for AM versus constant modulus signal types. The first step in these processes is called *feature extraction*. The computed features that are extracted in the first step are then provided to processors that perform *pattern recognition* that attempts to discriminate one signal type from another. Neural networks, being devices that do pattern recognition reasonably well and fast, are often applied to this form of signal classification.

The second general approach is based on probability theoretic notions, also referred to as *decision theoretic*. Likelihood ratios are determined and hypotheses are established based on probability density functions of random signals. Cost functions are determined and the minimum costs are achieved by the likelihood ratios. This makes these approaches optimum in that sense; costs are minimized.

Experimentation by simulation has shown that the latter of these approaches offers more of a possibility of operation at lower SNRs than the former does. The former approaches tend to fall off in performance when the SNR falls below about

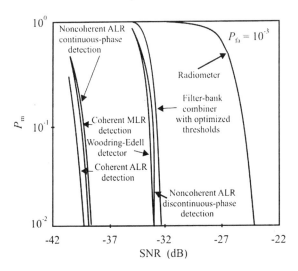

Figure 10.20 Probability of missing a detection for the detector shown in Figure 10.20 compared with other forms of slow frequency-hopping signal detectors. (*Source:* [18], © 1994, IEEE. Reprinted with permission.)

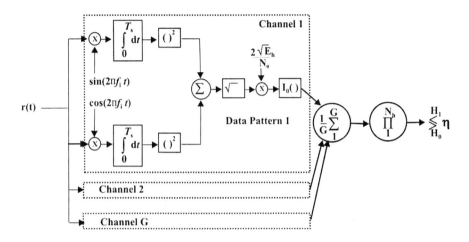

Figure 10.21 Architecture of an optimum detector for fast frequency-hopped signals. (*Source:* [17], © 1992, IEEE. Reprinted with permission.)

15 dB or so. The latter has, at least in simulations, successfully operated at SNRs less than zero. Low-SNR operation of such classifiers is important when long target ranges are involved. The more robust a classifier is, the more applicability it will have.

10.6.2.1 Pattern Recognition Approaches

When discussing signal detection, the notion of hypothesis testing was introduced in the context of a binary hypothesis problem: determining the presence of a signal in a channel. The signal classification problem can be cast in similar terms, although generalization to multiple-hypotheses is required. In that case, there are N hypotheses: selecting one of $N-1$ signal types and selecting the Nth choice, which is "none of the above."

As a simple example, suppose that the classes of interest are the following: AM DSB, AM SSB, Wideband FM (BW > 100 kHz), and Narrowband FM (BW < 100 kHz). Then the hypotheses are as follows.

- H_1: AM DSB;
- H_2: AM SSB;
- H_3: Wideband FM;
- H_4: Narrowband FM;

- H_5: Other (cannot determine or something other than the specified signal types).

The first step in these approaches is to extract relevant features. For this simple example, suppose that these features consist of the following

- p_1 = bandwidth of the predemodulated signal;
- p_2 = SNR of the output of an AM demodulator;
- p_3 = SNR of the output of an FM discriminator.

This is shown diagrammatically in Figure 10.22. It is tacitly assumed here that the width of the bandpass filter is adequate to pass the signal relatively undistorted. In this case, BW = 150 kHz might be about right.

The logic for selecting among the hypotheses using these features might be as follows. In this simple scheme, analog AM SSB would be very narrowband, so the bandwidth of the signal would be an important discriminator. Narrowband FM would have a narrower measure of bandwidth than wideband FM as well. These, coupled with the SNR at the output of the AM and FM demodulators, would determine the type of signal present. In those cases where there is potential confusion, the classifier indicates indeterminate results, which is probably the desired output in such circumstances. The threshold level for the SNR calculations in this example is 6 dB. In practice, this is a threshold that would be determined from experimentation. The resultant logic that implements this thought process is presented in Table 10.2.

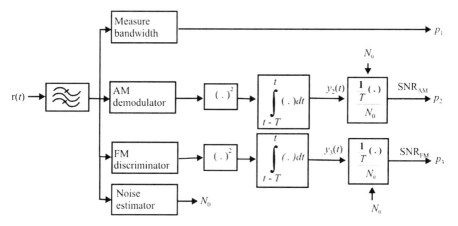

Figure 10.22 Simple pattern recognition signal classifier.

Table 10.2 Hypotheses Selections for the Example Classifier

p_1	p_2	p_3	Choose
< 5 kHz	< 6 dB	< 6 dB	H_5
< 5 kHz	< 6 dB	> 6 dB	H_4
< 5 kHz	> 6 dB	< 6 dB	H_2
< 5 kHz	> 6 dB	> 6 dB	H_5
> 5 kHz and < 100 kHz	< 6 dB	< 6 dB	H_5
> 5 kHz and < 100 kHz	< 6 dB	> 6 dB	H_4
> 5 kHz and < 100 kHz	> 6 dB	< 6 dB	H_1
> 5 kHz and < 100 kHz	> 6 dB	> 6 dB	H_5
> 100 kHz	< 6 dB	< 6 dB	H_5
> 100 kHz	< 6 dB	> 6 dB	H_3
> 100 kHz	> 6 dB	< 6 dB	H_1
> 100 kHz	> 6 dB	> 6 dB	H_5

The nemesis to accurate signal classification is the same as other communication signal processing problems—noise. This is manifest in the above example by the presence of N_0, the estimate of the noise floor used in the calculation of the SNR. This noise was discussed in Chapter 2. Since in all real instances of communication signals, noise is always present, the usual parameters calculated for signal classification are statistical. The various statistical moments of the signal over a short time interval, for example, are common parameters.

Nandi-Azzouz Classifier

Nandi and Azzouz have proposed an algorithm for the automatic classification of communication signals [20–22]. In their case, the problem was to distinguish among 13 analog and digital modulation types: AM, FM, multiple (= 2, 4) FSK, multiple (= 2, 4) ASK, DSB, LSB, USB, multiple (= 2, 4) PSK, *vestigial sideband* (VSB), and combined modulations. Combined in this case refers to a signal that has both substantial amplitude and phase modulations present. Nine parameters are used to perform the signal classification and these parameters are based on measurements of the signal in the time domain as well as the frequency domain. The flow diagram for this classifier is shown in Figure 10.23 while the parameter definitions are given in Table 10.3. In most cases, the data from which these parameters are obtained is normalized to some relevant value.

Each of these parameters was compared with threshold values that were derived by simulation. The thresholds are indicated in Table 10.4, where at each stage a parameter is compared to its threshold, thus dividing the classes into two groups at each such stage, gradually separating each modulation until only a single type remains in the group.

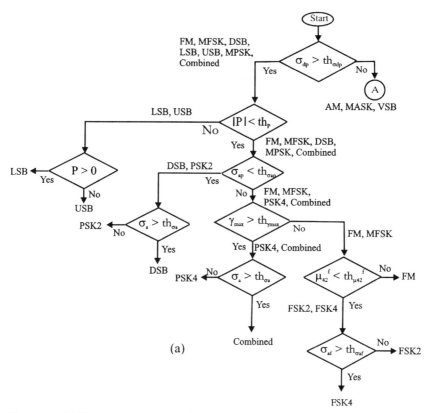

Figure 10.23(a) Flow diagram for the Nandi-Azzouz signal classifier; and (b) remainder of that flow diagram.

Results of this classifier against the target signals set are reported only for correct signal classification in [23], except for the digital modulations. These results are repeated in Figure 10.24. These parameters yielded very good performance in general against the specified target class types at the SNRs indicated. These parameters and thresholds could result in the classification of some signals into one of these 13 bins when the signal is actually something else altogether. Thus, it is necessary when designing a signal classification system to know as best as possible all of the signal types that are to be encountered in the environment. All of the signals may not be of interest, and therefore could be grouped into a class called "other" for practical purposes. Nevertheless, it is important to consider all of the signal types if a high-performance signal classifier is to be realized.

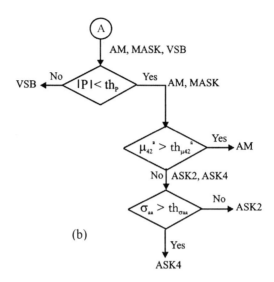

Figure 10.23 (Continued.)

Note that in this, as well as any other classifier that uses features from the power spectrum, it is important to know the center frequency involved. If it is not known, for example, it would be impossible to determine such things as the skewness and asymmetry characteristics of the PSD. The frequency of operation is sometimes tracked with a PLL that extracts the operating frequency over time, determining the moving average of frequency.

As a last note on the performance of this classifier, the SNRs considered are

Table 10.3 Parameters Used for the Nandi-Azzouz Pattern Recognition Signal Classifier

Parameter	Definition
γ_{max}	Maximum of the PSD of the normalized-centered instantaneous amplitude
σ_{ap}	Standard deviation of the absolute value of the instantaneous phase
σ_{dp}	Standard deviation of the instantaneous phase
P	Symmetry of the spectrum of the signal
σ_{aa}	Standard deviation of the absolute value of instantaneous amplitude
σ_{af}	Standard deviation of the absolute value of the instantaneous frequency
σ_a	Standard deviation of the instantaneous amplitude
μ_{42}^a	Kurtosis of the instantaneous amplitude
μ_{42}^f	Kurtosis of the instantaneous frequency

Table 10.4 Threshold Values of the Parameters for the Nandi-Azzouz Classifier

Parameter	Threshold Value (th$_i$)
γ_{max}	2.5
σ_{ap}	$\pi/5.5$
σ_{dp}	$\pi/6$
P	0.6
σ_{aa}	0.13
σ_{af}	2.15
σ_a	2.03
$\mu_{42}{}^a$	0.25
$\mu_{42}{}^f$	0.4

typical of those obtained from an airborne EW system, but are much too high to consider for ground applications. In fact, 15–20-dB SNRs are not unusual for even standoff airborne systems because signals propagate much better when one (or both) of the antennas is significantly elevated. For ground-to-ground applications however, the power decreases substantially more rapidly and SNRs of 15–20 dB are difficult to produce.

Assaleh-Farrell-Mammone Classifier

A technique was devised by Assaleh et al. [24] that classifies digitally modulated signals. It is based on calculating the autocorrelation coefficients of the predetected data. The IF (predemodulation, also referred to as predetected) signal is represented by

$$x(k) = s(k) + n(k) \tag{10.74}$$

where $s(k)$ are samples of the signal to be identified and $n(k)$ are samples of noise. The signal $s(k)$ is given by

$$s(k) = A\cos(2\pi f_c k + \phi(k) + \theta) \tag{10.75}$$

where f_c is the signal frequency, θ is the phase, and the parameter ϕ varies according to the signal type as given in Table 10.5.

The autocorrelation estimates of this signal are given by

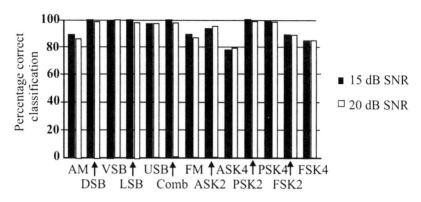

Figure 10.24 Results of the Nandi-Azzouz signal classifier described in the reference for two SNRs. (*Source:* [20], © 1998, IEEE. Reprinted with permission.)

$$\hat{R}_{rr}(k) = \sum_{n=0}^{M} r(n)r(n+k) \quad (10.76)$$

These autocorrelation estimates are the coefficients of a polynomial of dimension N given by

$$1 - a_1 z^{-1} - a_2 z^{-2} - \cdots - a_N z^{-N} \quad (10.77)$$

The *average instantaneous frequency* is related to the roots of this polynomial, z_i, by

Table 10.5 Values of Phase Shift According to the Type of Modulation

Modulation Type	$\phi(k)$
CW	0
BPSK	$0, \pi$
QPSK	$0, \pm\pi/2, \pi$
BFSK	$\pm 2\pi f_d k$
QFSK	$\pm \pi f_d k, \pm 2\pi f_d k$

$$f_i = \frac{f_s}{2\pi} \tan^{-1}\left(\frac{\text{Im}(z_i)}{\text{Re}(z_i)}\right) \tag{10.78}$$

and the *bandwidth* of this root is

$$BW_i = -\frac{f_s}{\pi} 10 \log_{10}\left[\frac{1}{\text{Im}^2(z_i) + \text{Re}^2(z_i)}\right] \tag{10.79}$$

In these expressions, f_s is the sample rate. It is these two parameters that the algorithm uses to classify the signals shown in Table 10.5. The classification algorithm is shown in Figure 10.25. The authors report that usually a second-order ($N = 2$) model is all that is necessary in this algorithm.

The simulated performance of this classifier against the signals considered is shown in Figure 10.26. For these results, the SNR was 15 dB. Almost perfect classification ensued under the conditions of the experiment. The algorithm was not tested, however, for robustness in noisy environments.

Whelchel-McNeill-Hughes-Loos Classifier

Implementation of a modulation classifier using neural networks was described by Whelchel et al. [25]. The signals of concern for their classifier were AM, AM SC, FM, CW, WGN, and QPSK. WGN is wideband Gaussian noise. They describe two experiments in which the SNR of the signals were 23 dB and 30 dB, representing relatively high values of SNR.

As with other pattern classification approaches, the first step is to extract the features from the incoming predetection signal. The features used in this classifier are based on several statistical moments and are given in Table 10.6. Two types of neural networks were trained: a back propagation network and a counterpropagation network [26]. Another variation that was tried was to connect (or not) the input layer of the neural network to the output layer. The performance results are shown in Figure 10.27 for the former network configuration to illustrate the type of performance achieved. These results are similar to the performance of the other pattern recognition approaches. The significant advantage of the neural network, however, is the processing speed.

Neural networks are massively parallel processors and compute results very rapidly. They emulate one model of how the human brain functions, with processing nodes that are interconnected with synapses. Many interconnects per node are implemented which facilitate the rapid computations. Neural networks are probably best known for their performance at pattern recognition.

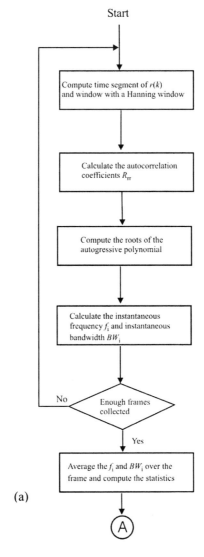

Figure 10.25 (a) Flow diagram for the Assaleh-Farrell-Mammone signal classifier; and (b) remainder of that classifier. (*Source:* [24], © 1992, IEEE. Reprinted with permission.)

10.6.2.2 Decision Theoretic Approaches

The decision theoretic signal classifiers establish likelihood ratios as discussed in Section 10.6.1.1 and compute the most likely signal type present by choosing the

Signal Processing 313

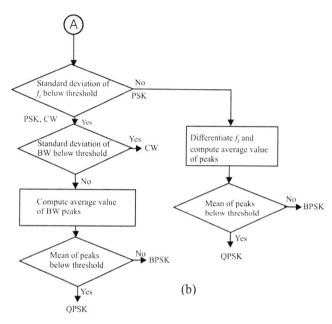

Figure 10.25 (Continued.)

signal corresponding to the largest likelihood ratio. Since the likelihood ratios are based on stastical properties of the signal, they produce optimal decisions, with optimality depending on the particular ratio used.

Table 10.6 Statistical Parameters Used in the Whelchel et al. Neural Network Signal Classifier

Amplitude variance
Amplitude skew
Amplitude kurtosis
Phase variance
Phase skew
Phase kurtosis
Frequency variance
Frequency skew
Frequency kurtosis

Source: [23], © 1992, IEEE. Reprinted with permission.

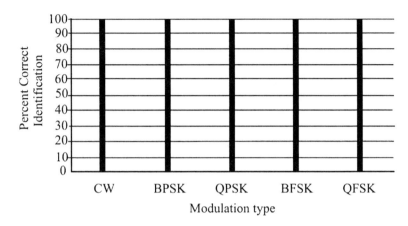

Figure 10.26 Simulated performance results of the Assaleh-Farrell-Mammone classifier. (*Source:* [23], © 1992, IEEE. Reprinted with permission.)

Kim-Polydoros Classifier

Kim and Polydoros describe a decision theoretic approach to discriminating between BPSK and QPSK [26]. The *quasi-likelihood ratio* (qLLR), which they determined for this discrimination, using their notation, is given by

$$\text{qLLR} = (\Sigma_I - \Sigma_Q)^2 + 4\Sigma_{IQ}^2 \tag{10.80}$$

where

$$\Sigma_I = \sum_{n=1}^{N} r_{I,n}^2 \qquad \Sigma_Q = \sum_{n=1}^{N} r_{Q,n}^2$$

$$\Sigma_{IQ} = \sum_{n=1}^{N} r_{I,n} r_{Q,n} \tag{10.81}$$

when the *complex envelope* of the output of a matched filter with input $r(t)$ is given by

$$\tilde{r}_n = r_{I,n} + j\, r_{Q,n} \tag{10.82}$$

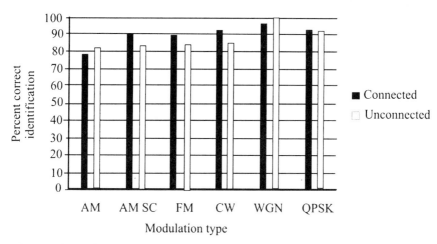

Figure 10.27 Performance of the Whelchel et al. neural network signal classifier. (*Source:* [24], © 1989, IEEE. Reprinted with permission.)

and the components given by

$$r_{I,n} = \int_{(n-1)T_s}^{nT_s} r(t)\cos(2\pi f_c t)\, dt$$

$$r_{Q,n} = \int_{(n-1)T_s}^{nT_s} r(t)\sin(2\pi f_c t)\, dt$$

(10.83)

$n = 1, 2, \ldots, N$ and N is the number of symbols processed.

The authors of [26] compared the performance of this classifier with two others. The *square law classifier* (SLC) to which they refer simply squares the input signal. Since $\cos^2 x = \frac{1}{2}(1 + \cos 2x)$, this effectively doubles the frequency of the signal. If a significant component is found at this twice frequency, then it is known that the input frequency was at x. For BPSK, the double frequency term shows up at twice the input data rate, while for QPSK it is at four times the input data rate. The other type of simple classifier discussed was the *phase-based classifier* (PBC), which operates directly on the phase changes from symbol to

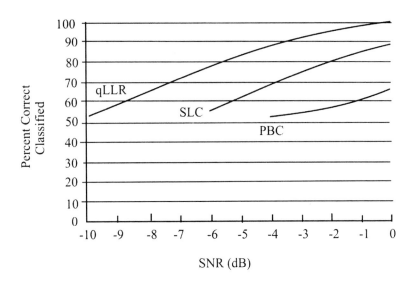

Figure 10.28 Performance of the Kim and Polydoros classifier. (*Source:* [26], © 1988, IEEE. Reprinted with permission.)

symbol. A histogram is generated of these phase changes and decisions about whether the signal is BPSK or QPSK are based on this histogram.

The results of this approach are shown in Figure 10.28. These are simulated results with the number of symbols included, $N = 100$. As seen, better than 80% correct classification is possible even with the SNR as low as –5 dB with the qLLR algorithm. That algorithm significantly outperforms the other two at the low SNRs considered here.

Sills Classifier

Sills compared the performance of coherent versus incoherent maximum likelihood modulation classification when the modulations of interest were three PSK and three QAM signals [27]. In particular, the modulations investigated are shown in Figure 10.29. The coherent maximum-likelihood classifiers were the traditional types that maximize the probability of a particular modulation type given the output of a coherent detector. For the coherent case, it was assumed that all the parameters of the signal were known, including the carrier phase. The results of the simulation are shown in Figure 10.30. All of the signals were classified correctly at least 90% of the time whenever the SNR was greater than 10 dB.

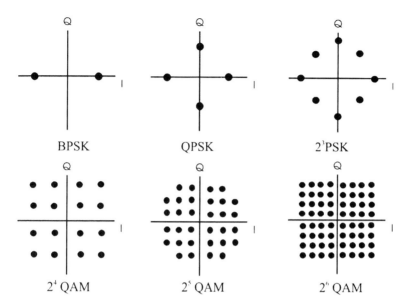

Figure 10.29 Constellations corresponding to standard modulations considered by Sills for signal classification. (*Source:* [27], © 1999, IEEE. Reprinted with permission.)

As aptly pointed out, however, measuring the carrier phase for QAM is very difficult at these signal levels. Because of this, a second simulation was performed where it was assumed that the carrier phase was not known. Again, maximum-likelihood detection was assumed. These results are shown in Figure 10.31. There is about a 3-dB loss in performance from the coherent case, a number not unlike that associated with other communication problems when comparing coherent versus incoherent detection.

10.6.3 Recognition/Identification

Several functions in communication EW systems can be automated, at least to some degree. Those functions that can recognize or identify entities are prime candidates for such processing; however, automating the processes can be difficult. Usually the entity sets involved are limited in size; otherwise, performance degrades.

The general flow of recognition processing is shown in Figure 10.32. The input to these processes, in general, is different depending on the type of processing to be performed. In the case of voice recognition, for example, the input would be speech.

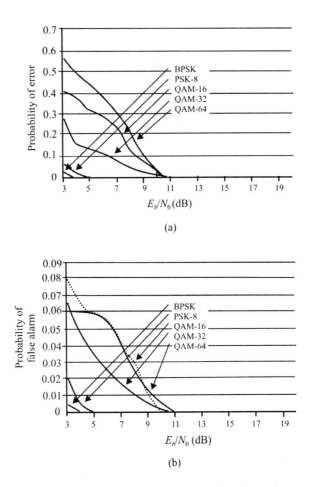

Figure 10.30 Sills maximum-likelihood classification results when coherent detection was assumed: (a) the probability of correct classification and (b) the probability of false alarm. (*Source:* [27], © 1999, IEEE. Reprinted with permission.)

In the flow diagram shown in Figure 10.32, features are first extracted from the input signal. Feature extraction is the process of determining from the input signal those parameters that can be used to discern between different entities and, conversely, are consistent with the same entity. These features are compared with the features stored in the local database. In one scenario (open identification) if there is a sufficiently close match of the new features to a database entry, then a successful match is declared; otherwise, the new features are added to the database. In another scenario (closed identification), if there is a match, then

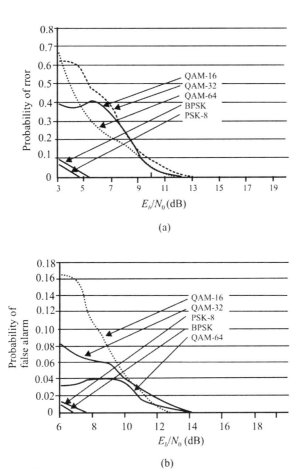

Figure 10.31 Sills incoherent maximum-likelihood classification. Shown in (a) are the probabilities of correct classification while (b) shows the probabilities of false alarm. (*Source:* [27], © 1999, IEEE. Reprinted with permission.)

success is declared, but if there is no match, then the conclusion is simply that the entity is not in the database—that is, the new features are not added to the database.

A flow diagram for the *verification* process is illustrated in Figure 10.33. It is somewhat different from the identification process in that only one comparison is required—the features are compared with the specific entity to be verified.

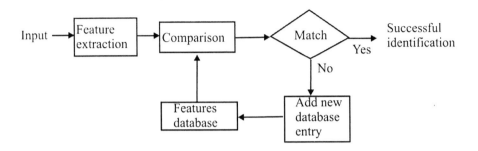

Figure 10.32 Flow diagram of the recognition/identification process.

Open identification is when the set of entities under consideration is not limited. Some form of identification would then be associated with all new entities encountered. *Closed identification* is when the set of entities is limited, and the question to be answered is whether the current entity belongs to the set.

10.6.3.1 Mathematical Modeling

Just as in other processing where entities are divided into groups based on features, the features form a space that must be segregated into regions. Each such region corresponds to a single decision. For just two features the feature space is illustrated in Figure 10.34. The dark circles represent the mean value of the two features in each region and the boundaries divide it into separate regions. Each feature must be sufficiently robust in order to be consistent with the same entity and sufficiently different from other entities. Each of the regions shown corresponds to a combination of features associated with each individual entity.

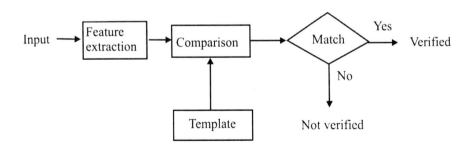

Figure 10.33 Flow diagram of the verification process.

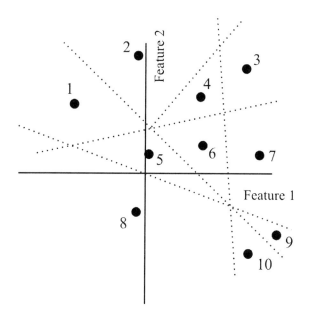

Figure 10.34 Segmenting a feature space for identification. In this example there are two features that make up the feature space and the regions are linearly separable.

There could be gaps between regions where undecided is the conclusion. Neural networks and fuzzy logic can be successfully applied to the recognition problem. In reality the space would be N-dimensional and the features would form a hyperspace.

Statistical Modeling

The distance between an arbitrary point x in feature space and a point μ representing a specific entity can be computed in several ways. Perhaps the most common is the Euclidean distance given by

$$d_E^2 = (x-\mu)^2 \tag{10.84}$$

An alternative is the *Mahalonobis distance* of $X = \{x_1, x_2, ..., x_n\}$ given by

$$d_M^2 = (\bar{x} - \mu)^T \Sigma^{-1} (\bar{x} - \mu) \tag{10.85}$$

where μ and Σ are the mean and covariance of the training features, respectively, and $\bar{x}$ is the average of the feature vectors.

The points of equi-Euclidean distance from a point $(x_1, x_2, ..., x_n)$ form a hyper-circle of radius d_E, with the point at its center. The points equi-Mahalonobis distant from this point form a hyperellipsoid. The axes are determined by the eigenvectors and eigenvalues of the covariance matrix Σ. There is an advantage of using the Mahalonobis distance over the Euclidean distance in some problems. Which to use depends on how the features are distributed in the feature space.

Probabilistic Modeling

An alternative way to use the features to determine a match is to use probability densities. If p_i are the continuous probability densities and $p_i(x)$ is the likelihood that a feature x is generated by the ith entity, then

$$P(\text{identity} = i|x) = \frac{p_i(x)}{p(x)} P_i \qquad (10.86)$$

where P_i is the probability that the input came from the ith entity and $p(x)$ is the probability of the feature x occurring from any entity. This is Bayes' rule, whose major shortcoming, as previously discussed, is that the a priori probabilities must be known (or assumed).

10.6.3.2 Source Recognition

This function attempts to identify the source of the information being transmitted. This source could be, for example, a person speaking, in the case of speaker identification, a person operating a Morse code key, or the computer providing digital data through a modem. Each component in a communication system has its own characteristics; however, they are not always distinguishable. The closer to the physical medium a component is, generally the easier it is to determine such characteristics. The performance of source recognition technologies depends on the type of recognition being attempted. Machine characteristics are generally more reliable than those imposed by human users of the systems.

Speaker Recognition

Speaker recognition is the process of estimating the identification of a speaker who is uttering voice signals [28]. The function can be divided into speaker identification and speaker verification as discussed above. Its use in the commercial sector is for automating such functions as bank transactions and voice mail.

Speaker identification can be classified as to whether it is text-dependent or text-independent. Text-dependent systems allow only a limited set of words to be used to perform the identification. Text-independent identification applies no such constraint. In general, text-dependent systems perform better than text-independent systems. Both requirements exist in communication EW systems. Text-independent capabilities apply when free and open speech is being processed, which applies in many circumstances. Text-dependent systems apply when the messages being processed are pro forma in character, such as artillery call for fire.

The language of the speaker may or may not be important. Humans can do a reasonably good job of recognizing other human voices, even over band-limited voice paths such as telephone channels. The objective behind automated speaker recognition technology is to have machines do the same thing.

In general, speaker identification applied to communication EW systems should be text-independent and robust. The latter means that performance must degrade smoothly as the SNR is decreased. There are cases that the former need not be true but those would be special situations.

Over relatively quiet channels, such as telephone lines, the problem is easier than over radio channels where noise and interference are prevalent. However, channel variability can be a problem over telephone lines as well. With the advent of digital telephony and fiber-optic telephone infrastructures, noise on telephone lines is rapidly becoming a problem of the past in the developed countries, anyway.

The speech spectrum is typically calculated with 20-ms time samples. Often the features used for speaker recognition are based on the *cepstrum*. The cepstrum contains information about time delays in speech. If *frame* represents a time segment of the voice signal, then

$$\text{cepstrum(frame)} = FFT^{-1}(\log|FFT(\text{frame})|) \tag{10.87}$$

Typically the first several cepstral coefficients are retained as features. The cepstrum is used to estimate the vocal tract parameters of the speaker. The subsequent computations after the cepstrum is obtained separate the parameters of the speech that are related to the vocal tract from those that are related to the pitch information.

One model of the vocal process in humans has periodic pulses generated by forcing air through the vocal chords. The vocal tract filters these pulses and it is the parameters of the vocal tract that are useful in identifying one individual from another. These processes can be viewed as (short term) linear time-invariant processes, so if $P(f)$ represents the Fourier transform of the pulses generated by and emitted from the vocal chords, and if $T(f)$ represents the filter function of the vocal tract, then

$$S(f) = P(f)T(f) \tag{10.88}$$

where $S(f)$ is the Fourier transform of the speech signal $s(t)$. Taking the logarithm of each side yields

$$\log(S(f)) = \log(P(f)) + \log(T(f)) \tag{10.89}$$

The periodic pulses produce peaks in the $\log(P(f))$ spectrum while the characteristics of the vocal tract are in the second term, which is represented by variables that are changing more slowly and are therefore represented by the shape of the cepstral spectrum, as represented by the lower cepstral component values. In this process, the higher frequencies of the log FFT are suppressed because it is felt that the lower frequencies of human speech produce more useful information. The resultant spectrum is called the mel-cepstrum. The features generated this way make up a point in feature space. Ideally a single, well-defined point in this space represents each speaker.

Morse Code Operator Recognition

Morse code is the transmission of dots and dashes that represent coding of some language, English, for instance. In addition to those features associated with a transmitter used to send the code, features of the person operating the key can sometimes be used to identify the operator. A particular operator may have consistent durations for the dots and dashes, which are differentiable from other such operators. Particular sequences of symbols may have repeatable characteristics that can be used for identification. The character for u (• • -) would almost always follow q (• • - •) in the English language and this sequence may always be sent in a unique way (such as speed) by a particular Morse code key operator.

The key itself may have measurable features that can be used. In that case, of course, one is identifying the key and not necessarily the operator but this could be just as useful from an information point of view.

The transients in the amplitude of the RF carrier at turn-on and turn-off may be unique for a particular transmitter and sufficiently different among transmitters

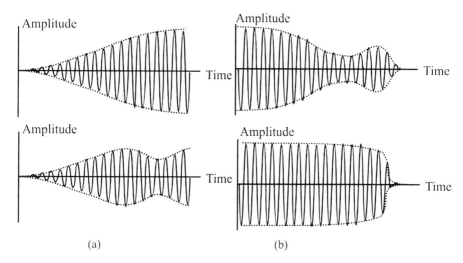

Figure 10.35 (a) Turn-on and (b) turn-off transients of a CW transmitter sending Morse code.

to be useful for identification. A few such cases are illustrated in Figure 10.35. The two turn-on transients are certainly different from one another as are the turn-off transients. If these characteristics are repeatable over time (the longer, the better) then they could be used to track targets. Frequently characteristics such as these are persistent until a repair is made to the system that changes whatever components are responsible for establishing this behavior.

Machine (Computer) Recognition

When computers are communicating via modems, the characteristics of the modems may present features that are usable for identification. The tone frequencies in FSK modems are a simple example. These tones are not the same each time they are sent, but exhibit statistical properties that can be measured.

Phase shifts in PSK modems or phase shifts and amplitude variations in QAM modems also have statistical properties. Two such modems might have the constellations shown in Figure 10.36, from which features could be extracted for identification. For example, one such possibility might simply be the amplitude in this case. Another may be the distance between the centers of the constellation points.

Even when modems are not being used and the communication media is driven directly with digital signals (V.90, for example), the source may contain

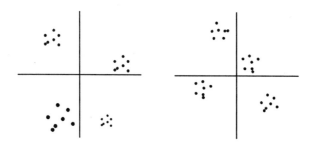

Figure 10.36 Two QAM or PSK constellations that exhibit different features that may be useful for modem identification.

identifiable features. Timing jitter on system clocks may be unique to an individual or a class of machines. Characteristic amplitude shifts may also contain features.

10.6.3.3 Language Identification

In those cases when the content of a transmission is important, and that transmission is a voice conversation, it may be necessary to first ascertain in which language the conversation is occurring [29]. Thus, automated language identification may be necessary. This is especially true in cases when automated translation is performed. Language identification as well as automated translation systems can be applied to written text as well. Herein this is not the case of interest.

In the commercial sector, language identification is important to, for example, phone companies so that calls can be routed to operators with the correct language. The same is true for international business transactions.

The features for language identification are usually based on the acoustic signature of that language. The *acoustic signature* is the unique characteristics of a language that make it different from all others. Possible sources of information to ascertain the acoustic signature are listed as follows:

1. Acoustic phonetics: The tones characteristic of a language, which differ among languages, and even when close, whose frequency of use varies;
2. Prosodics: The duration of phones, speech rate, and intonation;
3. Phonotactics: The rules that govern the use of phones in a language;
4. Vocabulary.

Recent results indicate that 60–80% of correct language classification is possible with phone line quality sources of audio using these parameters. Phone line quality audio is usually very good, with a characteristically high SNR. Unfortunately for many cases of interest, phone line quality speech is unavailable. RF transmitted voice is typically much noisier. Cellular phone and PCS phgone audio is somewhere in between these two extremes. Because there is more data to work with, the longer the utterance, the better.

10.6.3.4 Emitter Identification

Discovering properties of a transmitter in order to ascertain what type of radio it is and, perhaps, what specific transmitter it is, is sometimes possible. Identification of the type of transmitter can assist in determining the EOB. If possible, identifying a specific transmitter that has been intercepted in the past can assist in tracking battlespace units over time and space when that transmitter can be associated with a specific unit.

Type

Some of the parameters that can help identify the type of transmitter include the RF bandwidth, the frequency of operation, the type of modulation (e.g., AM, FM, and FSK), and power level. With the advent of software programmable radios, which emulate the waveforms of other radios, identification of the type of radio will likely become more difficult over time.

The frequency of operation can be an indication of the purpose for which the transmitter was built. For example, reception of a signal in the low VHF frequency range is an indication that the radio is probably a military system. Reception of a signal in the 130-MHz range would indicate that the radio is probably intended for aircraft traffic control.

Identification of the type of modulation can be used to segregate transmitters by type. The modulation index typically varies somewhat even in the same radio type due to variations in the components used to make the modulator, especially for inexpensive radios.

As for the transmitter sending Morse code above, the transients in *push-to-talk* (PTT) radio networks may also be usable as features. In the case of CW or AM transmitters, the features shown in Figure 10.35 may also apply to PTT networks. For the case of FM, the signal may have to be demodulated first but the demodulated output may exhibit features similar to those shown in Figure 10.35. Unintentional tones located anywhere in the spectrum may indicate particular oscillators in the transmitter, which are likely to be unique to a type of transmitter.

Specific

Once a transmitter has been detected and parameters extracted, it may be possible to identify that specific transmitter again if it is intercepted a second time. If an association can be made between that specific transmitter and the entity using it, then specific units or persons can be tracked over time.

The automatic extraction of features would be highly desirable, as there would be too many emitters in an MRC for operators to do this. Such features would likely include fine measurements of the parameters mentioned above for emitter type identification, as well as others.

For MFSK, the statistical variation of the frequency tones can be determined. The natural way to do this is via the FFT discussed earlier. Some pulsed transmitters exhibit unintentional modulation on the pulses. This modulation could be low levels of AM or FM signals generated in the transmitter by a wide variety of mechanisms.

Filters in transmitters frequently exhibit ringing effects. Most wideband transmitters implement tuned filters at the output of the final amplifier stage and before the antenna. This is to conjugate match the amplifier to the antenna for the purpose of transferring maximum power to the antenna. These filters are typically reasonably narrowband and therefore have a reasonably high Q. High Q filters tend to ring, or oscillate. This ringing may be detectable and useful as an identifying feature.

Background noise in voice PTT networks may contain artifacts of the surrounding area within which the radio is operating (inside a tank or inside a busy TOC as examples). While over extended time these characteristics are likely to change, background noise in a TOC, for example, for short temporal conditions they may be useful for tracking targets.

The spectral shape of the predetected signal is measurable and contains usable features. The selectivity of the radio is determined by how sharp the filters are in the modulation process. If the skirts on the spectrum of signals are sharp, it is an indication that the radio is selective as opposed to broad skirts, which might indicate another type of radio. In order to remove modulation effects, the spectrums used for this purpose would need to be averaged.

10.7 CONCLUDING REMARKS

This chapter introduces the reader to some fundamental concepts associated with signal processing in communication EW systems. Probably the most prolific signal processing task in these systems is calculating the Fourier transform, or its fast approximation. This transform is useful for determining the frequency content of signals and is fundamental to many of the signal detection functional requirements.

Noise is always present when processing communication signals. Some of the newer signal processing techniques such as the wavelet transform, cyclostationary processing, and high-order processing of signals can be used to reduce much of this noise, promising to make the signal detection function perform better.

Some of the many signal processing applications are discussed in this chapter as well. This list is certainly not all-inclusive, but it is presented to illustrate applications of some of the signal processing techniques.

References

[1] Stephens, J. P., "Advances in Signal Processing Technology for Electronic Warfare," *IEEE AES Magazine*, November 1996, pp. 31–38.
[2] Bracewell, R. N., *The Hartley Transform*, Oxford, U.K.: Oxford University Press, 1986, p. 10.
[3] Rioul, O., and M. Vetterli, "Wavelets and Signal Processing," *IEEE Signal Processing Magazine*, October 1991, pp. 14–38.
[4] Bruce, A., D. Donoho, and H. Gao, "Wavelet Analysis," *IEEE Spectrum*, October 1996, pp. 26–35.
[5] Guillemain, P., and R. Kronland-Martinet, "Characterization of Acoustic Signals Through Continuous Linear Time-Frequency Representations," *Proceedings of the IEEE*, Vol. 84, No. 4, April 1996, pp. 561–585.
[6] Ramchandran, K., M. Vetterli, and C. Herley, "Wavelets, Subband Coding, and Best Bases," *Proceedings of the IEEE*, Vol. 84, No. 4, April 1996, pp. 561–585.
[7] Wornell, G., "Emerging Applications of Multirate Signal Processing and Wavelets in Digital Communications," *Proceedings of the IEEE*, Vol. 84, No. 4, April 1996, pp. 561–585.
[8] Mallat, S., "Wavelets for Vision," *Proceedings of the IEEE*, Vol. 84, No. 4, April 1996, pp. 604–614.
[9] Schroder, P., "Wavelets in Computer Graphics," *Proceedings of the IEEE*, Vol. 84, No. 4, April 1996, pp. 561–585.
[10] Graps, A., "Introduction to Wavelets," accessed August 2001, http://www/arma.com/IEEEwave/IW_see_wave.html.
[11] Daubechies, I., "Orthonormal Bases of Compactly Supported Wavelets," *Communications on Pure and Applied Mathematics*, Vol. 41, 1998, pp. 909–996.
[12] Bock, R. K., "Fast Transforms," *The Data Analysis Briefbook*, accessed July 2001, http://ikpe1101.ikp.kfa-juelich.de/briefbook_data_analysis/node83.html.
[13] Gardner, W. A., "Signal Interception: A Unifying Theoretical Framework for Feature Detection," *IEEE Transactions on Communications*, Vol. 36, No. 8, August 1988, pp. 897–906.
[14] Mendel, J. M., "Tutorial on Higher-Order Stastics (Spectra) in Signal Processing and System Theory: Theoretical Results and Some Applications," *Proceedings of the IEEE*, Vol. 79, No. 3, March 1991, pp. 278–305.

[15] Torrieri, D. J., *Principles of Secure Communication Systems*, 2nd ed., Norwood, MA: Artech House, 1992, pp. 294–307.

[16] Woodring, D. G., *Performance of Optimum and Suboptimum Detectors for Spread Spectrum Waveforms*, NRL Report 8432, December 30, 1980.

[17] Levitt, B. K., and U. Cheng, "Optimum Detection of Frequency-Hopped Signals," *Proceedings of IEEE MILCOM*, 1992.

[18] Levitt, B. K., et al., "Optimum Detection of Slow Frequency-Hopping Signals," *IEEE Transactions on Communications*, Vol. 42, Nos. 2/3/4, February./March/April 1994, p. 1990.

[19] Beaulieu, N. C., W. L. Hopkins, and P. J. McLane, "Interception of Frequency-Hopped Spread-Spectrum Signals," *IEEE Journal on Selected Areas of Communications*, Vol. 8, No. 5, June 1990, pp. 854–855.

[20] Nandi, A. K., and E. E. Azzouz, "Algorithms for Automatic Modulation Recognition of Communication Signals," *IEEE Transactions on Communications*, Vol. 46, No. 4, April 1998, pp. 431–436.

[21] Azzouz, E. E., and A. K. Nandi, "Procedure for Automatic Recognition of Analogue and Digital Modulations," *IEE Proceedings – Communications*, Vol. 143, No. 5, October 1996, pp. 259–266.

[22] Azzouz, E. E., and A. K. Nandi, *Automatic Modulation Recognition of Communication Signals*, Boston, MA: Kluwer Academic Publishers, 1996.

[23] Assaleh, K., K. Farrell, and R. J. Mammone, "A New Method of Modulation Classification for Digitally Modulated Signals," *Proceedings of IEEE MILCOM*, 1992, pp. 712–716.

[24] Whelchel, J. E., et al., "Signal Understanding: An Artificial Intelligence Approach to Modulation Classification," *Proceedings of IEEE MILCOM*, 1984, pp. 231–236.

[25] Hush, D. R., and B. G. Horne, "Progress in Supervised Neural Networks," *IEEE Signal Processing Magazine*, January 1993.

[26] Kim, K., and A. Polydoros, "Digital Modulation Classification: The BPSK Versus QPSK Case," *Proceedings of IEEE MILCOM*, 1988, pp. 431–436.

[27] Sills, J. A., "Maximum-Likelihood Modulation Classification for PSK/QAM," *Proceedings of IEEE MILCOM*, 1999.

[28] Gish, H., and M. Schmidt, "Text-Independent Speaker Identification," *IEEE Signal Processing Magazine*, October 1994, pp. 18–32.

[29] Muthusamy, Y. K., E. Barnard, and R. A. Cole, "Reviewing Automatic Language Identification," *IEEE Signal Processing Magazine*, October 1994, pp. 33–41.

Chapter 11

Direction-Finding Position-Fixing Techniques

11.1 INTRODUCTION

Determining the geolocation of an emitting target, as mentioned previously, is an important aspect of ES systems. This is more commonly known as *position fixing*. There are several methods to accomplish this and some of them are described in this chapter. In all cases some parameters associated with the incoming signal, such as its time of arrival or differential time of arrival measured at two or more locations, are used to compute the location of the emitter.

11.2 BEARING ESTIMATION

One technique for position fixing determines the *angle of arrival* (AOA) or LOB of the incoming signal relative to magnetic or true north (they are not the same) at two or more locations. Where these bearings intersect is taken as the location of the target. This is sometimes referred to as triangulation. Herein, these bearings will be referred to as LOPs.

To determine the LOPs, several techniques can be used. The phase (or almost equivalently, time) difference can be measured between the signals at two or more antenna elements. Alternately if the antennas have directionality, the amplitude difference between these two signals can be measured. Frequency differences measured between two antennas can be used to determine the bearing, if one or more of the antennas are moving relative to the other or to the target. Some of these techniques are discussed in this section.

LOP systems normally do not operate at the frequencies of the signals. The frequency is typically converted to an IF and the phase or time measurements are made on this converted signal. Ignoring noise and other error sources, the IF version of the RF signals has the same information in it as the latter—it is just

11.2.1 Circular Antenna Array

One of the more common forms of antenna arrays for bearing determination is a circular array. An example of a four-element array mounted on a mast is illustrated in Figure 11.1. Other forms of this antenna include more or fewer number of elements. A sense antenna is sometimes included with a circular array, which is used to remove ambiguities in the bearing. This sense antenna can be formed virtually by combining in phase the output of the other antenna elements.

Let R be the radius of the circular array. Thus, R is the length of an "arm" of the array measured from the center to any one of the antenna elements. The quantity $2R/\lambda$ is sometimes referred to as the *aperture* of a circular array and it indicates the number of wavelengths presented by the array to a signal impinging orthogonal to the plane of the array which is the plane determined by the arms connecting the antenna elements.

An incoming signal $s(t)$ represented in phasor form is shown in Figure 11.2.

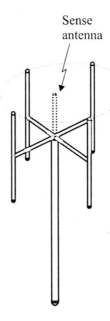

Figure 11.1 A square antenna array also forms a circular antenna array. The circle shown here is not part of the antenna but is included to show the circular nature of the square array.

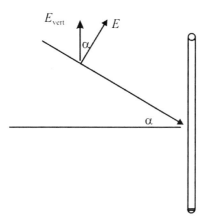

Figure 11.2 An E-field of arbitrary orientation incident on a vertically orientated antenna will only excite the antenna with the vertical component of the E-field.

The EM wave, as indicated previously, has an associated E-field $E(t)$ with a magnitude E. The vertical component of this field is the only portion that excites a vertically oriented antenna. Thus, E_{vert} is given by $E \cos \alpha$ and the corresponding signal amplitude must be adjusted by this factor.

11.2.2 Interferometry

One popular technique for measuring the AOA of an incoming signal, and thereby determining its LOP, is *interferometry*. In this case the phase difference $\Delta\theta_{12}$ or time difference Δt_{12} between two antennas is measured directly by some means.

The signal impinging on an antenna displaced from a second antenna as shown in Figure 11.3 must travel an extra distance of $d = D \sin \theta$ compared with that second antenna. This imposes an additional phase shift on the signal given by

$$\phi = \frac{2\pi D}{\lambda} \sin \theta \qquad (11.1)$$

To determine how errors in calculation of the phase angle relate to errors in calculating the azimuth of arrival of the signal, consider the following [1]. If U represents some system parameter, which is a function of independent

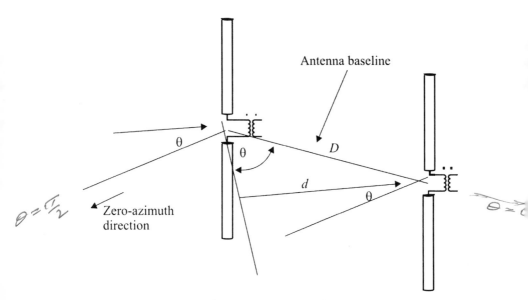

Figure 11.3 Definition of variables for an interferometer.

measurements represented by the variables, $u_1, u_2, \ldots, u_N$ and the u_i have probable errors denoted by $e(u_i)$, then the probable error of U is approximated by

$$e(U) = \sqrt{\left(\frac{\partial U}{\partial u_1}\right)^2 e^2(u_1) + \left(\frac{\partial U}{\partial u_2}\right)^2 e^2(u_2) + \cdots + \left(\frac{\partial U}{\partial u_N}\right)^2 e^2(u_N)} \qquad (11.2)$$

If the probable errors are from an independent statistical set with zero means, then the probable error is the same as the standard deviation $e(u_i) = \sigma_i$. In the simplified case under consideration here, there is one independent variable, θ. Thus in this expression, $U = \phi$ and $u_1 = \theta$. Assume that the mean of ϕ is zero so that

$$\sigma_\varphi = \sqrt{\left(\frac{\partial \phi}{\partial \theta}\right)^2 \sigma_\theta^2} = \frac{\partial \phi}{\partial \theta} \sigma_\theta \qquad (11.3)$$

The partial derivative in this expression can be calculated from (11.3) as

Direction-Finding Position-Fixing Techniques

$$\frac{\partial \phi}{\partial \theta} = \frac{2\pi D}{\lambda} \cos \theta \qquad (11.4)$$

which yields the expression for the standard deviation of the azimuth calculation as

$$\sigma_\varphi = \frac{2\pi D}{\lambda} \cos \theta \, \sigma_\theta \qquad (11.5)$$

Thus, the standard deviation of the azimuth measurement is maximum when $\theta = 0$, or off the ends of the baseline. It is minimum for $\theta = \pi/2$, or orthogonal to the baseline.

There are theoretical limits on how accurately parameters associated with signals can be measured when noise is present and taken into consideration. Herein the one that will be used is called the *Cramer-Rao* lower bound and it is a measure of parameter estimation accuracy in the presence of white noise. White noise is an approximation in most situations and can be a fairly good approximation. For example, the hissing noise that one hears sometimes from a radio when it is not tuned to a station, especially when the volume is turned up, is very close to white noise. The term "white" comes from the fact that white light contains all of the colors—white noise contains energy from all of the frequencies.

For an interferometric line of position system, the Cramer-Rao bound σ can be calculated to be [2]

$$\sigma^2 = \left(\frac{E}{N_0}\right)^{-1} \qquad (11.6)$$

where E is the signal energies at the two antennas (watt-seconds), assumed to be the same, and N_0 is the noise spectral density at the two antennas (watts per hertz), also assumed to be the same and independent of each other. The other assumption for this equation is that the angle of arrival, φ, does not vary over the measurement interval.

To convert this equation into a more usable form, recall that $E = S$ (W) T (seconds), and that N (W) $= N_0$ (W/Hz) B (Hz), where N is the noise power and B is the noise bandwidth, the latter assumed for convenience here to be the IF bandwidth. Thus, after some trivial algebra,

$$\sigma^2(\varphi) = \left(\frac{S}{N}TB\right)^{-1} \qquad (11.7)$$

with $\sigma(\varphi)$ in radians, which can be converted to degrees by multiplying by $180/\pi$. Therefore, the variance is reduced (accuracy improved) by increasing the SNR, the integration time, or the measurement bandwidth. Note, however, that increasing the bandwidth beyond the bandwidth of the signal of interest will decrease the SNR, so care must be used. With these assumptions, and an assumed noise bandwidth of 200 kHz, the instrumental accuracy, as given by the standard deviation of the measurement of the LOP, is shown versus (power) SNR and various measurement times in Figure 11.4.

11.2.2.1 Triple-Channel Interferometer

The antenna configuration shown in Figure 11.5 may be used as a three-channel interferometer. There are three baselines established by the three antenna elements, one for each antenna pair. Figure 11.6 shows the top view of this antenna arrangement. The AOA is related to the antenna parameters as follows.

The distance traveled by a signal impinging on this array from an azimuth angle denoted by φ and an elevation angle denoted by α, between antenna 0 and the center of the array, given by d_{0c} is calculated as

$$d_{0c} = R \cos\alpha \cos\varphi \qquad (11.8)$$

The $\cos\alpha$ term is required to reflect the signal perpendicular to the plane of

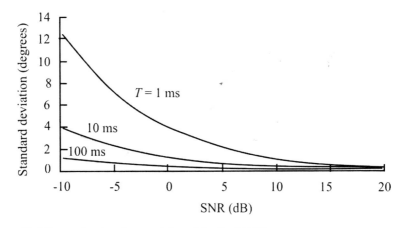

Figure 11.4 Cramer-Rao interferometric bound for $B = 200$ kHz.

Direction-Finding Position-Fixing Techniques

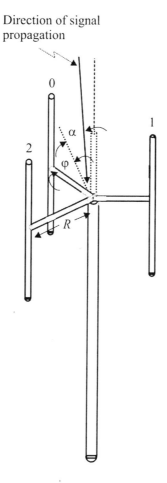

Figure 11.5 Definition of variables for a triple-channel interferometer.

the antenna array as mentioned above. The phase difference corresponding to this distance is given by

$$\theta_{0c} = 2\pi f \Delta t_{0c}$$
$$= 2\pi \frac{c}{\lambda} \Delta t_{0c}$$

(11.9)

where Δt_{0c} is the time it takes to traverse this distance. But $d_{0c} = c\Delta t_{0c}$ so

$$\begin{aligned}\theta_{0c} &= 2\pi \frac{d_{0c}}{\lambda} \\ &= \frac{2\pi R}{\lambda} \cos\alpha \cos\varphi\end{aligned} \quad (11.10)$$

Likewise, the distance traveled by this same signal from the center of the array to antenna 1 is given by

$$d_{c1} = R\cos\alpha \cos\left(\frac{\pi}{3} + \varphi\right) \quad (11.11)$$

with a corresponding phase difference given by

$$\theta_{c1} = -\frac{2\pi R}{\lambda} \cos\alpha \cos\left(\frac{\pi}{3} + \varphi\right) \quad (11.12)$$

The minus sign is required here because the phase at antenna 1 lags behind the phase at the center of the array.

The distance traveled between the center of the array and antenna 2 is given by

$$d_{c2} = R\cos\left(\frac{\pi}{3} - \varphi\right) \quad (11.13)$$

with the phase difference being

$$\theta_{c2} = -\frac{2\pi R}{\lambda} \cos\alpha \cos\left(\frac{\pi}{3} - \varphi\right) \quad (11.14)$$

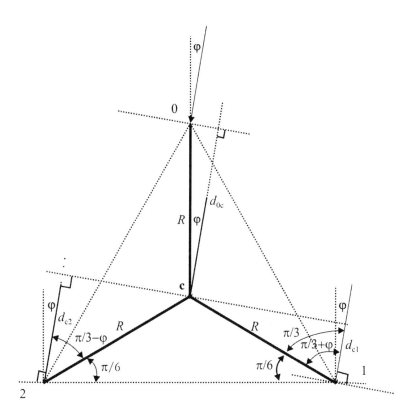

Figure 11.6 Top view of a triple-channel interferometer. Shown are the definitions for the angles and wavefronts as they pass through the array.

The signals at each of the antennas can thus be represented in terms of the AOA and elevation angle as follows.

$$s_0(t) = s(t) \cos\alpha \, e^{j\frac{2\pi R}{\lambda}\cos\alpha\cos\varphi}$$

$$s_1(t) = s(t) \cos\alpha \, e^{-j\frac{2\pi R}{\lambda}\cos\alpha\cos\left(\frac{\pi}{3}+\varphi\right)} \quad (11.15)$$

$$s_2(t) = s(t) \cos\alpha \, e^{-j\frac{2\pi R}{\lambda}\cos\alpha\cos\left(\frac{\pi}{3}-\varphi\right)}$$

340   Introduction to Communication Electronic Warfare Systems

To calculate the AOA, φ, and the elevation angle, α, first define $\Delta_1 = \theta_{c2} - \theta_{c1}$. Then

$$\Delta_1 = -\frac{2\pi}{\lambda} R \cos\alpha \left[\cos\left(\frac{\pi}{3} - \varphi\right) + \cos\left(\frac{\pi}{3} + \varphi\right) \right] \quad (11.16)$$

$$= -\frac{2\pi}{\lambda} R \cos\alpha \left[\begin{array}{l} \cos\left(\dfrac{\pi}{3}\right)\cos\varphi + \sin\left(\dfrac{\pi}{3}\right)\sin\varphi \\ + \cos\left(\dfrac{\pi}{3}\right)\cos\varphi - \sin\left(\dfrac{\pi}{3}\right)\sin\varphi \end{array} \right]$$

$$= -\frac{2\pi}{\lambda} R \cos\alpha \left[2\cos\left(\frac{\pi}{3}\right)\cos\varphi \right]$$

$$= -\frac{2\pi}{\lambda} R \cos\alpha \cos\varphi$$

Define $\Delta_2 = \theta_{0c} + \theta_{c1}$. Then

$$\Delta_2 = \frac{2\pi}{\lambda} R \cos\alpha \left[\cos\varphi - \cos\left(\frac{\pi}{3} + \varphi\right) \right]$$

$$= \frac{2\pi}{\lambda} R \cos\alpha \left[\cos\varphi - \cos\left(\frac{\pi}{3}\right)\cos\varphi + \sin\left(\frac{\pi}{3}\right)\sin\varphi \right] \quad (11.17)$$

$$= \frac{2\pi}{\lambda} R \cos\alpha \left[\frac{1}{2}\cos\varphi + \frac{\sqrt{3}}{2}\sin\varphi \right]$$

The AOA is given by

$$\varphi = \tan^{-1}\left(-\frac{\theta_{0c}}{\Delta_1} \right) \quad (11.18)$$

because

$$\tan^{-1}\left(-\frac{\theta_{0c}}{\Delta_1}\right) = \tan^{-1}\left(-\frac{\frac{2\pi}{\lambda}R\cos\alpha\sin\varphi}{-\frac{2\pi}{\lambda}R\cos\alpha\cos\varphi}\right) \quad (11.19)$$

$$= \tan^{-1}\left(\frac{\sin\varphi}{\cos\varphi}\right)$$

To obtain an expression for the elevation angle, expand

$$\left(\frac{\sqrt{3}}{2}\Delta_1\right)^2 + \left(\Delta_2 - \frac{1}{2}\Delta_1\right)^2 =$$

$$= \frac{3}{4}\left(\frac{2\pi R}{\lambda}\right)^2 \cos^2\alpha\cos^2\varphi + \left(\frac{2\pi R}{\lambda}\cos\alpha\right)^2$$

$$\times \left(\frac{1}{2}\cos\varphi + \frac{\sqrt{3}}{2}\sin\varphi - \frac{1}{2}\cos\varphi\right)^2$$

$$= \frac{3}{4}\left(\frac{2\pi R}{\lambda}\right)^2 \cos^2\alpha\cos^2\varphi + \left(\frac{2\pi R}{\lambda}\cos\alpha\right)^2\left(\frac{3}{4}\sin^2\varphi\right) \quad (11.20)$$

$$= \frac{3}{4}\left(\frac{2\pi R}{\lambda}\right)^2 \cos^2\alpha\left(\cos^2\varphi + \sin^2\varphi\right)$$

$$= \frac{3}{4}\left(\frac{2\pi R}{\lambda}\right)^2 \cos^2\alpha$$

so that

$$\alpha = \cos^{-1}\sqrt{\frac{4}{3}\left(\frac{\lambda}{2\pi R}\right)^2\left[\left(\frac{\sqrt{3}}{2}\Delta_1\right)^2 + \left(\Delta_2 - \frac{1}{2}\Delta_1\right)^2\right]} \quad (11.21)$$

Thus, by measuring the phase differences between the antenna elements, the AOA in the plane of the antenna as well as the elevation AOA of a signal can be determined with a triple channel interferometer. With a given realization of such an antenna array, these measurements are dependent on the array parameter, R, as well as the frequency (via the wavelength) of the signal. The larger the array

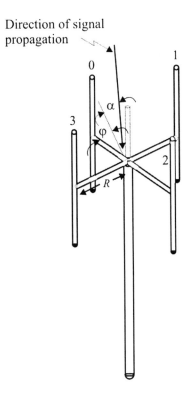

Figure 11.7 Defining the terms associated with a four-element circular array.

radius, R, relative to the wavelength of the signal, the better these AOAs can be measured.

There is a limit on this length, however. The baseline length, which in this case is along a diagonal of the periphery of the array, must be less than one-half of the wavelength to avoid ambiguous results. If it is longer than this, then multiple AOAs will generate the same time differences and therefore the same indicated angle. The wavelength is smallest at the highest operating frequency of the antenna array, so at the lowest operating frequency one would expect the poorest measurement accuracy.

11.2.2.2 Four-Element Interferometer

The geometric relationships in this antenna are as shown in Figure 11.7. As above, R is the radius of the array while the azimuth AOA of the signal $s(t)$ is given as φ while the arrival angle of the signal in the vertical dimension (zenith) is α relative to the plane of the array. The frequency of the signal is given as $f = c/\lambda$, the number of antenna elements $N = 4$, and c is the speed of light. It is assumed that $R < \lambda/4$ so that phase ambiguities do not arise.

If $s(t)$ represents the signal at the center of the array, then at each antenna element a replica of $s(t)$ exists but has shifted in phase somewhat. Thus

$$s_0(t) = s(t)\cos(\alpha)e^{j\cos(\alpha)\theta_0}$$
$$s_1(t) = s(t)\cos(\alpha)e^{j\cos(\alpha)\theta_1}$$
$$s_2(t) = s(t)\cos(\alpha)e^{j\cos(\alpha)\theta_2}$$
$$s_3(t) = s(t)\cos(\alpha)e^{j\cos(\alpha)\theta_3}$$

(11.22)

where θ_i's represent this phase shift. As above, the $\cos \alpha$ multiplying terms on $s(t)$ project the magnitude of the incoming signal $s(t)$ perpendicular to the array plane since the elevation angle of the signal relative to this plane is α. In Figure 11.8 the phase difference at antenna element 1 can be calculated as follows.

Now

$$\cos \alpha \sin \varphi = \frac{d_{1c}}{R} \qquad (11.23)$$

where d_{1c} is the distance traveled by the signal isophase line A-A' to isophase line B-B' in the array plane. Since

$$\theta_{1c} = 2\pi f \Delta t_{1c}$$
$$= \frac{2\pi c \Delta t_{1c}}{\lambda}$$
$$= \frac{2\pi \cos \alpha d_{1c}}{\lambda}$$
$$= \frac{2\pi R}{\lambda} \cos \alpha \sin \varphi$$

(11.24)

but

$$\sin \beta = \cos(\beta - \frac{\pi}{2}) \qquad (11.25)$$

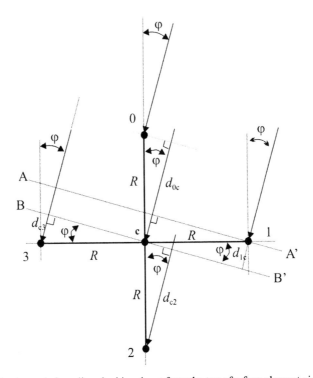

Figure 11.8 Angles and phase lines looking down from the top of a four-element circular array.

for any β so that

$$\theta_{1c} = \frac{2\pi R}{\lambda} \cos\alpha \cos(\varphi - \frac{\pi}{2}) \qquad (11.26)$$

The distance the signal must travel between the center of the array and antenna 2 is given by

$$d_{c2} = R \cos\alpha \cos\varphi \qquad (11.27)$$

The corresponding phase shift associated with this distance is

$$\theta_{c2} = -\frac{2\pi R}{\lambda}\cos\alpha\cos\varphi \qquad (11.28)$$

The minus sign is required because the phase at antenna 2 lags that at the center of the array, whereas at antenna 1 above, the phase led the phase at the center. These signs could be reversed as long as consistency is maintained.

Likewise, the distance the signal must travel between the center of the array and antenna 3 is given by

$$d_{c3} = R\cos\alpha\sin\varphi \qquad (11.29)$$

while the corresponding phase shift is given by

$$\begin{aligned}\theta_{c3} &= -\frac{2\pi R}{\lambda}\cos\alpha\sin\varphi \\ &= -\frac{2\pi R}{\lambda}\cos\alpha\cos\left(\varphi - \frac{\pi}{2}\right)\end{aligned} \qquad (11.30)$$

where, again, the phase at antenna 3 lags that at the center of the array, so that the minus sign is required.

The distance the signal must travel between antenna 0 and the center of the array is given by

$$d_{0c} = R\cos\alpha\cos\varphi \qquad (11.31)$$

so that

$$\theta_{0c} = \frac{2\pi R}{\lambda}\cos\alpha\cos\varphi \qquad (11.32)$$

Equations (11.22–11.32) can be summarized with (11.33) and (11.34) as:

$$s_i(t) = s(t)\cos\alpha \; e^{j\frac{2\pi R}{\lambda}\cos\alpha \; \cos\left(-\frac{2\pi i}{N}+\varphi\right)} \qquad i = 1,2,3,4 \qquad (11.33)$$

and

$$\theta_i = \frac{2\pi R}{\lambda} \cos(-\frac{2\pi i}{N} + \varphi) \qquad i = 1,2,3,4 \qquad (11.34)$$

To determine the bearing calculate

$$\frac{\theta_{1c}}{\theta_{0c}} = \frac{\frac{2\pi R}{\lambda} \cos\alpha \sin\varphi}{\frac{2\pi R}{\lambda} \cos\alpha \cos\varphi} \qquad (11.35)$$

so that

$$\varphi = \tan^{-1}\left(\frac{\theta_{1c}}{\theta_{0c}}\right) \qquad (11.36)$$

This calculation yields $0 \leq \varphi < \pi$ so θ_{c2} and θ_{c3} are used to determine from which side of the array the signal is coming.

To determine the elevation formulate

$$\theta_{0c}^2 + \theta_{1c}^2 = \left(\frac{2\pi R}{\lambda}\right)^2 \cos^2\alpha \cos^2\varphi + \left(\frac{2\pi R}{\lambda}\right)^2 \cos^2\alpha \cos^2\varphi$$

$$= \left(\frac{2\pi R}{\lambda}\right)^2 \cos^2\alpha \left(\cos^2\varphi + \sin^2\varphi\right) \qquad (11.37)$$

yielding

$$\alpha = \sqrt{\cos^{-1}\left[\left(\frac{\lambda}{2\pi R}\right)^2 \left(\theta_{0c}^2 + \theta_{1c}^2\right)\right]} \qquad (11.38)$$

So, again, like the triple-channel interferometer, a four-element circular antenna array can be used to measure the AOA of a signal, and thus facilitate

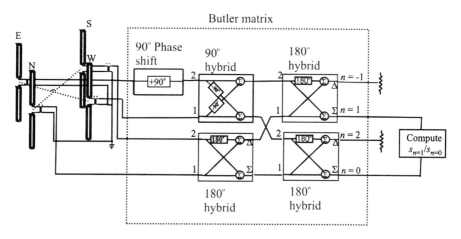

Figure 11.9 Monopulse direction finder using a Butler matrix for phase shifting.

calculation of the line of position to the target. It also allows calculation of the elevation AOA.

11.2.3 Monopulse Direction Finder

By shifting the phase of a signal that has been received by several antennas in specified ways and then comparing the phase relationship of these signals, the phase information at the antenna array is converted into amplitude differences with which it is possible to calculate the AOA of the signal. One such configuration to accomplish this is shown in Figure 11.9 [3]. In this case the amplitude relationship between the mode 1 output is compared with the mode 0 output of a *Butler matrix*.

One of the most significant advantages of this type of system is its speed of operation. All of the necessary computations are performed in the hardware and are instantaneously available. In fact, the appellation monopulse is based on this instantaneous operation and is borrowed from radar systems, where the angle of arrival is measured on a single pulse.

The devices used to perform the necessary phase shifts are called hybrids. These components shift the phase of the inputs in one of several possible ways. In this circuit both 90° hybrids and 180° hybrids are used as shown in Figure 11.9. For an 180° hybrid typically the phase-shifted inputs are summed for one of the outputs, while the other output sums one input with an 180°-shifted version of the other. The 90° hybrid shifts both inputs by –90° as shown.

By identifying the north antenna with antenna 0, west as antenna 1, south as antenna 2, and east as antenna 3 in the previous derivation then the operation of the monopulse direction finder with a Butler matrix can be analyzed as follows. By simply following the signals through the Butler matrix, the following results ensue.

$$s_N(t) = s_0(t) = s(t)\cos\alpha e^{j\frac{2\pi R}{\lambda}\cos\alpha\cos\varphi} \tag{11.39}$$

$$\begin{aligned} s_W(t) = s_1(t) &= s(t)\cos\alpha e^{j\frac{2\pi R}{\lambda}\cos\alpha\cos(-\frac{2\pi}{4}+\varphi)} \\ &= s(t)\cos\alpha e^{j\frac{2\pi R}{\lambda}\cos\alpha[\cos(-\frac{\pi}{2})\cos\varphi - \sin(-\frac{\pi}{2})\sin\varphi]} \\ &= s(t)\cos\alpha e^{j\frac{2\pi R}{\lambda}\cos\alpha\sin\varphi} \end{aligned} \tag{11.40}$$

$$\begin{aligned} s_E(t) = s_3(t) &= s(t)\cos\alpha e^{j\frac{2\pi R}{\lambda}\cos\alpha\cos(-\frac{6\pi}{4}+\varphi)} \\ &= s(t)\cos\alpha \\ &\quad \times e^{j\frac{2\pi R}{\lambda}\cos\alpha[\cos(-\frac{3\pi}{2})\cos\varphi - \sin(-\frac{3\pi}{2})\sin\varphi]} \\ &= s(t)\cos\alpha\, e^{-j\frac{2\pi R}{\lambda}\cos\alpha\sin\varphi} \end{aligned} \tag{11.41}$$

$$\begin{aligned} s_S(t) = s_2(t) &= s(t)\cos\alpha e^{j\frac{2\pi R}{\lambda}\cos\alpha\cos(-\pi+\varphi)} \\ &= s(t)\cos\alpha e^{j\frac{2\pi R}{\lambda}\cos\alpha[\cos(-\pi)\cos\varphi-\sin(-\pi)\sin\varphi]} \\ &= s(t)\cos\alpha e^{-j\frac{2\pi R}{\lambda}\cos\alpha\cos\varphi} \end{aligned} \tag{11.42}$$

The mode zero output is therefore

$$\begin{aligned} s_{n=0}(t) &= s_N(t) + s_S(t) + s_W(t) + s_E(t) e^{j\frac{\pi}{2}} e^{-j\frac{\pi}{2}} \\ &= s(t)\cos\alpha\, e^{j\frac{2\pi R}{\lambda}\cos\alpha\cos\varphi} + s(t)\cos\alpha\, e^{-j\frac{2\pi R}{\lambda}\cos\alpha\cos\varphi} \\ &\quad + s(t)\cos\alpha\, e^{j\frac{2\pi R}{\lambda}\cos\alpha\sin\varphi} + s(t)\cos\alpha\, e^{-j\frac{2\pi R}{\lambda}\cos\alpha\sin\varphi} \\ &= s(t)\cos\alpha \left[\begin{array}{l} e^{j\frac{2\pi R}{\lambda}\cos\alpha\cos\varphi} + e^{-j\frac{2\pi R}{\lambda}\cos\alpha\cos\varphi} \\ + e^{j\frac{2\pi R}{\lambda}\cos\alpha\sin\varphi} + e^{-j\frac{2\pi R}{\lambda}\cos\alpha\sin\varphi} \end{array} \right] \end{aligned} \tag{11.43}$$

$$s_{n=0}(t) = s(t)\cos\alpha \left[2\cos\left(\frac{2\pi R}{\lambda}\cos\varphi\right) + 2\cos\left(\frac{2\pi R}{\lambda}\sin\varphi\right) \right] \quad (11.44)$$

The term inside the brackets represents the phase shift induced on the signals from the antennas by the Butler matrix. The effects are perhaps most clearly seen by considering this term as a phasor, with amplitude given by

$$\text{Amplitude}_{n=0} = \sqrt{\left[2\cos\left(\frac{2\pi R}{\lambda}\cos\alpha\cos\varphi\right) + 2\cos\left(\frac{2\pi R}{\lambda}\cos\alpha\sin\varphi\right) \right]^2} \quad (11.45)$$

Likewise, the $n = 1$ output of the Butler matrix is calculated as follows.

$$\begin{aligned} s_{n=1}(t) &= s_N(t) + s_S(t)e^{j\pi} + s_W(t)e^{-j\frac{\pi}{2}} + s_E(t)e^{j\frac{\pi}{2}} \\ &= s(t)\cos\alpha\, e^{j\frac{2\pi R}{\lambda}\cos\alpha\cos\varphi} - s(t)\cos\alpha\, e^{-j\frac{2\pi R}{\lambda}\cos\alpha\cos\varphi} \\ &\quad - js(t)\cos\alpha\, e^{j\frac{2\pi R}{\lambda}\cos\alpha\sin\varphi} + js(t)\cos\alpha\, e^{-j\frac{2\pi R}{\lambda}\cos\alpha\sin\varphi} \end{aligned} \quad (11.46)$$

The amplitude of $s_{n=1}$ is

$$\text{Amplitude}_{n=1} = \sqrt{4\sin^2\left(\frac{2\pi R}{\lambda}\cos\alpha\sin\varphi\right) + 4\sin^2\left(\frac{2\pi R}{\lambda}\cos\alpha\cos\varphi\right)} \quad (11.47)$$

The amplitude response patterns for the monopulse array for $n = 0$ and $n = 1$ are shown in Figure 11.10.

In a similar fashion the output at the $n = -1$ port is calculated to be

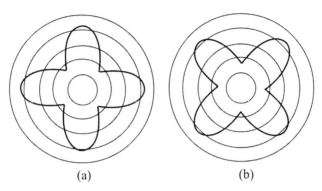

Figure 11.10 Amplitude response patterns for a monopulse array: (a) $n = 0$ and (b) $n = 1$.

$$s_{n=-1}(t) = s(t)\cos\alpha \left[\begin{array}{c} e^{j\frac{2\pi R}{\lambda}\cos\alpha\cos\varphi} - e^{-j\frac{2\pi R}{\lambda}\cos\alpha\cos\varphi} \\ -je^{j\frac{2\pi R}{\lambda}\cos\alpha\sin\varphi} + je^{-j\frac{2\pi R}{\lambda}\cos\alpha\sin\varphi} \end{array} \right]$$

$$= s(t)\cos\alpha \left[\begin{array}{c} 2j\sin\left(\frac{2\pi R}{\lambda}\cos\alpha\cos\varphi\right) \\ -2j^2\sin\left(\frac{2\pi R}{\lambda}\cos\alpha\sin\varphi\right) \end{array} \right]$$

$$= s(t)\cos\alpha \left[\begin{array}{c} 2\sin\left(\frac{2\pi R}{\lambda}\cos\alpha\sin\varphi\right) \\ +j2\sin\left(\frac{2\pi R}{\lambda}\cos\alpha\cos\varphi\right) \end{array} \right] \quad (11.48)$$

$$= s(t)\cos\alpha \left[\begin{array}{c} -2\sin\left(\frac{2\pi R}{\lambda}\cos\alpha\sin\varphi\right) \\ +j2\sin\left(\frac{2\pi R}{\lambda}\cos\alpha\cos\varphi\right) \end{array} \right]$$

with amplitude

$$\text{Amplitude}_{n=-1} = \sqrt{4\sin^2\left(\frac{2\pi R}{\lambda}\cos\alpha\sin\varphi\right) + 4\sin^2\left(\frac{2\pi R}{\lambda}\cos\alpha\cos\varphi\right)} \quad (11.49)$$

Note that this amplitude is the same as the amplitude of the $n = 1$ port. It is the phase shifts that are different at these ports, not the amplitudes. Therefore, the amplitudes at either the $n = 1$ or the $n = -1$ port can be used to compare with the $n = 0$ port to determine the angle of arrival.

Finally, the signal at the $n = 2$ port is given by

$$s_{n=2}(t) = s(t)\cos\alpha\left[2\cos\left(\frac{2\pi R}{\lambda}\cos\alpha\cos\varphi\right) - 2\cos\left(\frac{2\pi R}{\lambda}\cos\alpha\sin\varphi\right)\right] \quad (11.50)$$

The amplitude is given by

$$\text{Amplitude}_{n=2} = \sqrt{\left[2\cos\left(\frac{2\pi R}{\lambda}\cos\alpha\cos\varphi\right) - 2\cos\left(\frac{2\pi R}{\lambda}\cos\alpha\sin\varphi\right)\right]^2} \quad (11.51)$$

Note that there is a phase progression around the array, as one considers one antenna compared with another $\pi/2$ radians from it. For $n = 1$ and $n = -1$, the phase shift is in increments of 90°. For $n = 2$, this phase shift is in increments of 180° [0, 180; 0, 360; and 0, 540].

By taking the ratio of the signal at the $n = 1$ port to that of the $n = 0$ port, one obtains

$$\frac{s_{n=1}}{s_{n=0}} = \frac{2s(t)\cos\alpha\left[\sin\left(\frac{2\pi R}{\lambda}\cos\alpha\sin\varphi\right) + j\sin\left(\frac{2\pi r}{\lambda}\cos\alpha\cos\varphi\right)\right]}{2s(t)\cos\alpha\left[\cos\left(\frac{2\pi R}{\lambda}\cos\alpha\cos\varphi\right) + \cos\left(\frac{2\pi R}{\lambda}\cos\alpha\sin\varphi\right)\right]} \quad (11.52)$$

Separating this into the real and imaginary parts and canceling appropriate terms yields

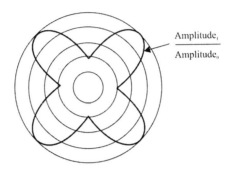

Figure 11.11 Amplitude ratio when $R = 0.2$m and $f = 100$ MHz.

$$\text{Amplitude}_{1/0} = \left\{ \left[\frac{\sin\left(\frac{2\pi R}{\lambda}\cos\alpha\sin\varphi\right)}{\cos\left(\frac{2\pi R}{\lambda}\cos\alpha\cos\varphi\right) + \cos\left(\frac{2\pi R}{\lambda}\cos\alpha\sin\varphi\right)} \right]^2 + \left[\frac{\sin\left(\frac{2\pi R}{\lambda}\cos\alpha\cos\varphi\right)}{\cos\left(\frac{2\pi R}{\lambda}\cos\alpha\cos\varphi\right) + \sin\left(\frac{2\pi R}{\lambda}\cos\alpha\sin\varphi\right)} \right]^2 \right\}^{1/2} \quad (11.53)$$

This amplitude ratio function is plotted in Figure 11.11 for $R/\lambda = 0.067$. This ratio is clearly dependent on the AOA of the signal and will yield different amplitudes depending on this angle. Note, however, that there are ambiguities. The ratio will yield the same result for signals arriving at several different angles around the array. Thus, additional information is required to resolve these ambiguities. There are no ambiguities if the signals are assumed to arrive from only a limited set of directions—for example, from 0 to $\pi/4$. If this is not the case, these ambiguities are usually resolved by using a sense antenna that is located at the center of the array. It provides a reference phase difference of zero. Similar results ensue when taking the ratio of the signal at $n = -1$ to that at $n = 0$—in fact, they are negatives of each other.

11.2.4 Amplitude Direction Finding

The antenna response patterns of directional antennas can be used to measure the direction of arrival of signals. Consider the antenna response pattern shown in

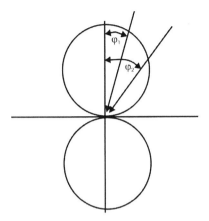

Figure 11.12 Amplitude response pattern for the Adcock antenna configuration shown in Figure 11.13.

Figure 11.12 of two dipoles configured in an Adcock configuration as shown in Figure 11.13. The amplitude of the signal arriving at the φ_1 shown would have a greater value than one arriving at φ_2 and that amplitude difference can be used as an indication of the AOA.

The output of this antenna configuration is given by

$$v(t) = s_0(t) - s_2(t)$$
$$= s(t)\cos\alpha \left[e^{j\frac{2\pi R}{\lambda}\cos\alpha\cos\varphi} - e^{-j\frac{2\pi R}{\lambda}\cos\alpha\cos\varphi} \right]$$

$$=s(t)\cos\alpha \left[\begin{array}{l} \cos\left(\dfrac{2\pi R}{\lambda}\cos\alpha\cos\varphi\right) + j\sin\left(\dfrac{2\pi R}{\lambda}\cos\alpha\cos\varphi\right) \\ -\cos\left(\dfrac{2\pi R}{\lambda}\cos\alpha\cos\varphi\right) + j\sin\left(\dfrac{2\pi R}{\lambda}\cos\alpha\cos\varphi\right) \end{array} \right] \quad (11.54)$$

$$= s(t)\cos\alpha \left[j2\sin\left(\frac{2\pi R}{\lambda}\cos\alpha\cos\varphi\right) \right]$$

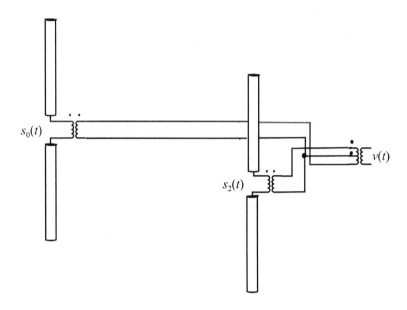

Figure 11.13 Adcock antenna array configuration.

The amplitude of this signal is

$$Amplitude_{v(t)} = 2\sin\left(\frac{2\pi R}{\lambda}\cos\alpha\cos\varphi\right) \tag{11.55}$$

Of course, in this simple case ambiguities can arise since there are other bearings that would produce the same amplitude. Another pair of dipoles arranged orthogonal to this antenna baseline could be incorporated to remove such ambiguities.

Since it is difficult to build antennas where the amplitude response is known exactly at all azimuths and all frequencies, especially in the lower-frequency ranges, amplitude comparison direction finding is only used when precise measurements of bearings are not important. Again, this antenna configuration is also a way to convert the phase of signals similar to the hybrids previously discussed.

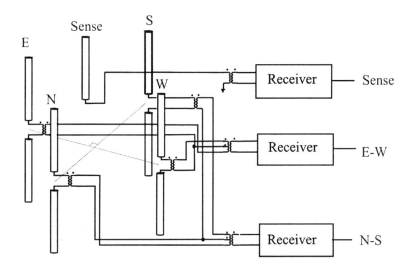

Figure 11.14 Watson-Watt direction-finding architecture.

11.2.4.1 Watson-Watt Direction Finder

This form of direction finder employs the amplitude measurement technique described above. It typically uses a circular array of dipoles, usually an orthogonal arranged set of the dipoles shown in Figure 11.14. The outputs of all four of these antennas are combined in phase to form a sense antenna or a separate sense antenna is incorporated as shown.

Orthogonally arranged loop antennas can also be used in the Watson-Watt direction finding architecture. The receiver arrangement is the same and, like the dipole arrangement, the orthogonal antennas outputs can be combined to generate the sense output without a separate antenna.

In this arrangement, the AOA is determined as follows. The signal from the south antenna, $s_S(t)$ is subtracted from the signal from the north antenna $s_N(t)$ yielding

$$s_{NS}(t) = s_N(t) - s_S(t) = s(t)\cos\alpha\left[e^{j\frac{2\pi R}{\lambda}\cos\alpha\cos\varphi} - e^{-j\frac{2\pi R}{\lambda}\cos\alpha\cos\varphi}\right]$$

$$= s(t)\cos\alpha\begin{bmatrix}\cos\left(\frac{2\pi R}{\lambda}\cos\alpha\cos\varphi\right) \\ + j\sin\left(\frac{2\pi R}{\lambda}\cos\alpha\cos\varphi\right) \\ -\cos\left(\frac{2\pi R}{\lambda}\cos\alpha\cos\varphi\right) \\ + j\sin\left(\frac{2\pi R}{\lambda}\cos\alpha\cos\varphi\right)\end{bmatrix} \quad (11.56)$$

$$= s(t)\cos\alpha\left[j2\sin\left(\frac{2\pi R}{\lambda}\cos\alpha\cos\varphi\right)\right]$$

Likewise, the signal from the west antenna, $s_W(t)$, is subtracted from the signal at the east antenna, $s_E(t)$.

$$s_{EW}(t) = s_E(t) - s_W(t) = s(t)\cos\alpha\left[e^{j\frac{2\pi R}{\lambda}\cos\alpha\sin\varphi} - e^{-j\frac{2\pi R}{\lambda}\cos\alpha\sin\varphi}\right]$$

$$= s(t)\cos\alpha\left[-j2\sin\left(\frac{2\pi R}{\lambda}\cos\alpha\sin\varphi\right)\right] \quad (11.57)$$

Next the ratio of the amplitude of these two quantities is formed.

$$\frac{s_{EW}(t)}{s_{NS}(t)} = \frac{-\sin\left(\frac{2\pi R}{\lambda}\cos\alpha\sin\varphi\right)}{\sin\left(\frac{2\pi R}{\lambda}\cos\alpha\cos\varphi\right)} \quad (11.58)$$

The Watson-Watt principle is based on approximating the AOA by the arctan of this ratio.

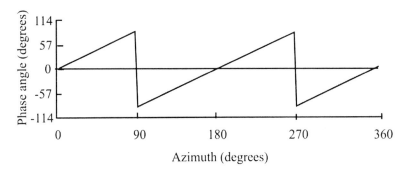

Figure 11.15 Indicated phase angle given by $\tan^{-1} s_{EW}(t)/s_{NS}(t)$.

$$\varphi \approx \tan^{-1} \frac{s_{EW}(t)}{s_{NS}(t)} \tag{11.59}$$

This ratio is plotted versus the azimuth in Figure 11.15 for $R = 0.3$m and $f = 200$ MHz.

The N-S and E-W antenna patterns are shown in Figure 11.16 for the values of $R = 0.1$m and $f = 200$ MHz ($\lambda = 1.5$m) yielding $R/\lambda = 0.067$, which are fairly typical parameters. The R/λ values are critical. Shown in Figure 11.17 are these patterns for several other values of R/λ.

The degree to which this approximation is not accurate is called the *spacing error*. Thus, plotted in Figure 11.18 is the true φ versus the function given by (11.59). For small values of R/λ this error is not too large, but it can get significant as R gets larger (or the wavelength gets shorter in relation to R). This characteristic is similar to the array response patterns discussed above. In fact, it

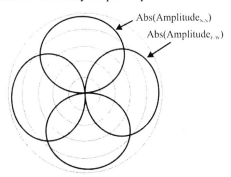

Figure 11.16 N-S and E-W antenna patterns for a Watson-Watt Adcock array. In this case $R = 0.1$m and $f = 200$ MHz. The absolute values are plotted for plotting convenience only.

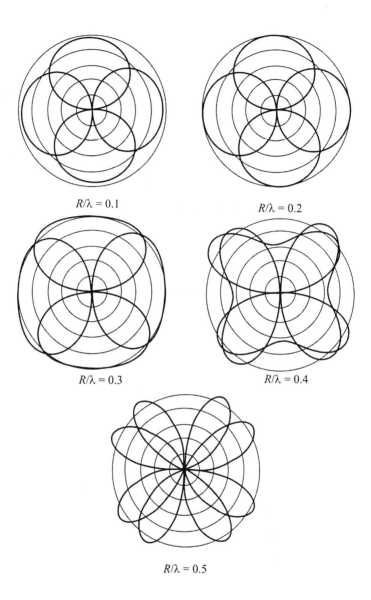

Figure 11.17 N-S and E-W response patterns for various values of R/λ.

is caused by the same effect. Note that the spacing error is a *systematic error* (caused by the characteristics of the system and is present all the time) and so can

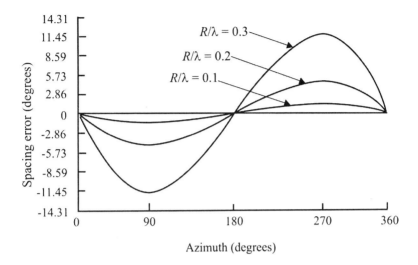

Figure 11.18 Spacing error caused by approximations to $\tan^{-1} s_{\text{E-W}}/s_{\text{N-S}}$.

be removed by calibration in many cases. Such calibration usually involves sending signals toward the array from known directions, one at a time. The array response is measured and compared with the ground truth. The difference is stored in the calibration table. It may be necessary to do this at more than one elevation.

One of the advantages of this system is that it can be instantaneous. That is particularly important when trying to locate signals that have short duration. In order to be instantaneous, however, three receiver channels are necessary. Alternately, if minimization of hardware is required, the antenna array patterns can be electronically scanned and a single receiver channel is adequate.

Shown in Figure 11.19 are the antenna response patterns when combined with the sense data. Dominant directional patterns evolve, yielding unambiguous bearings to be determined even from a single receiver. Also note that combining the signals, as here, is simply another way to accomplish the phase-shifting operation in the monopulse method.

11.2.5 Doppler Direction Finder

If an antenna is moving toward the source of a signal, then the frequency of the signal at the antenna is somewhat higher than if the antenna were still. If the antenna were moving away from the source of the signal, then the frequency

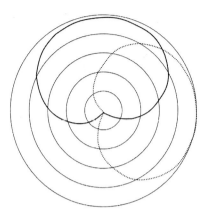

Figure 11.19 Antenna response patterns when the N-S and E-W patterns are combined with the sense antenna data.

would be somewhat lower than the true frequency. These frequency differences are an effect called *Doppler shift*, and are the same as when one listens to a passing train where, as the train is coming toward you, it sounds higher in frequency than after it passes and is moving away, when the audio signal is lower in frequency.

These effects can be exploited to determine arrival angles of signals as well. An antenna can be rotated in an EM field, as shown in Figure 11.20, and the output frequency measured. The frequency difference between the signal received at the rotating antenna and the true frequency measured at the stationary antenna determines the AOA.

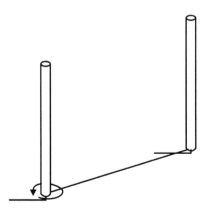

Figure 11.20 Doppler antenna configuration.

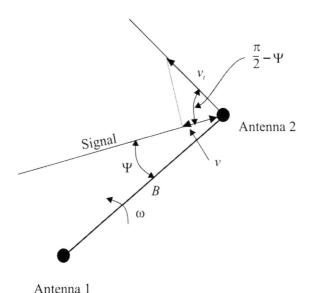

Figure 11.21 Top view of the Doppler rotating antenna array. Antenna 1 is stationary while antenna 2 rotates around it.

Consider the diagram shown in Figure 11.21, which is a top view of Figure 11.20. The Doppler shift is given approximately by (assuming that $v \ll c$)

$$\Delta f = \frac{v}{c} f_0 \qquad (11.60)$$

where v is the rotating antenna velocity in the direction of the signal, f_0 is the signal frequency, and c is the speed of propagation. The tangential velocity, v_t, shown in Figure 11.21 is given by

$$v_t = B\omega \qquad (11.61)$$

where B is the baseline length and ω is the angular velocity of the rotating antenna. The velocity v is related to the tangential velocity by

$$v = v_t \cos\left(\frac{\pi}{2} - \Psi\right) = v_t \sin \Psi \qquad (11.62)$$

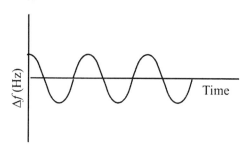

Figure 11.22 Output of the comparison of the two signals in a Doppler direction-finding system.

Therefore, the Doppler shift is given by

$$\Delta f = \frac{B\omega}{c} f_0 \sin \Psi \qquad (11.63)$$

A sense antenna located in the center of the circle determines the true frequency. The response would be as shown in Figure 11.22. When the rotating antenna is moving toward the signal source, the difference frequency is positive, and when moving away, it is negative. In most implementations, a circular array of fixed antennas would be used as opposed to actually mechanically rotating an antenna. The effects of rotation can be achieved by electronically switching each antenna in the array in turn. The sense antenna is not really needed since the output of all the antennas in the circular array can be combined in phase to accomplish this function.

11.2.6 Array-Processing Bearing Estimation

In operational direction-finding applications, frequently there is more than one signal present at the frequency at which one wishes to determine an LOP. This is the case when two separate transmitters are using the same channel and it is also the case when one signal arrives at the antenna from two or more directions. The former case is called cochannel interference and the latter case is called multipath interference (see Chapter 4).

Many of the techniques described above do not correctly compute the bearing to any of the sources of the signals in these situations. Typically some other value is computed, sometimes being the vector sum of the signals, sometimes not, depending on the methods used. Therefore, some other way of finding the bearings is required.

The algorithms derived to deal with multiple signals are collectively called *array processing* direction finding, or sometimes called *superresolution* direction finding. Gething called this "multicomponent wave fields" [4]. There are three criteria that are normally associated with these algorithms. The first is their ability to resolve two signals that arrive at closely spaced azimuths. This is the criterion from which the term "superresolution" is derived. The closer in azimuth that two signals arrive, the more difficult it is to separate them—to calculate the two distinct AOAs. The second criterion is whether there is a statistical bias in the computed azimuths. This bias is reflected in incorrectly computing the azimuths, irrespective of how well they are resolved. Since the measurements of AOAs in real situations are statistical calculations, this bias is a bias of the mean of the density function away from the true values. These two criteria are often traded for one another, as increasing one can deleteriously affect the other. The third criterion is the variability, corresponding to the standard deviation of the probability density function, which represents the range of azimuths over which the calculations are expected to vary when noise is present.

Most of the high-resolution techniques require searching over the spaces spanned by the steering vectors. That is, it is assumed that signals are arriving from a particular direction and the steering vectors are calculated with that assumption. The correct answer is assumed to be the bearings where the largest spectrum results. Thus, for incremental azimuths around the antenna array, a set of linear equations are solved. The two techniques for calculating the azimuth angles of signals impinging on an antenna array discussed here are MUSIC and beamforming.

The *multiple signal classification* (MUSIC) technique for spectral estimation is sensitive to fully coherent signals since then the covariance matrix is singular (described subsequently). Performance of the algorithm then depends on the degree of coherency in the signals. One advantage of MUSIC is that it can accommodate a nonuniform antenna array and no a priori knowledge of the number of signals is required.

There are various ways to compute the angular spectrum associated with an antenna and a particular algorithm, and each way has individual characteristics and limitations. These techniques vary in the assumptions about the signals. For example, it may be assumed that the two signals are not correlated with one another (correlated signals usually imply that one signal is a multipath reflection of the other). There are also assumptions about the configuration of the antenna array. One technique might require a circular array and another a linear array, for example. It should be noted that while this discussion talks to measuring the azimuth angle of arrival, the vertical angle of arrival could be calculated in the same way with an appropriately oriented antenna. This is particularly useful when computing the locations of HF targets, which could be arriving at a nonzero elevation angle, reflected off the ionosphere.

11.2.6.1 MUSIC

A method of superresolution direction finding based on eigen-decomposition of the signal correlation matrix has been devised, originally by Schmidt [5]; it is called MUSIC. In this technique the eigenvalues and eigenvectors of the signal correlation matrix are determined. The largest eigenvalues are assumed to be associated with signal vectors and the smallest eigenvalues are associated with noise vectors. Successive samples of antenna data are collected. Each of these samples corresponds to a single *frame*. Since, in general, RF signals are stochastic in nature, due to random noise as well as other reasons such as signal fading and modulation effects, the samples from frame to frame in general will be different.

Let μ_{jk} denote the contribution of signal k of *unit* amplitude to antenna j and if the amplitude of signal k in frame i is ψ_{ki}, the signal v_{jk} at antenna j without considering noise is determined by summing over all these signals. Thus, for S signals and N antennas

$$
\begin{aligned}
v_{1i} &= \mu_{11}\psi_{1i} + \mu_{12}\psi_{2i} + \cdots + \mu_{1R}\psi_{Si} \\
v_{2i} &= \mu_{21}\psi_{1i} + \mu_{22}\psi_{2i} + \cdots + \mu_{2R}\psi_{Si} \\
&\vdots \\
v_{Ni} &= \mu_{N1}\psi_{1i} + \mu_{N2}\psi_{2i} + \cdots + \mu_{NR}\psi_{Si}
\end{aligned}
\tag{11.64}
$$

or

$$
v_{ji} = \sum_{k=1}^{S} \mu_{jk}\psi_{ki} \tag{11.65}
$$

In this expression, the product $\mu_{jk}\psi_{ki}$ represents the component of the signal at antenna j due to signal k during frame i. Note that if there is only one signal present, then there is only one term on the right side of these equations.

With noise components η_{ji} included, this becomes

$$
v_{ki} = \sum_{k=1}^{S} \mu_{jk}\psi_{ki} + \eta_{ji} \tag{11.66}
$$

that is,

$$\Lambda_i = \mathbf{M}\Psi_i + \mathbf{N}_i \qquad (11.67)$$

In (11.66) Λ_i, $\mathbf{N}_i$, and Ψ_i vary from frame to frame while matrix $\mathbf{M}$ is constant. $\mathbf{M}$ is a function of the constant antenna geometry and of the signal arrival time differences at the two antennas on a baseline (and thus the AOAs), assumed constant over the duration of interest. Again, if there is only one signal present, then $\mathbf{M}$ is a column vector. Such column vectors herein will be denoted by $\mathbf{m}^{<i>}$.

The columns of $\mathbf{M}$ are the steering vectors associated with the ith antenna and are given by

$$\mathbf{m}_i = e^{j2\pi \bar{k} \cdot \bar{r}_i} \qquad (11.68)$$

This steering vector is the response of the ith antenna if a signal were coming from direction k.

$\mathbf{N}_i$ varies because of the random characteristics of Gaussian noise; Ψ_i varies too due to random variations in the signals caused by effects other than noise. Such effects include fading and random modulation on the antenna signals used to derive Ψ_i.

It is assumed that the noise is uncorrelated from one tap to the next as well as one frame to the next, and Gaussian, with zero mean and standard deviation σ. Using the notation $E\{x\}$ for the expected value of x, we then have

$$E\{\eta_{ji}\} = 0$$
$$E\{\eta_{ji}\eta_{ki}\} = \sigma^2 \delta_{jk} \qquad (11.69)$$

where δ_{jk} is the Kronecker delta function. The expected value of a matrix is the matrix of the element expected values, namely,

$$E\left\{\begin{bmatrix} x_1 & x_2 \\ x_3 & x_4 \end{bmatrix}\right\} = \begin{bmatrix} E\{x_1\} & E\{x_2\} \\ E\{x_3\} & E\{x_4\} \end{bmatrix} \qquad (11.70)$$

The conjugate transpose of a matrix $\mathbf{G} = [g_{ij}]$ is that matrix denoted by $\mathbf{G}^H$ with entries $g_{ij} = \bar{g}_{ji}$, where the overbar denotes conjugation. $\mathbf{G}^H$ is sometimes called the *Hermitian* of **G**. Also given matrix **G** such that $g_{ij} = \bar{g}_{ji}$, then **G** is said to be a Hermitian matrix.

The sample covariance matrix Γ is generated by forming matrix

$$\mathbf{Y} = [\boldsymbol{\Lambda}_1 \quad \boldsymbol{\Lambda}_2 \quad \cdots \quad \boldsymbol{\Lambda}_F] \tag{11.71}$$

where F is the number of frames collected. Each column of this matrix corresponds to a set of data samples from the antennas.

Then

$$\begin{aligned}
\Gamma &= E\{\mathbf{YY}^H\} \\
&= E\left\{[\boldsymbol{\Lambda}_1 \quad \boldsymbol{\Lambda}_2 \quad \cdots \quad \boldsymbol{\Lambda}_F]\begin{bmatrix}\boldsymbol{\Lambda}_1^H \\ \boldsymbol{\Lambda}_2^H \\ \vdots \\ \boldsymbol{\Lambda}_F^H\end{bmatrix}\right\} \\
&= E\{\boldsymbol{\Lambda}_1\boldsymbol{\Lambda}_1^H + \boldsymbol{\Lambda}_2\boldsymbol{\Lambda}_2^H + \cdots + \boldsymbol{\Lambda}_F\boldsymbol{\Lambda}_F^H\} \\
&= [E\{\boldsymbol{\Lambda}_1\boldsymbol{\Lambda}_1^H\} + E\{\boldsymbol{\Lambda}_2\boldsymbol{\Lambda}_2^H\} + \cdots + E\{\boldsymbol{\Lambda}_F\boldsymbol{\Lambda}_F^H\}]
\end{aligned} \tag{11.72}$$

Consider an arbitrary element of this matrix.

$$\begin{aligned}
E\{\boldsymbol{\Lambda}_t\boldsymbol{\Lambda}_t^H\} &= E\{(\mathbf{M}\boldsymbol{\Psi}_t + \mathbf{N}_t)(\mathbf{M}\boldsymbol{\Psi}_t + \mathbf{N}_t)^H\} \\
&= E\{(\mathbf{M}\boldsymbol{\Psi}_t + \mathbf{N}_t)(\boldsymbol{\Psi}_t^H\mathbf{M}^H + \mathbf{N}_t^H)\} \\
&= E\{\mathbf{M}\boldsymbol{\Psi}_t\boldsymbol{\Psi}_t^H\mathbf{M}^H + \mathbf{M}\boldsymbol{\Psi}_t\mathbf{N}_t^H + \mathbf{N}_t\boldsymbol{\Psi}_t^H\mathbf{M}^H + \mathbf{N}_t\mathbf{N}_t^H\} \\
&= E\{\mathbf{M}\boldsymbol{\Psi}_t\boldsymbol{\Psi}_t^H\mathbf{M}^H\} + E\{\mathbf{M}\boldsymbol{\Psi}_t\mathbf{N}_t^H\} + E\{\mathbf{N}_t\boldsymbol{\Psi}_t^H\mathbf{M}^H\} + E\{\mathbf{N}_t\mathbf{N}_t^H\} \\
&= \mathbf{M}E\{\boldsymbol{\Psi}_t\boldsymbol{\Psi}_t^H\}\mathbf{M}^H + \mathbf{M}E\{\boldsymbol{\Psi}_t\mathbf{N}_t^H\} + E\{\mathbf{N}_t\boldsymbol{\Psi}_t^H\}\mathbf{M}^H + E\{\mathbf{N}_t\mathbf{N}_t^H\} \\
&= \mathbf{MHM}^H + \boldsymbol{\Phi}
\end{aligned} \tag{11.73}$$

The last step follows because the signals are assumed to be uncorrelated with the noises.

In this expression

$$\mathbf{H} = \mathrm{E}\{\mathbf{\Psi}_\iota \mathbf{\Psi}_\rho^H\} = \mathrm{E}\left\{\begin{bmatrix} \psi_{1\iota} \\ \psi_{2\iota} \\ \psi_{3\iota} \\ \vdots \\ \psi_{S\iota} \end{bmatrix} \begin{bmatrix} \overline{\psi}_{1\rho} & \overline{\psi}_{2\rho} & \overline{\psi}_{3\rho} & \cdots & \overline{\psi}_{S\rho} \end{bmatrix}\right\}$$

$$= \mathrm{E}\left\{\begin{bmatrix} \psi_{1\iota}\overline{\psi}_{1\rho} & \psi_{1\iota}\overline{\psi}_{2\rho} & \psi_{1\iota}\overline{\psi}_{3\rho} & \cdots & \psi_{1\iota}\overline{\psi}_{S\rho} \\ \psi_{2\iota}\overline{\psi}_{1\rho} & \psi_{2\iota}\overline{\psi}_{2\rho} & \psi_{2\iota}\overline{\psi}_{3\rho} & \cdots & \psi_{2\iota}\overline{\psi}_{S\rho} \\ \psi_{3\iota}\overline{\psi}_{1\rho} & \psi_{3\iota}\overline{\psi}_{2\rho} & \psi_{3\iota}\overline{\psi}_{3\rho} & \cdots & \psi_{3\iota}\overline{\psi}_{S\rho} \\ \vdots & \vdots & \vdots & \ddots & \vdots \\ \psi_{S\iota}\overline{\psi}_{1\rho} & \psi_{S\iota}\overline{\psi}_{2\rho} & \psi_{S\iota}\overline{\psi}_{3\rho} & \cdots & \psi_{S\iota}\overline{\psi}_{S\rho} \end{bmatrix}\right\}$$

(11.74)

and

$$\mathbf{\Phi} = \mathrm{E}\{\mathbf{N}_\iota \mathbf{N}_\rho^H\} = \mathrm{E}\left\{\begin{bmatrix} \eta_{1\iota} \\ \eta_{2\iota} \\ \eta_{3\iota} \\ \vdots \\ \eta_{N\iota} \end{bmatrix} \begin{bmatrix} \eta_{1\rho} & \eta_{2\rho} & \eta_{3\rho} & \cdots & \eta_{N\rho} \end{bmatrix}\right\}$$

(11.75)

$$= \mathrm{E}\left\{\begin{bmatrix} \eta_{1\iota}\overline{\eta}_{1\rho} & \eta_{1\iota}\overline{\eta}_{2\rho} & \eta_{1\iota}\overline{\eta}_{3\rho} & \cdots & \eta_{1\iota}\overline{\eta}_{N\rho} \\ \eta_{2\iota}\overline{\eta}_{1\rho} & \eta_{2\iota}\overline{\eta}_{2\rho} & \eta_{2\iota}\overline{\eta}_{3\rho} & \cdots & \eta_{2\iota}\overline{\eta}_{N\rho} \\ \eta_{3\iota}\overline{\eta}_{1\rho} & \eta_{3\iota}\overline{\eta}_{2\rho} & \eta_{3\iota}\overline{\eta}_{3\rho} & \cdots & \eta_{3\iota}\overline{\eta}_{N\rho} \\ \vdots & \vdots & \vdots & \ddots & \vdots \\ \eta_{N\iota}\overline{\eta}_{1\rho} & \eta_{N\iota}\overline{\eta}_{2\rho} & \eta_{N\iota}\overline{\eta}_{3\rho} & \cdots & \eta_{N\iota}\overline{\eta}_{N\rho} \end{bmatrix}\right\}$$

$\mathbf{\Gamma}$, $\mathbf{H}$, and $\mathbf{\Phi}$ are square matrices of order $(N \times N)$, $(S \times S)$, and $(N \times N)$, respectively. $\mathbf{M}$ and $\mathbf{M}^H$ are of order $(N \times S)$ and $(S \times N)$, respectively.

When there is no noise, measurement errors, or other random effects, then when R is less than $(N-1)$, the system must be overdetermined and one or more linear relationships will be satisfied by the measured elements of $\mathbf{\Lambda}_i$ on each frame and $\mathbf{\Gamma}$ will be of reduced rank r. It is said to have *nullity* $(N-r)$.

With the above noise model applied to a large sample of frames, the off-diagonal elements of Φ approach zero and (11.72) assumes the asymptotic form

$$\Gamma = \mathbf{MHM}^H + \sigma^2 \mathbf{I} \tag{11.76}$$

For completely uncorrelated noise, the off-diagonal elements of **H** also tend to zero. The off-directional elements of Γ contain the directional information about the arriving signals and do not tend towards zero, but the leading diagonal elements are real and tend toward equality; the physical interpretation is that all antennas are expected to sense approximately the same power over a long sampling period.

A common way of estimating Γ is to average the frame data, namely,

$$\hat{\Gamma} = \frac{1}{F} \sum_{i=1}^{F} \Lambda_i \Lambda_i^H$$

$$= \frac{1}{F} \sum_{i=1}^{F} \begin{bmatrix} v_{1i} v_{1i}^H & v_{1i} v_{2i}^H & \cdots & v_{1i} v_{Ni}^H \\ v_{2i} v_{1i}^H & v_{2i} v_{2i}^H & \cdots & v_{2i} v_{Ni}^H \\ \vdots & \vdots & \ddots & \vdots \\ v_{Ni} v_{Ni}^H & v_{Ni} v_{2i}^H & \cdots & v_{Ni} v_{Ni}^H \end{bmatrix} \tag{11.77}$$

For the above approach to work, the frame data must be samples of a zero-mean statistical process. The reason for this is that the *time difference of arrival* (TDOA) information is in the off-diagonal entries in $\hat{\Gamma}$. If the frame data is not based on a zero mean process, the off-diagonal elements of the frame covariance matrices $\Lambda_i \Lambda_i^H$ will not tend to zero with averaging. They thus contribute to the off-diagonal elements of $\hat{\Gamma}$ causing erroneous results.

Calculating the TDOA

The goal of this analysis is to ascertain the TDOAs of multiple signals arriving at a pair of antennas. The above discussion was a prelude establishing the basics of the processes to do just this.

An *eigenvector* of a square matrix **G** ($N \times N$) is a nonzero column vector e_i that satisfies

$$\mathbf{G} e_i = \lambda_i e_i \tag{11.78}$$

Direction-Finding Position-Fixing Techniques 369

for an associated scalar λ_i known as its *eigenvalue*. The eigenvalues of **G** are found from the *characteristic equation* of **G** given by

$$\det[\mathbf{G} - \lambda \mathbf{I}] = 0 \tag{11.79}$$

where **I** is the identity matrix. Matrix **G** has N eigenvalues, not all of which are necessarily distinct. The eigenvectors associated with any λ form an independent set of vectors and form a basis for the linear manifold associated with λ. This linear manifold is referred to as the *eigenvector manifold associated with λ*. Once the eigenvectors are found, the associated eigenvectors are determined by solving (11.77).

When the eigenvalues of $\hat{\Gamma}$ are found, they fall into two groups. The first group is of larger values than the second group which, when there is no noise or measurement error, is equal to each other and is of a small value. The number of larger eigenvalues is a measure of the number of constituent waveforms present. In most cases of interest Γ is singular since normally $S < N - 1$. MUSIC is based on the singular value decomposition on the covariance matrix Γ.

Singular Value Decomposition

Any $m \times n$ matrix **Y** can be decomposed into

$$\mathbf{Y} = \mathbf{U}\mathbf{\Sigma}\mathbf{V}^H = \begin{bmatrix} \mathbf{U}_s & \mathbf{U}_o \end{bmatrix} \begin{bmatrix} \mathbf{\Sigma}_s & \mathbf{0} \\ \mathbf{0} & \mathbf{0} \end{bmatrix} \begin{bmatrix} \mathbf{V}_s^H \\ \mathbf{V}_o^H \end{bmatrix} \tag{11.80}$$

where **U** is an $m \times n$ orthogonal matrix, **V** is an $n \times m$ orthogonal matrix and $\mathbf{\Sigma}_s$ is an $r \times r$ diagonal matrix with real, nonnegative elements σ_i, $i = 1, 2, \ldots, r = \min(m, n)$ arranged in descending order

$$\sigma_1 \geq \sigma_2 \geq \cdots \geq \sigma_r > \sigma_{r+1} = \sigma_{r+2} = \cdots = 0 \tag{11.81}$$

These σ_i are called the *singular values* of **Y**. The first r columns of **U** are the left singular vectors of **Y** and the first r columns of **V** are the right singular vectors of **Y**. The structure of Σ is

$$\Sigma = \begin{bmatrix} \sigma_1 & & & & & \\ & \sigma_2 & & & \mathbf{0} & \\ & & \ddots & & & \\ & & & \sigma_r & & \\ & & & & 0 & \\ & \mathbf{0} & & & & 0 \\ & & & & & & \ddots \\ & & & & & & & 0 \end{bmatrix} \tag{11.82}$$

$$= \begin{bmatrix} \Sigma_s \\ \mathbf{0} \end{bmatrix} \quad \text{if } m \geq n$$

$$= \begin{bmatrix} \Sigma_s & \mathbf{0} \end{bmatrix} \quad \text{if } m < n$$

where r is the rank of $\mathbf{Y}$. Thus, $\mathbf{U}_s$ consists of the left singular vectors associated with the nonzero singular values and $\mathbf{U}_o$ consists of the left singular vectors associated with the zero singular values. These $\mathbf{U}_s$ vectors span the subspace consisting of the vectors in $\mathbf{M}$. The vectors in $\mathbf{U}_o$ span the orthogonal subspace so

$$m^{<i>H} \mathbf{U}_o = \mathbf{0} \tag{11.83}$$

Also, the singular vectors are normalized so that $\mathbf{U}^H \mathbf{U} = \mathbf{I}$.

$\mathbf{U}$ is the $m \times m$ matrix of orthonormalized row eigenvectors of $\mathbf{Y}\mathbf{Y}^T$ while $\mathbf{V}$ is the $n \times n$ matrix of orthonormalized column eigenvectors of $\mathbf{Y}^T\mathbf{Y}$. The singular values of $\mathbf{Y}$ are defined as the nonnegative square roots of the eigenvalues of $\mathbf{YY}^T$.

The singular value decomposition is perhaps most known for solutions to the least squares problem. For the linear system defined by the set of equations

$$\mathbf{Ax} = \mathbf{b} \tag{11.84}$$

where $\mathbf{A}$ is not square, but of dimensions $m \times n$, and $\mathbf{x}$ and $\mathbf{b}$ are vectors, then the least squares solution is that value of x where

$$\min_x \|\mathbf{Ax} - \mathbf{b}\| \tag{11.85}$$

which is

$$\hat{\mathbf{x}} = \mathbf{V}\mathbf{\Sigma}^{\Diamond}\mathbf{U}^T\mathbf{b} \tag{11.86}$$

Matrix $\mathbf{\Sigma}^{\Diamond}$ is

$$\mathbf{\Sigma}^{\Diamond} = \begin{bmatrix} 1/\sigma_1 & & & & 0 & 0 & \cdots & 0 \\ & \ddots & & & & & & \\ & & 1/\sigma_r & & & \vdots & & \vdots \\ & & & 0 & & & & \\ & & & & \ddots & & & \\ & & & & & 0 & 0 & \cdots & 0 \end{bmatrix} \tag{11.87}$$

then matrix

$$\mathbf{A}^{\Diamond} = \mathbf{V}\mathbf{\Sigma}^{\Diamond}\mathbf{U}^T \tag{11.88}$$

is known as the *pseudoinverse* of $\mathbf{A}$.

Note that if the zeros are removed from $\mathbf{\Sigma}^{\Diamond}$ and $\mathbf{U}$ and $\mathbf{V}$ are appropriately reduced in size, rank($\mathbf{A}$) = r, and the solution reduces to

$$\mathbf{x} = \left(\mathbf{A}^T\mathbf{A}\right)^{-1}\mathbf{A}^T\mathbf{b} \tag{11.89}$$

which, with $\mathbf{A} = \mathbf{U}\mathbf{\Sigma}\mathbf{V}^T$

$$\mathbf{x} = \mathbf{V}\mathbf{\Sigma}^{-1}\mathbf{U}^T\mathbf{b} \tag{11.90}$$

Because Γ is positive definite and Hermitian, (11.79) can be expressed as

$$\Gamma = \mathbf{U}\mathbf{\Sigma}\mathbf{U}^H \tag{11.91}$$

where **U** is unitary. Note that if a vector **x** is orthogonal to $\mathbf{M}^H$, then it is an eigenvector of Γ with eigenvalue σ^2 because

$$\Gamma\mathbf{x} = \mathbf{MHM}^H\mathbf{x} + \sigma^2\mathbf{x} = \sigma^2\mathbf{x} \tag{11.92}$$

The eigenvector of Γ with eigenvalue σ^2 lies in $\mathcal{N}(\mathbf{M}^H)$, the null space of $\mathbf{M}^H$. Thus, the smallest $(N-L)$ nonzero eigenvalues are

$$\lambda_{L+1} = \lambda_{L+2} = \cdots \lambda_N = \sigma^2 \tag{11.93}$$

It is therefore possible to partition the eigenvectors into noise eigenvectors and signal eigenvectors and the covariance matrix Γ can be written as

$$\Gamma = \mathbf{U}_s \Sigma_s \mathbf{U}_s^H + \mathbf{U}_n \Sigma_n \mathbf{U}_n^H \tag{11.94}$$

Letting $\mathcal{R}(\mathbf{X})$ denote the range space of **X**, the range of **Q** is the orthogonal complement to the range of **M**, because

$$\mathcal{R}(Q) = \mathcal{N}(\mathbf{M}^H) = {}^\perp\mathcal{R}(M) \tag{11.95}$$

where $\mathcal{N}(\mathbf{A})$ is the null space of **A**. Therefore,

$$\begin{aligned}\mathcal{R}(\mathbf{U}_s) &= \mathcal{R}(\mathbf{M}) \\ \mathcal{R}(\mathbf{U}_n) &= {}^\perp \mathcal{R}(\mathbf{M}^H)\end{aligned} \tag{11.96}$$

$\mathcal{R}(\mathbf{U}_s)$ is called the *signal subspace* and $\mathcal{R}(\mathbf{U}_n)$ is called the *noise subspace*. The projection operators onto these signal and noise subspaces are defined as

$$\begin{aligned}\mathbf{P}_\mathbf{M} &= \mathbf{MM}^\Diamond = \mathbf{U}_s\left(\mathbf{U}_s^H \mathbf{U}_s\right)^{-1}\mathbf{U}_s^H = \mathbf{U}_s\mathbf{U}_s^H \\ \mathbf{P}_\mathbf{M}^\perp &= \mathbf{I} - \mathbf{MM}^\Diamond = \mathbf{U}_n\left(\mathbf{U}_n^H \mathbf{U}_n\right)^{-1}\mathbf{U}_n^H = \mathbf{U}_n\mathbf{U}_n^H\end{aligned} \tag{11.97}$$

Determining the TDOAs

The eigenstructure of the covariance matrix Γ can be analyzed to determine the TDOAs. The normal nonsingular nature of Γ can cause numerical problems in some cases, however. In this technique the N-dimensional space is partitioned into the aforementioned signal subspace and the orthogonal noise subspace. The computation approach, starting from the covariance matrix estimate $\hat{\Gamma}$, requires the determination of the eigenvalues and eigenvectors of this matrix. As previously mentioned, the eigenvalues form two sets, one consisting of larger eigenvalues than the other. Starting with $k = 0$, and going until $k = N$, where k is the estimated number of eigenvalues, a logical dividing point is selected for the number of signals present. The eigenvectors corresponding to the signal subspace eigenvalues in the larger set (larger in value, not cardinality), denoted $\mathbf{R}_s$, span the signal space while the eigenvectors corresponding to the smaller, denoted $\mathbf{R}_n$, set span the orthogonal noise subspace.

The matrix $\mathbf{R}_n$ of eigenvectors spanning the noise subspace is used to calculate the function

$$P_{\text{MUSIC}} = \frac{1}{\Omega^H \mathbf{R}_n \mathbf{R}_n^H \Omega} \qquad (11.98)$$

since the denominator in this expression tends to zero due to orthogonality. The peaks in this functional are determined, and the S highest peaks are an estimate of where the time differences occur. A typical plot of P_{MUSIC} might look like that in Figure 11.23. In this example there are two signals impinging on the antenna array, one at about 90° and another at about 270°. The abscissa on a chart like this corresponds to the angle around the antenna array, while the ordinate corresponds to the amplitude response of the array.

11.2.6.2 Beamforming

One of the oldest techniques for dealing with the cochannel problem is *beamforming*. This technique is not normally considered to be in the superresolution family since historically the techniques used did not separate two signals, but suppressed the antenna response in all but a given direction. In this approach, the outputs of an array of K antennas are multiplied by weights, which are usually complex, and are delayed. The amplitude of the weight affects the amplitude of the signal from that particular antenna element while the phase of the weight changes the phase of the signal. The outputs of these multipliers and delay elements are then summed to form the array pattern. By properly adjusting the

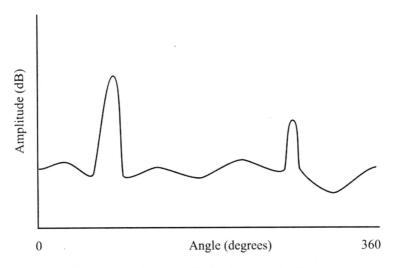

Figure 11.23 Azimuthal response of an array with signals at two azimuths.

weights and time delays, the antenna's "beam" can be steered in a desired direction, enhancing the reception in that direction while suppressing the reception in others. The structure is shown in Figure 11.24.

If $x_j(t)$ represents the output of the jth antenna element and $s(t)$ is the signal that is transmitted then

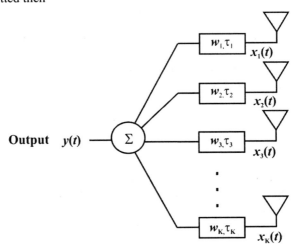

Figure 11.24 One of many types of beamforming network.

$$x_j(t) = s(t + t_j) + n_j(t) \tag{11.99}$$

where $n_j(t)$ represents the additive noise at antenna element j. These signals are multiplied by weights with amplitudes w_j, delayed by a time τ_j, and then summed. The output of the beamformer is

$$y(t) = \sum_{j=1}^{K} w_j x_j(t-\tau_j) \tag{11.100}$$

If the time delays are adjusted so that $t_j = \tau_j$, and assuming the weights are set to 1 then

$$\begin{aligned} y(t) &= Ks(t) + \sum_{j=1}^{K} n_j(t-\tau_j) \\ &= \sum_{j=1}^{K} w_j s(t+t_j-\tau_j) + w_j n_j(t-\tau_j) \end{aligned} \tag{11.101}$$

which under appropriate conditions will maximize $y(t)$. Search over a range of azimuths is accomplished by using several values of τ_j for each antenna path. Those directions where the response is maximum are assumed to be the direction of the arrival of signals.

11.2.7 Line-of-Bearing Optimization

The least squares range difference location problem has been investigated by Schmidt [6] and others. The range difference from a target to two sensors creates a hyperbolic line of position as shown in Figure 11.25. Thus, there are ambiguous

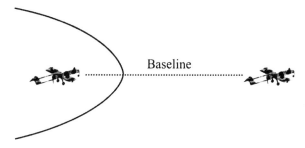

Figure 11.25 Two sensors creating a single baseline form a hyperbolic LOP upon which the target is located.

answers for where the target is located. In order to ascertain where on this LOP the target lies, either one (or both) of the sensors must move to another location and another measurement taken, or a third sensor must be added creating a second (and unnecessary third) baseline. The resulting position fix determined by the intersection of the LOPs is then unique.

In the case shown, the sensors are widely separated. That need not be the case, however. The sensors could be two antennas in the same array, for example. The range difference in such an array is still a hyperboloid, but to calculate the target location this way is difficult. The LOPs intersect at too large a range from the array. Small errors in calculating the LOPs result in large errors in calculating the target location because the LOPs are essentially parallel in that region.

Another technique to obtain a PF on a target is to measure the LOP from two or more sensors to the target. Where two or (preferably) more LOBs intersect is where the target is presumed to be located. One way of determining such lines of bearing is with phase interferometry where the phase difference is measured between $N > 2$ antennas and the LOB is computed as shown above in Section 11.2.2.

11.2.7.1 Phase Difference Averaging

Schmidt [6] showed that the TDOA averaging process produced the geolocation that was the closest feasible one in a least-squared sense based on measured range differences. Although in those results the calculations produced geolocations based on range differences, herein they are somewhat extended to show that for an N-channel interferometer, calculating the signal phase differences in the same way produces bearings that are optimum in the least squares sense.

TDOA averaging is the process of removing the average TDOA, τ_{avg}, or equivalently the average range difference, Δ_{avg}, from the measured TDOAs or range differences. This produces feasible TDOAs or range differences that in a least squares sense are the closest to the measured data.

The top view of a triple channel interferometer is shown in Figure 11.26. The phase angle of a planer signal crossing the array orthogonal to both the plane of the array and the antenna elements varies modulo 2π as the signal passes the array. Herein it is assumed that the baseline length of the array is small enough to avoid ambiguities at the highest operational frequency. The phase differences between the signals at the antennas taken a pair at a time are measured by one of several means not of consequence here. These phase angle differences are denoted θ_{ij}. They are related to the frequency and TDOAs at the antenna pairs τ_{ij} by $\theta_{ij} = 2\pi f \tau_{ij}$. Furthermore, since $\tau_{ij} = \Delta_{ij}/c$, where Δ_{ij} is the path distance traversed by the signal between antennas i and j and c is the speed of propagation of the wave in the medium, then $\theta_{ij} = 2\pi f \Delta_{ij}/c$. It is tacitly assumed that the sensor

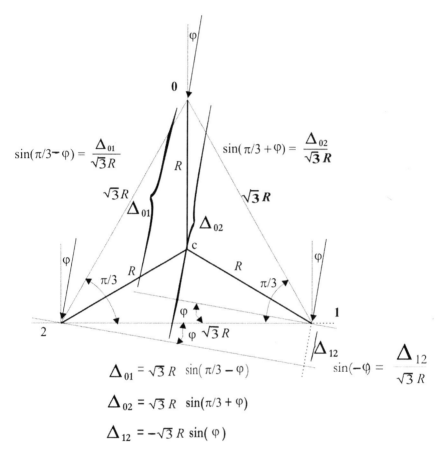

Figure 11.26 Top view of an $N = 3$ channel antenna array forming an interferometer.

system is adequately far from the target that the signal is appropriately represented as a vertically polarized plane wave and that the AOA of the signal at each antenna is the same.

Ideally, the measured phase differences around any closed set of antennas sum to zero. In practice the sum will normally be nonzero due to measurement errors and noise. The process described by Schmidt [6] removes the measurement errors associated with measuring the range differences assuming that the sensors are widely spaced (i.e., not the antenna array shown in Figure 11.26, but separate sensors). The procedure described here assumes the sensors consist of the antenna array shown, with all antennas at the same geographic location, albeit separated somewhat by the baseline length shown.

In any case, the measured phase differences will not necessarily correspond to any feasible AOA due to the noise in the measurements. Therefore, as for the determination of the closest feasible location by Schmidt, the AOA based on the closest feasible phase angles is sought here. This will be called *phase difference averaging*. The resulting phase difference measurements are over-determined since there is more than one equation and only one dependent variable φ, thus yielding to least squares optimization.

The procedure described by Schmidt subtracts the average distances, or equivalently, the average TDOA, from each of the measurements, thereby calculating the closest feasible set of range differences. Thus,

$$\Delta_{avg} = \frac{1}{3}(\Delta_{01} + \Delta_{12} + \Delta_{20}) \tag{11.102}$$

and Δ_{avg} is subtracted from each measurement. Since $\Delta_{ij} = c\theta_{ij}/2\pi f$, then the average range difference can be expressed as

$$\Delta_{avg} = \frac{c}{2\pi f} \frac{1}{3}(\theta_{01} + \theta_{12} + \theta_{20}) \tag{11.103}$$

but

$$\frac{1}{3}(\theta_{01} + \theta_{12} + \theta_{20}) = \theta_{avg} \tag{11.104}$$

is the average of the measured phase differences. Therefore, subtracting the average phase difference from the measured phase differences will produce a set of adjusted phase differences that are feasible and least squares optimum.

As shown in Figure 11.26, the path-length differences for a signal with an AOA of φ are given by

$$\begin{aligned}\Delta_{01} &= d_1 - d_0 = \sqrt{3}R\sin\left(\frac{\pi}{3}+\varphi\right) \\ \Delta_{02} &= d_2 - d_0 = \sqrt{3}R\sin\left(\frac{\pi}{3}-\varphi\right) \\ \Delta_{12} &= d_2 - d_1 = \sqrt{3}R\sin(-\varphi)\end{aligned} \tag{11.105}$$

The relevant distances are Δ_{01}, Δ_{12} and $\Delta_{20} = -\Delta_{02}$. Care must be exercised when defining the signs of the angles involved. The d_i here represent the distance of antenna i from the target. Of course, over 2π radians these equations are ambiguous, each yielding two possible answers for the AOA. Any two taken at the same time, however, yield a unique answer.

The procedure for calculating the AOA is then:

1. Measure the phase differences between all pairs of antennas. This will yield $\binom{N}{2}$ phase difference measurements.
2. Determine the average path-length difference from (11.103).
3. Calculate the AOA φ using any two of (11.105).

This procedure works for any interferometric antenna array as long as the number of elements $N > 2$. When $N > 3$, the subarrays consisting of three elements each are taken one at a time and the procedure is applied to them.

11.3 POSITION-FIXING ALGORITHMS

The previous discussions focused on calculating an LOP, upon which an emitting target is assumed to lay, based on measuring an azimuth and perhaps elevation AOAs of the EM wavefront at an antenna array. The target could lie at any point on the LOP, and therefore its location has not as yet been determined. Knowing just the LOP in some situations is useful—as an azimuth upon which to home a missile, for example. The direction in which to point a jammer antenna is another situation where just the LOP is useful information. It is generally not enough for most ES applications, however. Two or more such LOPs are usually combined to geolocate a target. If only two LOPs are available, the resultant fix calculated is called a *cut*. When three or more are used, it is called a *fix*.

A cut is less reliable than a fix. Two LOPs will always intersect at a single point, unless, of course, they are parallel. Three or more real LOPs will rarely intersect at the same point. Since determining the geolocation of a target is a statistical process, calculating the absolute location (as represented by a cut) is unreliable. Also, in general, for any statistical process the more relevant data that can be brought to bear on the solution, the better because independent statistical results typically improve in accuracy inversely proportional to some power of the number of measurements made.

There are many algorithms available to facilitate the calculation of the location of an emitter based on multiple lines of position. This calculated location estimate is called the *best point estimate* (BPE). A few of these are presented in

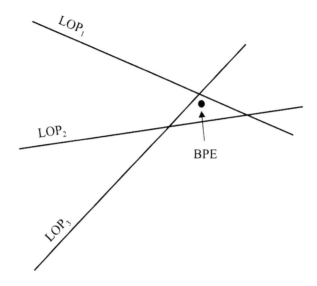

Figure 11.27 Determining the BPE. It will be some function of the individual LOPs.

this section. The problem is illustrated in Figure 11.27. Normally, along with the BPE, a region is also calculated within which the emitter lies with a specified probability, usually 50% or 90%. The most common form of region is an ellipse, in which case it is called an EEP. Other forms are a circle, called a CEP, and a rectangle.

In these calculations, the a priori variance of the bearing for a system would typically be assumed to be the instrumental accuracy of the system, if the real accuracy is not known (by measurement, for example). This would typically be determined by experimentation and test, sometimes called *array calibration,* or just *calibration*. For ground-based systems, a turntable is typically used for this purpose. For larger aircraft, the aircraft are flown and data collected against the known location of emitters. For wide bandwidth systems, such calibration tables can become quite large.

11.3.1 Eliminating Wild Bearings

There is a technique for eliminating wild bearings when a sufficient number of bearings are available [4]. Such wild bearings, when included in BPE calculations, tend to yield erroneous results. If it can be determined which bearings are "wild," they can be eliminated when the BPE is calculated.

Once the BPE has been calculated, determine the dispersion factor, given by

$$E = \sum \frac{\theta_i^2}{\sigma_i^2} \qquad (11.106)$$

where σ_i is the standard deviation (σ_i^2 is the variance) of the ith bearing and θ_i is the difference between the ith AOA and the ith line to the BPE. If there are wild bearings present, this number will be large. To eliminate these bearings, eliminate one bearing at a time and recalculate the dispersion factor. The best estimate will be the one associated with the smallest dispersion factor, and the bearing(s) excluded can be assumed wild.

A minimum of four LOPs is required in order to apply this technique. With three, eliminating one results in only two lines of position remaining; thus, a cut. It is impossible to tell which one of these BPEs is correct based on only two bearings. Therefore, a minimum of three LOPs at a time is required, necessitating at least four initially.

Care must be taken when eliminating bearings from the calculation of the BPE. The technique described here would apply if there were enough LOPs to start with, but it does not if there are not. In general, it is not known which bearings accurately correspond to those from the target and which might not. There is no statistical basis normally for eliminating bearings from the BPE calculation. There can be exceptions to this such as if the system operator "hears" interference while the bearing is being calculated. Another case would be if it is known that the system was inoperative when the bearing was measured.

11.3.2 Stansfield Fix Algorithm

One of the first algorithms developed for the purpose of calculating the location of an emitter based on multiple lines of bearing was due to Stansfield [7]. In that algorithm it is assumed that the bearing errors of the ES systems are normally (Gaussian) distributed. The joint probability density function of multiple lines of bearings is then a multivariate Gaussian probability density function. A maximum likelihood estimator for the BPE ensues by maximizing the exponent in the equation for the joint probability density function (which will minimize the total probability of error because the exponent is negative).

Using Figure 11.28 to define the variables, then the joint probability of the miss distances between the LOPs and the true target location, denoted as p_i, is

$$P(p_1, p_2, \cdots, p_N) = \frac{1}{(2\pi)^{N/2} \sum_{i=1}^{N} \sigma_{p_i}} \exp\left(-\frac{1}{2} \sum_{i=1}^{N} \frac{p_i^2}{\sigma_{p_i}^2}\right) \qquad (11.107)$$

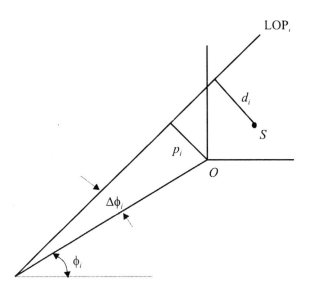

Figure 11.28 Definition of the terms for the derivation of the position fixing algorithm of Stansfield. (*Source:* [7], © 1947, IRE. Reprinted with permission.)

where N = number of LOPs.

It is assumed that the target lies at point $S = (x_T, y_T)$, so that

$$d_i = p_i + x_T \sin \phi_i - y_T \cos \phi_i \qquad (11.108)$$

where d_i is the distance from point S to the LOP$_i$. It is also assumed that the angular error is small so that $p_i \approx \Delta\phi_i D_i$, and thus $\sigma_{pi} \approx D_i \sigma_\phi$.

The joint probability of the d_i's is given by an expression similar to that above.

$$P(d_1, d_2, \cdots, d_N) = \frac{1}{(2\pi)^{N/2} \sum_{i=1}^{N} \sigma_{p_i}} \exp\left(-\frac{1}{2} \sum_{i=1}^{N} \frac{d_i^2}{\sigma_{p_i}^2}\right) \qquad (11.109)$$

$$= \frac{1}{(2\pi)^{N/2} \sum_{i=1}^{N} \sigma_{p_i}}$$

$$\times \exp\left(-\frac{1}{2} \sum_{i=1}^{N} \frac{(p_i + x_T \sin\phi_i - y_T \cos\phi_{ii})^2}{\sigma_{p_i}^2}\right)$$

Direction-Finding Position-Fixing Techniques

The values of x_T and y_T that maximize the exponent in this expression are the coordinates with the highest probability of being those of the target—that is, the coordinates of the BPE. This yields the following expression for the target coordinates [7].

$$x_T = \frac{1}{ab-c^2} \sum_{i=1}^{N} p_i \frac{c\cos\phi_i - b\sin\phi_i}{\sigma_{p_i}^2}$$

$$y_T = \frac{1}{ab-c^2} \sum_{i=1}^{N} p_i \frac{a\cos\phi_i - c\sin\phi_i}{\sigma_{p_i}^2}$$

(11.110)

where

$$\tan\phi_i = \frac{y' - y_i}{x - x_i}$$

$$\cos\phi_i = \frac{d_i}{y' - y_T}$$

$$y' = (x_T - x_i)\tan\phi_i + y_i$$

$$d_i = [(x_T - x_i)\tan\phi_i + y_i - y_T]\cos\phi_i$$

$$d_i = a_i x_T + b_i y_T - c_i$$

(11.111)

$$a = \sum_{i=1}^{N} \frac{\sin^2\phi_i}{\sigma_{p_i}^2 D_i^2}$$

$$b = \sum_{i=1}^{N} \frac{\cos^2\phi_i}{\sigma_{p_i}^2 D_i^2}$$

$$c = \sum_{i=1}^{N} \frac{\sin\phi_i \cos\phi_i}{\sigma_{p_i}^2 D_i^2}$$

and D_i is the distance between the true position and the system i.

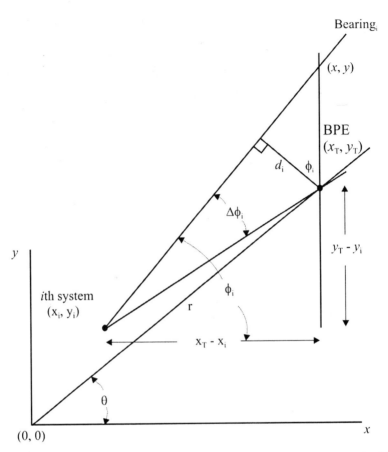

Figure 11.29 Definitions of the terms for derivation of Brown's mean-squared distance algorithm. (*Source:* [9].)

11.3.3 Mean-Squared Distance Algorithm

An algorithm developed by Brown [8] based on an earlier algorithm by Legendre [9] will be presented in this section for calculating the BPE, which is based on minimizing the square of the miss distance of the BPE from the measured lines of position. Refer to Figure 11.29. The direction-finding systems are located at coordinates (x_i, y_i) and the bearing from the ith system is ϕ_i and is shown in Figure 11.29. The BPE is calculated to be (x_T, y_T) and the bearing error for the ith system is $\Delta\phi_i$.

To minimize the sum of the squares of the total miss distance, formulate

$$D = \sum_{i=1}^{N} d_i^2$$
$$= \sum_{i=1}^{N} a_i^2 x_T^2 + \sum_{i=1}^{N} 2a_i b_i x_T y_T - \sum_{i=1}^{N} 2a_i c_i x_T \qquad (11.112)$$
$$+ \sum_{i=1}^{N} b_i^2 y_T^2 - \sum_{i=1}^{N} 2b_i c_i y_T + \sum_{i=1}^{N} c_i^2$$

where

$a_i = \sin \phi_i$
$b_i = -\cos \phi_i$
$c_i = x_i \sin \phi_i - y_i \cos \phi_i$
N = number of LOPs

Setting the first partial derivative of D with respect to x_T and then y_T equal to zero will find the values of x_T and y_T for which the total squared distance is minimized.

$$\frac{\partial D}{\partial x_T} = 0 = 2x_T \sum_{i=1}^{N} a_i^2 + 2y_T \sum_{i=1}^{N} a_i b_i - 2\sum_{i=1}^{N} a_i c_i \qquad (11.113)$$
$$\frac{\partial D}{\partial y_T} = 0 = 2x_T \sum_{i=1}^{N} a_i b_i + 2y_T \sum_{i=1}^{N} b_i^2 - 2\sum_{i=1}^{N} b_i c_i$$

which yield

$$x_T = \frac{\sum_{i=1}^{N} b_i^2 \sum_{i=1}^{N} a_i c_i - \sum_{i=1}^{N} a_i b_i \sum_{i=1}^{N} b_i c_i}{\sum_{i=1}^{N} a_i^2 \sum_{i=1}^{N} b_i^2 - \left(\sum_{i=1}^{N} a_i b_i\right)^2} \qquad (11.114)$$

$$y_T = \frac{\sum_{i=1}^{N} a_i^2 \sum_{i=1}^{N} b_i c_i - \sum_{i=1}^{N} a_i b_i \sum_{i=1}^{N} a_i c_i}{\sum_{i=1}^{N} a_i^2 \sum_{i=1}^{N} b_i^2 - \left(\sum_{i=1}^{N} a_i b_i\right)^2}$$

Hertel extended these results using statistical estimation arguments (Hertel, R., personal communication, 1982) based on linear system theory. The above miss distance for system i is expressed as

$$d_i = a_i x_T + b_i y_T - c_i \tag{11.115}$$

where i is the ith measurement of a line of bearing and a_i, b_i, and c_i are as given above. In matrix form this is

$$\mathbf{D} = \mathbf{HP} - \mathbf{C} \tag{11.116}$$

In this expression

$$\mathbf{C} = \begin{bmatrix} c_0 \\ c_1 \\ c_2 \\ \vdots \\ c_N \end{bmatrix} \quad \mathbf{P} = \begin{bmatrix} x_T \\ y_T \end{bmatrix} \quad \mathbf{H} = \begin{bmatrix} a_0 & b_0 \\ a_1 & b_1 \\ a_2 & b_2 \\ \vdots & \vdots \\ a_N & b_N \end{bmatrix} \quad \mathbf{D} = \begin{bmatrix} d_0 \\ d_1 \\ d_2 \\ \vdots \\ d_N \end{bmatrix}$$

The least-squared error estimator for the target location vector $\mathbf{P}$ is given by [10]

$$\hat{\mathbf{P}} = \left[\mathbf{H}^T \mathbf{R}^{-1} \mathbf{H}\right]^{-1} \mathbf{H}^T \mathbf{R}^{-1} \mathbf{C} \tag{11.117}$$

where, as usual, -1 denotes inverse and T denotes transpose. In this equation, $\mathbf{R}$ is a weighting matrix that is selected to optimize the calculation in some sense. The variance of this estimator is given by [10]

$$\mathbf{Q} = \left[\mathbf{H}^T \mathbf{R}^{-1} \mathbf{H}\right]^{-1} = \begin{bmatrix} \sigma_x^2 & \rho \sigma_x \sigma_y \\ \rho \sigma_x \sigma_y & \sigma_y^2 \end{bmatrix} \tag{11.118}$$

which is a covariance matrix [11]. The EEP parameters are related to the elements of this covariance matrix as follows:

$$L_A = \text{Semimajor axis} = \frac{2(\sigma_x^2\sigma_y^2 - \rho^2\sigma_x^2\sigma_y^2)C^2}{\sigma_x^2 + \sigma_y^2 - \left[(\sigma_y^2 - \sigma_x^2)^2 + 4\rho^2\sigma_x^2\sigma_y^2\right]^{1/2}}$$

(11.119)

$$L_1 = \text{Semiminor axis} = \frac{2(\sigma_x^2\sigma_y^2 - \rho^2\sigma_x^2\sigma_y^2)C^2}{\sigma_x^2 + \sigma_y^2 + \left[(\sigma_y^2 - \sigma_x^2)^2 + 4\rho^2\sigma_x^2\sigma_y^2\right]^{1/2}}$$

$$\tan 2\phi = \frac{2\rho\sigma_x\sigma_y}{\sigma_y^2 - \sigma_x^2}$$

$C = -2\ln(1 - P_e)$

$P_e = $ Probability of being inside

(11.120)

Here, ϕ is the tilt angle of the semimajor axis of the ellipse relative to the x-axis.

The weighting matrix $\mathbf{R}^{-1}$ in the above expression is used to optimize the performance. In one application of this algorithm $\mathbf{R}^{-1}$ is given by

$$\mathbf{R}^{-1} = \frac{1}{\sum_i QF_i} \begin{bmatrix} QF_0 & 0 & 0 & \cdots & 0 \\ 0 & QF_1 & 0 & \cdots & 0 \\ 0 & 0 & QF_2 & \cdots & 0 \\ \vdots & \vdots & \vdots & \ddots & \vdots \\ 0 & 0 & 0 & \cdots & QF_N \end{bmatrix}$$

$$\times \begin{bmatrix} \frac{1}{\sigma_{d_0}^2} & 0 & 0 & \cdots & 0 \\ 0 & \frac{1}{\sigma_{d_1}^2} & 0 & \cdots & 0 \\ 0 & 0 & \frac{1}{\sigma_{d_2}^2} & \cdots & 0 \\ \vdots & \vdots & \vdots & \ddots & \vdots \\ 0 & 0 & 0 & \cdots & \frac{1}{\sigma_{d_N}^2} \end{bmatrix}$$

(11.121)

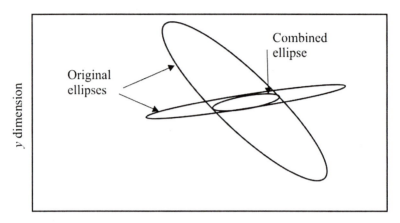

Figure 11.30 Combined error ellipses.

where QF_i is some quality factor associated with measurement i. It might be the variance of the measurement, it could be some higher-order statistic as it makes sense, or it could be a measure of the SNR, for example.

11.3.4 Combining Error Contours

The above algorithms for calculating the BPE of the location on an emitter also calculate an ellipse within which the emitter lies with a specified probability. If there are multiple BPEs, with associated EEPs, for a given emitter it is possible to combine these location estimates and their associated EEPs to refine the BPE and EEP. This is true assuming that the random variables involved have symmetric density functions, which includes the Gaussian case. An example of this is shown in Figure 11.30 where two original EEPs are combined. The resultant EEP is the smallest one shown in Figure 11.30. The original EEPs must be based on the same probability, for example, 50% and 90%. If this is not the case, adjustment of the original contours is necessary—fortunately such computations are simple, as given above. Blachman [12] derived the method for combining the EEPs based on two dimensions. These results were extended by Roecker [13] to include as many dimensions as desired using statistical estimation theory.

If the equations for the original EEPs are given by

Direction-Finding Position-Fixing Techniques

$$a_i x^2 + b_i x y + c_i y^2 - d_i x - e_i y + f_i = 1 \qquad i = 1, 2, \cdots, N \qquad (11.122)$$

with the centers of the ellipses denoted as α_i and β_i, the length of the semimajor axis denoted as L_{Ai}, the semiminor axis length denoted by L_{li}, and the angle of the semimajor axis with the x-axis given by θ_i, the parameters in this equation are:

$$a_i = \frac{\cos^2 \theta_i}{L_{Ai}^2} + \frac{\sin^2 \theta_i}{L_{li}^2} \qquad b_i = 2\cos\theta_i \sin\theta \left(\frac{1}{L_{Ai}^2} - \frac{1}{L_{li}^2} \right)$$

$$d_i = 2a_i \alpha_i + b_i \beta_i \qquad (11.123)$$

$$e_i = b_i \alpha_i + 2 c_i \beta_i \qquad c_i = \frac{\sin^2 \theta_i}{L_{Ai}^2} + \frac{\cos^2 \theta_i}{L_{li}^2}$$

$$f_i = a_i \alpha_i^2 + b_i \alpha_i \beta_i + c_i \beta_i^2$$

The parameters for the composite ellipse are determined from these parameters as follows. The equation for the composite ellipse is

$$a x^2 + b x y + c y^2 - d x - e y + f = 1 \qquad (11.124)$$

with

$$a = \sum_{i=1}^{N} a_i \qquad e = \sum_{i=1}^{N} e_i$$

$$b = \sum_{i=1}^{N} b_i \qquad \alpha = \frac{2cd - be}{4ac - b^2} \qquad (11.125)$$

$$c = \sum_{i=1}^{N} c_i \qquad \beta = \frac{2ae - bd}{4ac - b^2}$$

$$d = \sum_{i=1}^{N} d_i \qquad \theta = \frac{1}{2} \arctan \frac{-b}{c - a}$$

$$f = a\alpha^2 + b\alpha\beta + c\beta^2$$

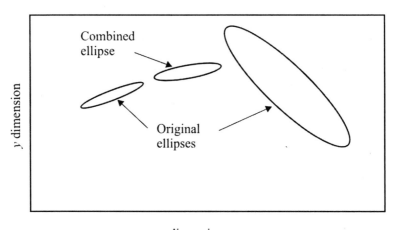

Figure 11.31 Combining error ellipses when the two original ellipses do not overlap can yield unexpected results as shown here.

The major and minor axes of the combined ellipse are given by

$$L_A = \sqrt{\frac{2}{a+c-\sqrt{(a-c)^2+b^2}}}$$
$$L_1 = \sqrt{\frac{2}{a+c+\sqrt{(a-c)^2+b^2}}}$$
(11.126)

When the original ellipses overlap considerably as shown in Figure 11.30, the composite ellipse will be largely contained within the region of overlap of the original ellipses. If not, however, the combined ellipse can fall outside the confines of the original EEPs as shown in Figure 11.31. The combined ellipse, however, always has a smaller major and minor axis than the original EEPs. Therefore, as long as the above conditions are met, combining ellipses will always produce better (smaller) error contours for a given probability.

11.4 SINGLE-SITE LOCATION TECHNIQUES

The source of HF signals propagating via sky-wave paths can be located by triangulation using any of the techniques discussed previously. However, it is also possible to use a single ES system for this purpose under some circumstances. If the elevation of the ionospheric layer that the HF signal is refracting through, or, more accurately, the equivalent height of the layer that is reflecting the signal is known, as shown in Figure 11.32, then the emitter can be located since we know the range to the target and its angle of arrival. The wave is assumed to be reflected at the midway point between the transmitter and the ES system. The elevation angle is related to the range and ionospheric height by

$$\tan \theta = \frac{R/2}{h} \qquad (11.127)$$

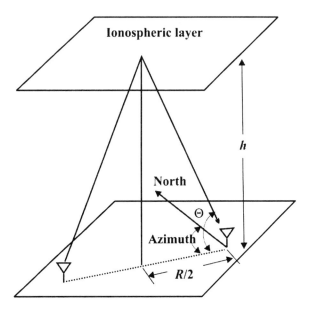

Figure 11.32 Single-site location of HF emitters.

So

$$R = 2h\tan\theta \qquad (11.128)$$

Thus, by measuring the elevation AOA, the range to the target can be estimated. Although this derivation was given using planes for the ionosphere and the Earth, it can also be derived using spherical surfaces, and is more accurate.

The devices used to measure the height of the ionosphere versus frequency are called *ionospheric sounders*. Typically a swept signal is radiated straight up for *vertical sounders* or at an angle for *oblique sounders*. The time of the return reflection is compared to when the signal was transmitted to ascertain the ionospheric height. The resultant display of the sounder results are called *ionograms*.

Vertical sounders, of course, measure the ionospheric height directly overhead. It is frequently assumed that the ionosphere is homogeneous, and the heights are the same throughout the region. The height is usually measured at the site of the ES system and it is assumed to be the same height at the reflection point. This is rarely true, so such techniques for measuring the location of transmitters are not that accurate. EEPs with axes that are 10% of the range to the target are typical. Thus, if the target is 100 km from the ES system, the major axis of the EEP can be 10 km or more. It should also be noted that although the reflecting surface discussed here was the ionosphere, any reflecting surface could also be used in some circumstances as long as the distance between that surface and the direction-finding system is known.

11.5 FIX ACCURACY

Like any system that measures some physical phenomena, direction-finding systems always exhibit some degree of error. This error is typically both systematic and operational. The first represents errors that normally can, for the most part, be removed by calibration. The latter are caused by some characteristic of the environment, and normally cannot be removed by calibration. These problems are solvable only by careful selection of the siting for the system.

The problem is to compute a fix on the emitter based on the LOPs computed at several separate physical locations. These LOPs could be obtained from different systems deployed on the ground or from an aircraft that is moving. Regardless, the bearings do not normally cross at a point, but describe a region. Stansfield [7] called these regions "cocked hats."

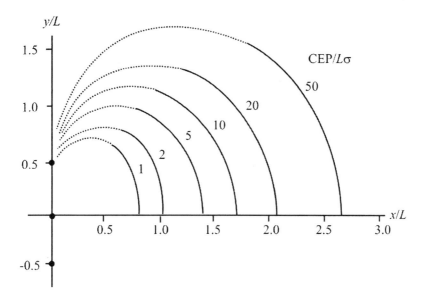

Figure 11.33 Contours of constant CEP/$L\sigma$ for a linear baseline. (*Source:* [14], © 1984, IEEE. Reprinted with permission.)

The configuration of the ES system baseline can affect the accuracy of the fixes computed by triangulation. This is referred to as *geometric dilution of precision* (GDOP). The effects can be seen by considering Figure 11.33 [14]. In this case a linear baseline is shown, with three ES systems placed at $y/L = -0.5, 0,$ and 0.5, where L refers to the length of the baseline in consistent units. For this chart it is assumed that the ES system is located close to the Earth's surface so that $1/R^4$ propagation characteristics predominate. Note that because of the baseline configuration, it is only necessary to show the first quadrant—the fourth quadrant is symmetric.

Where x/L is approximately 1.0 and y/L is approximately 0.5, the CEP/$L\sigma = 2$, so if the baseline length $L = 20$ km and the system bearing standard deviation $\sigma = 5° = 0.09$ radians, then the accuracy of the fix is CEP = 2×20 (km) $\times 0.09$ (radians) = 3,600m. As the target moves further off the perpendicular bisector of the baseline, the accuracy degrades—the GDOP effect. Off the end of the baseline all of the bearings are the same and therefore do not intersect at a single point. In that case there is no solution and the CEP or EEP is unbounded.

The dotted portion of these curves represent regions where the CEP is not a very good measure of the accuracy. In that case the two eigenvalues of the covariance matrix, λ_1 and λ_2 (for a two-dimensional PF calculation) are such that one is too large compared with the other, $\lambda_2/\lambda_1 < 0.01$. The calculation is changed to

$$\frac{L_A}{3.552\sqrt{\kappa L\sigma}} = \text{constant} \qquad (11.129)$$

where

$$L_A = 2\sqrt{\kappa\lambda_1} \qquad (11.130)$$

is the length of the major axis of the EEP and

$$\kappa = -2\ln(1-P_e) \qquad (11.131)$$

when P_e is the probability of being inside the EEP or CEP [14].

To illustrate the significance of these error bounds, shown in Figure 11.34 is a chart of the CEP of an emitter position fix when the emitter is located on the perpendicular bisector to the baseline. In this case the baseline is linear as shown in Figure 11.33, and it is 30 km long from end to end. Shown is the 90% CEP as a function of the distance from the baseline and the accuracy of the direction-finding systems.

The effects of the GDOP can be seen in Figure 11.35, where a similar chart as that in Figure 11.34 is shown except that in this case the radial from the baseline is at a 45° angle from the center EW system. It is clear that off axis, the accuracy is degraded.

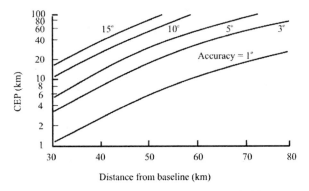

Figure 11.34 CEP versus range from the baseline on a perpendicular bisector.

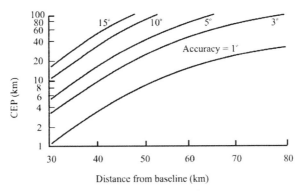

Figure 11.35 CEP versus range from the baseline on a 45° locus from the center of the baseline.

A set of curves for a concave baseline is shown in Figure 11.36. The effects of baseline extensions are clear in this case as well. Extensions of the individual baselines made up by any pair of ES systems causes unbounded CEPs. Off the end of these baselines, even close fixes can have very large errors.

The accuracy of the position fixes depends on the SNR at the ES system, as discussed previously. Elevating the ES system therefore normally will tend to improve the fix accuracy because the received power, given by (2.43), increases

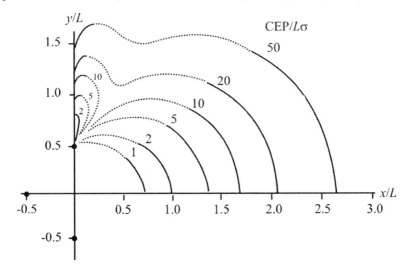

Figure 11.36 Contours of constant CEP/$L\sigma$ where the baseline is in the shape of a V. (*Source*: [14], © 1984, IEEE. Reprinted with permission.)

with antenna height while the noise tends to be the same or less. The noise can be less because some of it is caused by EMI generated by other electronic devices in the EW system. Raising the antenna moves it further away from these devices.

11.6 FIX COVERAGE

Deployment of both ground-based and airborne EW systems is limited by several, but normally different, factors. Ground system deployment is largely dictated by the terrain available for deployment, with higher ground normally being desired so the coverage area is maximized. Airborne systems are limited by the weather as well as flight restrictions to avoid other air traffic.

High ground is not exclusively the province of EW systems. Other operational systems need such terrain as well, such as communication relays and ground surveillance radars. Therefore, planning on interference from other systems sharing the same frequency bands is wise.

It is important to keep in mind that, as opposed to determining the coverage area for signal intercept where the area is given by the disjunction of the coverage area of each individual system, the coverage area for geolocation is determined by the conjunction of the coverage area of two or more systems. The movement of one system to a different location can also achieve two or more system locations, of course. Thus, the geolocation coverage area of a suite of EW systems will always be less than the intercept coverage area.

Deployment of EW systems can influence the accuracy that can be achieved for locations of targets. The area to be covered can define better baselines for the bearing collection. In general, it is best for location accuracy purposes to keep the target region of interest along the perpendicular bisector of the pseudobaseline which the EW systems make. This is obvious from Figures 11.33 and 11.36, which show the effects of GDOP. The best achievable accuracy is on this bisector, while the worst is off the ends of the pseudobaseline.

When computing the BPE using LOPs, it is best to get the LOPs to cross at as close to a $\pi/4$ radian angle as possible. At angular crossings less than that, the targets are farther away than desirable. Equivalently, the baseline is shorter than desired. At angular crossings greater than that, the target is too close to the baseline and the GDOP will increase (in the limit the target lies on the pseudo-baseline and the accuracy goes to zero, which means the error goes to infinity). To achieve these angular specifications, an airborne system should not fly in a straight line if it is the only ES system being considered. It should fly in a semicircular pattern around the geographical region of interest as illustrated in Figure 11.37. The equivalent situation when ground systems are being considered is shown in

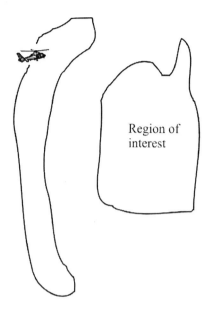

Figure 11.37 A low-flying aircraft flying a path that, as much as possible, surrounds the area of interest.

Figure 11.38 where the baseline is shaped more or less in a semicircular arrangement.

For airborne systems that obviously are moving, a single platform can collect several bearings in time succession. If targets are emitting for enough time, then such approaches produce enough bearings to compute a fix. For frequency agile targets this may not be the case. When UAVs are employed as EW sensor systems, then overflight of the target area is possible. This not only improves coverage of especially low-power targets, it also improves direction-finding geometry if employed correctly.

For ground-based situations, the collection systems are not necessarily moving when collecting data. Therefore, more than one EW system is typically necessary to collect bearings simultaneously with other systems. The most general configuration would consist of both ground systems and airborne systems operating simultaneously. (For ship applications at sea and line-of-sight EW, only the airborne case would normally apply since if EW is executed from a ship, the target is much closer than the ship commander would likely tolerate. For HF, these descriptions do apply since HF signals can propagate much further.)

Irrespective of whether the EW system is airborne or ground, if the area of interest is broad and shallow, then the best baseline configuration is one that is

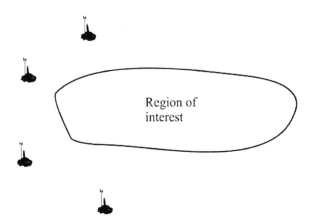

Figure 11.38 If the target area is narrow and deep, the EW systems should be configured in a concave configuration to maximize position fix accuracy.

convex, as shown in Figure 11.39. This configuration provides better coverage from three of the four EW systems on each side while sacrificing the bearing from the fourth system for the width of coverage. In essence this configuration tries to place a straight baseline against the targets in one part of the coverage area.

On the other hand, if the area to be covered is narrow and deep, then a concave baseline is best as shown in Figure 11.38. Here all four bearings are used, sacrificing the width of coverage.

When there are only three ground systems available and the coverage area is nonlinear, then the quality of coverage is going to be somewhat less than that available with standoff sensor systems. The term *nonlinear* in this case refers to when friendly forces are deployed intermixed with both neutral population and enemy forces. Lacking any information about specific areas to be covered it can be shown that the best arrangement to cover the whole, more or less circular, area of interest is an equilateral triangle as shown in Figure 11.40. However, due to the GDOP effects described above, fix accuracy in line with any of the baselines formed with the systems will degrade as shown in the figure. Augmenting this arrangement with a direction-finding system on a UAV linked with the three ground systems will significantly improve the coverage accuracy as well as the coverage area.

Bordering on optimal direction-finding geometries can be obtained with two or more standoff aircraft, be they manned fixed wing, manned rotary winged, or UAV with UAV augmentation. The UAV in this case is assumed to be capable of overflight of the target area and the geometry shown in Figure 11.41 is possible. Coverage areas are extended due to the overflight and therefore close proximity of

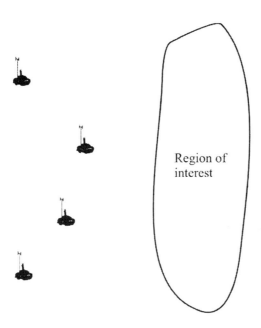

Figure 11.39 If the area of interest is shallow and long, the best configuration for EW systems is a convex baseline as shown here.

the UAV. In addition, since it is closer to the targets, whatever LOP inaccuracy is exhibited is somewhat ameliorated due to its close range to the targets. Of course, the same GDOP concerns as discussed above are still present on the baselines between any two of the systems shown.

11.7 CONCLUDING REMARKS

Several techniques for computing an LOP were presented in this chapter. Most of the techniques measure the phase difference at two or more antennas, the time of arrival differences at these antennas, or the amplitude comparisons at two or more antennas. Alternately the Doppler frequency shift of a rotating antenna compared to a stationary antenna is measured. Whatever the technique, the bearing angle to the target is estimated.

Systematic accuracy of practical implementations of these approaches can be as good as 1° or 2° or less when calibration of the antennas is performed. Operational accuracies are typically somewhat worse than this due to the various

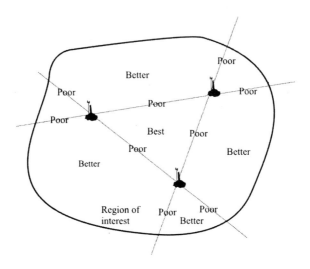

Figure 11.40 Expected quality of direction-finding coverage with three ground systems in a nonlinear battlespace.

operational scenarios involved. Three position-fixing algorithms were also presented that utilize the LOPs from two or more collection sites to compute the fix of the target.

In the HF frequency range where long-range communications relies on reflections of the signals from the ionosphere, it is possible to compute a location

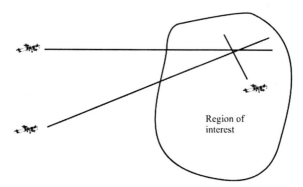

Figure 11.41 Configuration of three airborne direction-finding systems when two are standoff and one is overflight of the region of interest.

from a single system as long as the height of the reflection point in the ionosphere is known or estimated. This technique measures the range and azimuth to the target. Accuracy of doing so is about 10% of range. Alternately or in addition, two or more of these systems can be netted just as in the higher frequency ranges.

References

[1] Pender, H., and K. McIlwain, *Electrical Engineers Handbook: Electric Communication and Electronics,* 4th ed., New York: John Wiley & Sons, 1963, p. 1-16.

[2] Torrerri, D. J., *Principles of Secure Communication Systems,* 2nd ed., Norwood, MA: Artech House, 1992, p. 353.

[3] Wiley, R. G., *Electronic Intelligence: The Interception of Radar Signals,* Dedham, MA: Artech House, Inc., 1985, p. 102.

[4] Gething, P. J. D., *Radio Direction Finding and Superresolution,* London: Peter Peregrinus Ltd., 1991.

[5] Schmidt, R. O., "Multiple Emitter Location and Signal Parameter Estimation," *Proceedings of the RADC Spectrum Estimation Workshop,* Griffiths Air Force Base, Rome, NY, 1979, pp. 243–258; republished in *IEEE Transactions on Antennas and Propagation,* Vol. AP-34, No. 3, March 1986, pp. 276–280.

[6] Schmidt, R. O., "Least Squares Range Difference Location," *IEEE Transactions on Aeropace and Electronic Systems,* Vol. 32, No. 1, Jan 1996, pp. 234–242.

[7] Stansfield, R. G., "Statistical Theory of D.F. Fixing," *J. IEE.,* Vol. 94, Part IIIa, 1947, pp. 762–770.

[8] Brown, R. M., *Emitter Location Using Bearing Measurements from a Moving Platform,* NRL Report 8483, Naval Research Laboratory, Washington, D.C., June 1981.

[9] Legendre, A., "Nouvelles Methods Pour la Determination des Orbites des Cometes," Paris, 1805, pp. 72–75.

[10] Sage, A. P., and J. L. Melsa, *Estimation Theory with Applications to Communications and Control,* New York: John Wiley & Sons, 1983, pp. 237–239 and pp. 244–245.

[11] Foy, W. H., "Position Location Solutions by Taylor Series Estimation," *IEEE Transactions on Aerospace and Electronic Systems,* March 1976, pp. 187–193.

[12] Blachman, N. M., "On Combining Target-Location Ellipses," *IEEE Transactions on Aerospace and Electronic Systems,* Vol. AES-27, No. 2, March 1989, pp. 284–287.

[13] Roecker, J. A., "On Combining Multidimensional Target Location Ellipsoids," *IEEE Transactions on Aerospace and Electronic Systems,* Vol. 27, No. 1, January 1991, pp. 175–176.

[14] Torreiri, D. J., "Stastical Theory of Passive Location Systems," *IEEE Transactions on Aerospace and Electronic Systems,* Vol. AES-20, No. 2, March 1984, pp. 183–197.

Chapter 12

Hyperbolic Position-Fixing Techniques

12.1 INTRODUCTION

When the *time of arrival* (TOA), the TDOA (denoted here by τ), and/or the frequency difference of arrival ($\dot{\tau}$), or differential Doppler by another appellation, measured at two or more widely dispersed and moving sensors, are used to geolocate emitters, then quadratic LOPs result. Generally, the intersection of these curves is taken as the emitter location. In order for there to be frequency differences caused by movement, either the sensor (one or more) or the target, or both, must be in motion. Therefore, such sensor systems are typically mounted on airborne platforms. For those interested in more details about hyperbolic PF processing, [1–19] are recommended.

12.2 TIME DIFFERENCE OF ARRIVAL

The arrival time of a signal at two or more dispersed sensors can be used to estimate the geographic location of the emitting target 20]. The geometry is shown in Figure 12.1, where, for simplicity, only two receiving systems are shown. Since the transmitter and/or the receiving systems can be elevated, r_1 and r_2 are slant ranges between the transmitter and the sensor systems.
These distances can be expressed as

$$r_i = c\, t_i \qquad i = 1,2 \tag{12.1}$$

where c is the speed of propagation of the signal, normally for communication signals in the ether assumed to be the speed of light, and t_i is the time between

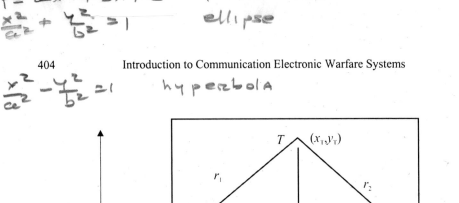

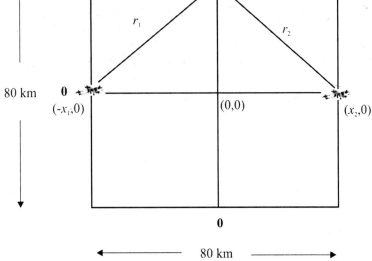

Figure 12.1 Example scenario for illustrating the concepts of τ and $\dot{\tau}$ emitter location.

when the signal leaves the transmitter and when it arrives at the sensor. The τ is the time difference between when the signal arrives at one receiving site and the other, namely,

$$\tau = t_2 - t_1 = \frac{r_2}{c} - \frac{r_1}{c} = \frac{1}{c}(r_2 - r_1) = \text{CONSTANT} \qquad (12.2)$$

From Figure 12.1,

$$r_i = \sqrt{(x_T - x_i)^2 + y_T^2} \qquad i = 1,2 \qquad (12.3)$$

so that

$$\tau = \frac{1}{c}\left[\sqrt{(x_T - x_1)^2 + y_T^2} - \sqrt{(x_T - x_2)^2 + y_T^2}\right] \qquad (12.4)$$

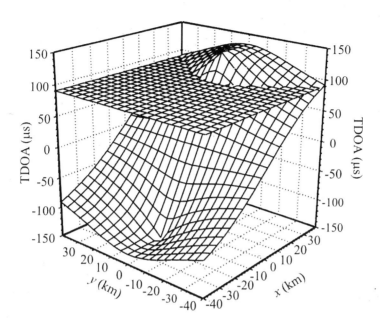

Figure 12.2 Intersection of the τ curve with the plane defined by a constant τ curve of 90 μs forms a parabolic curve.

The curve defined by this expression is a parabola. Since t_2 can be less than or greater than t_1, the τ can be greater than or less than zero. The τ contour for the scenario shown in Figure 12.1 is shown in Figure 12.2. The τ varies between about −130 μs to +130 μs. Also shown is a constant τ curve which is a plane; in this case $\tau = 90$ μs. The intersection of these curves forms a parabola as shown. As visualized from above, several of these curves are shown in Figure 12.3.

Clearly only two sensor systems cannot produce a unique geolocation instantaneously. To accomplish this at least three sensor systems are required. Suppose there are S sensors available for computation of the position fix [21]. As above, the governing equations are given by

$$d_i - d_j = \|\mathbf{r}_i - \mathbf{r}_T\| - \|\mathbf{r}_j - \mathbf{r}_T\| = c(t_i - t_j) = ct_{i,j} \qquad i, j = 0, 1, \cdots, S-1 \qquad (12.5)$$

for all pairs of sensors, (i, j).

Without loss of generality, assume that all of the arrival times are compared with the arrival time at a sensor located at coordinates (0,0) as shown in Figure

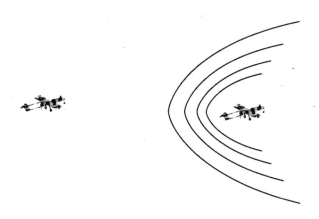

Figure 12.3 Lines of constant τ are parabolas.

12.4. Thus, the time differences of arbitrary (i, j) are not used, just $(i, 0)$ for all i. The equation in this case reduces to (because $r_j = 0$)

$$d_i = \|\mathbf{r}_i - \mathbf{r}_T\| = c(t_{i,0} + t_0) \qquad (12.6)$$

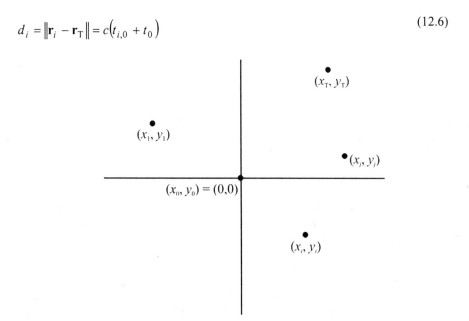

Figure 12.4 Sensor grid and target in two dimensions.

Squaring (12.6) yields

$$(x_i - x_T)^2 + (y_i - y_T)^2 + (z_i - z_T)^2 = c^2(t_{i,0} + t_0)^2 \qquad (12.7)$$

Expanding

$$\begin{aligned} x_i^2 - 2x_i x_T + x_T^2 + y_i^2 - 2y_i y_T + y_T^2 + z_i^2 - 2z_i z_T + z_T^2 \\ = c^2\left(t_{i,0}^2 + 2t_{i,0} + t_0^2\right) \end{aligned} \qquad (12.8)$$

At the reference site at (0,0)

$$\|\mathbf{x}\|^2 = \mathbf{x}^T \mathbf{x} = x_T^2 + y_T^2 + z_T^2 = c^2 t_0^2 \qquad (12.9)$$

Subtracting (12.9) from (12.8)

$$x_i^2 + y_i^2 + z_i^2 - 2(x_i x_T + y_i y_T + z_i z_T) = c^2 t_{i,0}^2 + 2c^2 t_{i,0} \qquad (12.10)$$

or

$$\begin{bmatrix} x_i & y_i & z_i \end{bmatrix} \begin{bmatrix} x_T \\ y_T \\ z_T \end{bmatrix} + ct_{i,0}\sqrt{x_T^2 + y_T^2 + z_T^2} = \\ -\frac{1}{2}c^2 t_{i,0}^2 + \frac{1}{2}\left(x_i^2 + y_i^2 + z_i^2\right) \qquad (12.11)$$

which utilized the identity

$$ct_0 = \sqrt{x_T^2 + y_T^2 + z_T^2} \qquad (12.12)$$

Putting this in matrix form

$$\mathbf{p}_i \mathbf{x}_T + ct_{i,0}\|\mathbf{x}_T\| = -\frac{1}{2}ct_{i,0}^2 + \frac{1}{2}\|\mathbf{p}_i\|^2 \qquad (12.13)$$

where $\mathbf{p}_i$ is the position vector of sensor i.

Extending this result for all of the $i = 1,\ldots, S$ sensors yields

$$\mathbf{P}\mathbf{x}_T + \mathbf{c}\|\mathbf{x}_T\| = \mathbf{d} \tag{12.14}$$

where

$$\mathbf{P} = \begin{bmatrix} x_1 & y_1 & z_1 \\ x_2 & y_2 & z_2 \\ \vdots & \vdots & \vdots \\ x_{S-1} & y_{S-1} & z_{S-1} \end{bmatrix} \qquad \mathbf{x}_T = \begin{bmatrix} x_T \\ y_T \\ z_T \end{bmatrix}$$

$$\mathbf{c} = c\mathbf{t} \qquad \mathbf{t} = \begin{bmatrix} t_{1,0} \\ t_{2,0} \\ \vdots \\ t_{S-1,0} \end{bmatrix} \tag{12.15}$$

$$\mathbf{d} = \frac{1}{2}\operatorname{diag}\left(\mathbf{P}\mathbf{P}^T - c^2 \mathbf{t}\mathbf{t}^T\right)$$

When $S = 4$ and $\mathbf{P}$ is nonsingular, then $\mathbf{x}_T$ can be determined as

$$\mathbf{x}_T = \mathbf{P}^{-1}(\mathbf{d} - \mathbf{c}\|\mathbf{x}_T\|) \tag{12.16}$$

Substituting (12.3) into (12.1) yields

$$\|\mathbf{x}_T\|^2 = \mathbf{x}^T\mathbf{x} = \left\{\mathbf{P}^{-1}(\mathbf{d} - \mathbf{c}\|\mathbf{x}_T\|)\right\}^T \left\{\mathbf{P}^{-1}(\mathbf{d} - \mathbf{c}\|\mathbf{x}_T\|)\right\} \tag{12.17}$$

Let $a = \|\mathbf{x}_T\|$ and $\mathbf{Q} = (\mathbf{P}\mathbf{P}^T)^{-1}$; then this expression reduces to

$$\left(\mathbf{c}^T\mathbf{Q}\mathbf{c} - 1\right)a^2 - 2\mathbf{d}^T\mathbf{Q}\mathbf{c}a + \mathbf{d}^T\mathbf{Q}\mathbf{d} = 0 \tag{12.18}$$

which is a quadratic equation that can easily be solved for a which is the range of the target from the origin. Substituting this range back into (12.14) will solve the problem for the target location. Two results ensue, so other information must be used to determine which is correct. Frequently one answer is negative and since a represents the range to the target, the positive root is the correct one. Otherwise, other information must be employed.

In general $S > 4$, so this is an overdetermined system of equations. Instead of the inverse, then, the pseudoinverse

$$\mathbf{P}^\Diamond = (\mathbf{P}^T\mathbf{P})^{-1}\mathbf{P}^T \tag{12.19}$$

above instead of the regular inverse. The pseudoinverse solves (12.14) in the least squares sense (see Section 11.2.6.1).

The Cramer-Rao bound on parameter estimation is a frequently used measure of how well such a parameter can be measured. Under some reasonable assumptions it represents the best that can be obtained under those assumptions. The Cramer-Rao bound for estimating the TOA of a signal at a sensor is given by [22]

$$\sigma_t^2 \geq \left(\frac{2E}{N_0}\beta^2\right)^{-1} \tag{12.20}$$

where σ_t^2 is the variance of the measurement, and is given in seconds squared; in this case, E is the energy in the signal (watt-seconds, for example), N_0 is the level of noise per unit bandwidth present (watts per hertz, for example), and β is a measure of the bandwidth occupied by the signal, measured in radians. This equation reduces to

$$\sigma_t = \frac{1}{\beta}\frac{1}{\sqrt{BT\gamma}} \tag{12.21}$$

where B is the noise bandwidth of the receivers, T is the integration time, and γ is the effective input SNR at the two sensor sites.

β is the rms radian frequency, which is a measure of the bandwidth of the signal and is given by

$$\beta = 2\pi\left[\frac{\int_{-\infty}^{\infty} f^2|S(f)|^2\,df}{\int_{-\infty}^{\infty}|S(f)|^2\,df}\right]^{1/2} \tag{12.22}$$

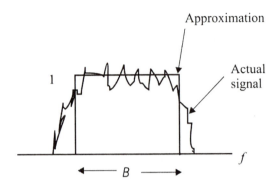

Figure 12.5 Example of an ideal signal that approximates a communication signal.

where $S(f)$ is the spectrum of the signal. For an ideal signal with sharp edges in the frequency spectrum as shown in Figure 12.5, $\beta = 2\pi B/\sqrt{3}$ (where B is in hertz), for example.

Thus, γ is a composite SNR at the two sensors. If γ_1 and γ_2 are the SNRs at the two sensors, then γ is given by

$$\frac{1}{\gamma} = \frac{1}{2}\left[\frac{1}{\gamma_1} + \frac{1}{\gamma_2} + \frac{1}{\gamma_1\gamma_2}\right] \tag{12.23}$$

12.3 DIFFERENTIAL DOPPLER

The frequency difference of arrival, or differential Doppler, for all practical cases of interest herein, where all sensor and/or target velocities are much smaller than the speed of light, is given by

$$\dot{\tau} = \frac{v_2}{c}f_0 - \frac{v_1}{c}f_0 = \frac{f_0}{c}(v_2 - v_1) \tag{12.24}$$

where v_i represents the instantaneous velocity of the sensors relative to the transmitter in the radial direction and f_0 is the frequency of the transmitted signal. Thus,

$$\dot{\tau} = \frac{f_0}{c}\left(\frac{dr_2}{dt} - \frac{dr_1}{dt}\right) \tag{12.25}$$

where r_i is the range between the transmitter and the sensor, as shown in Figure 12.1; thus, dr_i/dt is the rate of change of the range in the radial direction. From (12.3),

$$\frac{dr_i}{dt} = \frac{d\left[(x_T - x_i)^2 + y_T^2\right]^{1/2}}{dt}$$

$$= \frac{(x_T - x_i)}{\left[(x_T - x_i)^2 + y_T^2\right]^{1/2}} \frac{dx_i}{dt} \qquad i = 1, 2 \tag{12.26}$$

where it is assumed that the aircraft are flying at the same speed parallel to the x-axis so that $dy_i/dt = 0$. Denoting $v = v_1 = dx_1/dt = v_2 = dx_2/dt$, then

$$\dot{\tau} = \frac{f_0 v}{c}\left\{\frac{(x_T - x_1)}{\left[(x_T - x_1)^2 + y_T^2\right]^{1/2}} - \frac{(x_T - x_2)}{\left[(x_T - x_2)^2 + y_T^2\right]^{1/2}}\right\} \quad \text{Hz} \tag{12.27}$$

The surface formed by this expression for the example here is shown in Figure 12.6. Note that the largest differential Doppler frequency is approximately 6.5 Hz or so. This is not much of a frequency difference to try to measure out of 100 MHz. The example considered here uses parameters that are typical for slow flying applications. That is part of the reason that the differential Doppler values are so low—the velocity of the EW system is small (recall that for zero velocity the differential Doppler is also zero). So the lower the velocity, the lower the differential Doppler component. Faster-moving EW systems improve the measurable values of differential Doppler.

The differential Doppler curves are complex quadratic functions. Their form can be seen by examining Figure 12.7 where the above surface is viewed from almost directly overhead and the intersection of the $\dot{\tau} = 4.5$ Hz with the surface contour is shown. When visualized from above, the $\dot{\tau}$ curves look like those in Figure 12.8. Again, the $\dot{\tau}$ from only two sensor platforms do not yield unique fixes. The Cramer-Rao bound for estimating $\dot{\tau}$ is given by

$$\sigma_f = \frac{1}{T_e}\frac{1}{\sqrt{BT\gamma}} \tag{12.28}$$

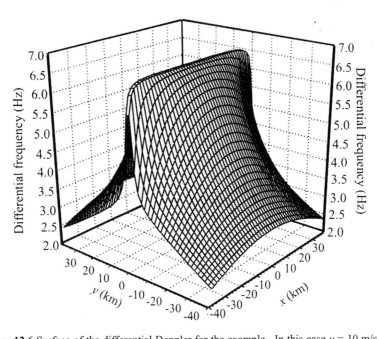

Figure 12.6 Surface of the differential Doppler for the example. In this case $v = 10$ m/s and $f = 100$ MHz ($\lambda = 3$m).

where B, T and γ are as above, σ_f is the standard deviation of the $\dot{\tau}$ measurement, and T_e is the rms integration time given by

$$T_e = 2\pi \left[\frac{\int_{-\infty}^{\infty} t^2 |u(t)|^2 dt}{\int_{-\infty}^{\infty} |u(t)|^2 dt} \right]^{1/2} \tag{12.29}$$

where $u(t)$ is the probability density function of the integration time. Again, for example, if the actual integration time is T, then $T_e = 2\pi T/\sqrt{3}$.

Hyperbolic Position-Fixing Techniques 413

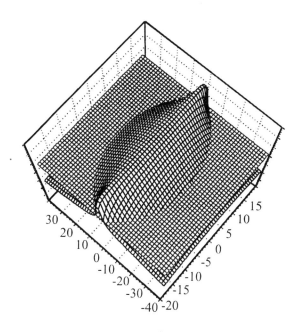

Figure 12.7 Top view of the $\dot{\tau}$ contour. The shape resembles an ellipse, but actually it is a complex quadratic.

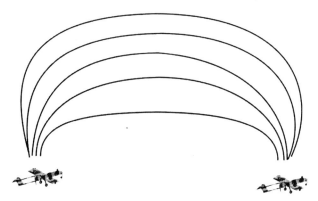

Figure 12.8 Lines of constant $\dot{\tau}$ are complex quadratic curves.

12.4 CROSS-AMBIGUITY FUNCTION PROCESSING

An alternative way to compute the τ and $\dot{\tau}$ for communication targets is with the *cross-ambiguity function* (CAF). This computation simultaneously yields the τ and $\dot{\tau}$ for two sensors, and the CAF must be computed for each pair in order to yield sufficient information to calculate a PF. The CAF is a generalization of the cross-correlation function and is given by

$$\text{CAF}(\tau,t) = \int_0^T s_1(t) s_2(t+\tau) e^{-j\omega t} \, dt \qquad (12.30)$$

A three-dimensional graphical plot of the magnitude of the ambiguity function shows the amplitude of the spectrum on one axis, as the dependent variable, versus the two independent variables on the other two axes, given by the τ and $\dot{\tau}$. With calculation of the ambiguity function, highly precise geolocation of targets can be obtained. A sketch of an example ambiguity function is shown in Figure 12.9. The magnitude of the ambiguity function is searched to ascertain where the highest peak lies. The resultant values of τ and $\dot{\tau}$ where this peak occurs is assumed to be those corresponding to the emitter. The width of the largest peak are measures of the standard deviations of the τ and $\dot{\tau}$ as illustrated in Figure 12.9.

Neither the τ nor the $\dot{\tau}$ yields unique solutions with two sensors. There are an infinite number of points that lead to the same τ and $\dot{\tau}$ values. This is

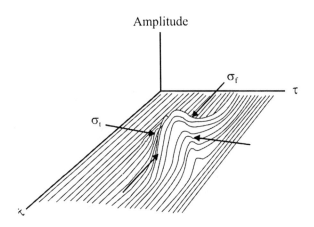

Figure 12.9 Example of the magnitude of a CAF.

illustrated in Figures 12.3 and 12.8. When the τ and $\dot{\tau}$ are combined, there no longer is an infinite number of points, but there are still more than one. Shown in Figure 12.10 is an example solution of computing an emitter location using $\dot{\tau}$ and τ combined. With just the two platforms shown, the emitter could be at either point 1 or 2 and the computed solutions would be the same.

There are ways to resolve this ambiguity, one of which is to add a third sensor. Alternately, as long as the emitter stays transmitting long enough, one or both of the two sensors could move to another location and the process repeated, yielding a unique solution as illustrated in Figure 12.11. The contours computed with the second baseline would cross at either point 1 or 2, but not both. The second sensor moved to another position, or there is a third sensor in the formation, so that a second baseline can be formed. In this example, the emitter is at location 1, and that is where the contours intersect for both baselines, whereas for position 2, the intersection of the contours is different. The tradeoffs here are time or a third sensor platform.

Of course, adding a third system to the configuration adds more than one more baseline. Only two baselines are needed, however, for the process just described. Operationally, the best baselines to use would be determined by the configuration relative to the location of the target. Some pairs of baselines would produce better results than others.

Another way to resolve the ambiguities is to use other *direction-finding* (DF) techniques to indicate which computed location corresponds to the target. Such

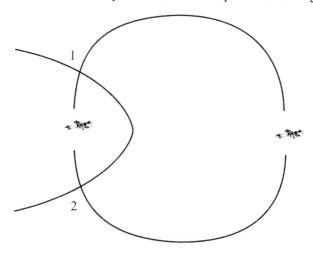

Figure 12.10 Combining the τ solution with the $\dot{\tau}$ solution does not yield a unique location with two platforms.

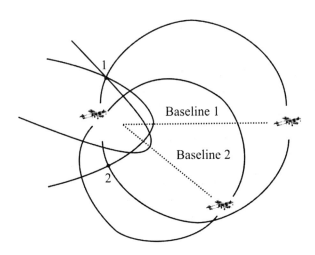

Figure 12.11 Adding a second baseline removes the ambiguity inherent with only a single baseline.

techniques are typically less accurate than $\tau/\dot{\tau}$ and would therefore not normally be used in the actual calculation of the BPE.

12.4.1 Position Fix Accuracy

As an example, suppose $B = 25$ kHz, the channel width in the low VHF frequency band. Further suppose that $\gamma_1 = 10$ dB (10) and $\gamma_2 = 15$ dB (32). Let $T = 100$ ms Then $\beta = 90.7 \times 10^3$ radians per seconds, $T_e = 0.363$ second and $\gamma = 12$. Therefore, $\sigma_t = 63$ ns and $\sigma_f = 16.1$ mHz. To put this in perspective, signals propagate at about the speed of light, 3×10^8 m/s. Thus σ_t of this magnitude equates to an area of uncertainty in the τ dimension of about 20m. In the Doppler direction the interpretation is not quite so straightforward.

While subject to the same detrimental factors as interferometers and other DF systems, especially the degradation due to multipath reflections, particularly in the ground applications, τ and $\dot{\tau}$ systems have been demonstrated to achieve considerably better accuracy than typical DF systems. This is accomplished, however, at the expense of the time necessary to obtain results. The integration time is a factor in the denominator in both of the above equations. Thus, increasing this time will decrease the standard deviation of the resulting position fix. It is not unusual for the integration time to increase by two orders of magnitude to obtain such precision results.

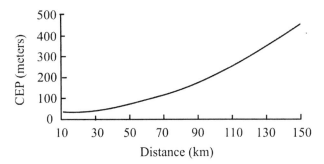

Figure 12.12 CEP as a function of distance from the sensor baseline on the perpendicular bisector.

The standard deviation of the measurements of the fix coordinates, calculated on the perpendicular bisector of the baseline shown in Figure 12.1, depends on

$$\sigma_x = \frac{c\sigma_t \sqrt{\left(\frac{b}{2}\right)^2 + D^2}}{D} \tag{12.31}$$

and

$$\sigma_y = \frac{\lambda \sigma_f \left[\left(\frac{b}{2}\right)^2 + D^2\right]^{3/2}}{vbD} \tag{12.32}$$

where λ is the wavelength, v is the velocity of the EW systems, b is the baseline length, and D is the distance from the baseline.

The standard deviations of the fix calculation form the major and minor axes of the elliptical error probable contour. As given previously, an approximate (to within 10%) CEP contour can be determined from these two values as

$$\text{CEP} = 0.75\sqrt{\sigma_x^2 + \sigma_y^2} \tag{12.33}$$

To get a sense of the geolocation accuracy possible with τ and $\dot{\tau}$ processing, consider the following example. Suppose that the target transmits enough power so that an SNR of 10 dB is received at each of two airborne EW systems. Further suppose that the transmitted frequency is 50 MHz, the signal bandwidth is 25 kHz, the integration time is 0.3 second, the distance between the systems is 50 km, and the velocity of the airborne systems is 32 m/s laterally to the target. The geolocation accuracy possible on the perpendicular bisector to the baseline is shown in Figure 12.12.

As can be seen, geolocation accuracy of less than 500m is possible with these parameters out to a range of 150 km and more (assuming a constant SNR of 10 dB). Note, however, that these results are along the perpendicular bisector of the baseline. Off the baseline the GDOP decreases the accuracy as discussed in Chapter 11.

12.5 TIME OF ARRIVAL

Geolocation of a target based on measuring the τ and $\dot{\tau}$ of an EM wave emanating from the target was presented Section 12.4. Presented in this section is an algorithm for computing the geolocation of a target based on the measurement of the TOAs of the signal wavefront at four (or more) EW systems. This derivation follows that in [23] closely. Let the location of the target be denoted by $r_T = (x_T, y_T, z_T)$ and the location of the systems be denoted as $r_i = (x_i, y_i, z_i)$, $i = 1, 2, 3, 4$. Further let the time of transmission of a signal be denoted by t_0, and t_i, $i = 1, 2, 3, 4$ be the arrival time of the signal at each of the receiving systems. The Euclidean distances between the systems and the target are given by

$$d_i = \|r_i - r_T\| = \sqrt{(x_i - x_T)^2 + (y_i - y_T)^2 + (z_i - z_T)^2} \tag{12.34}$$

and d_i is related to the arrival times t_i by

$$d_i = c(t_i - t_0) \tag{12.35}$$

where c is the speed of light. Therefore, by equating these expressions and squaring, one obtains

$$c^2 t_1^2 - 2c^2 t_1 t_0 + c^2 t_0^2 - x_1^2 + 2x_1 x_T - x_T^2 - y_1^2 + \\ 2y_1 y_T - y_T^2 - z_1^2 + 2z_1 z_T - z_T^2 = 0 \tag{12.36a}$$

$$c^2 t_2^2 - 2c^2 t_2 t_0 + c^2 t_0^2 - x_2^2 + 2x_2 x_T - x_T^2 - y_2^2 + \\ 2y_2 y_T - y_T^2 - z_2^2 + 2z_2 z_T - z_T^2 = 0 \tag{12.36b}$$

$$c^2 t_3^2 - 2c^2 t_3 t_0 + c^2 t_0^2 - x_3^2 + 2x_3 x_T - x_T^2 - y_3^2 +$$
$$2y_3 y_T - y_T^2 - z_3^2 + 2z_3 z_T - z_T^2 = 0 \qquad (12.36c)$$
$$c^2 t_4^2 - 2c^2 t_4 t_0 + c^2 t_0^2 - x_4^2 + 2x_4 x_T - x_T^2 - y_4^2 +$$
$$2y_4 y_T - y_T^2 - z_4^2 + 2z_4 z_T - z_T^2 = 0 \qquad (12.36d)$$

By subtracting these equations two at a time, t^2 can be eliminated from all of them, thereby making them a linear set of equations in t_0. This must be done, however, so that the resulting equations are independent.

Subtracting (12.36b) from (12.36a), then (12.36c) from (12.36b), and finally (12.36d) from (12.36c) yields

$$c^2\left(t_1^2 - t_2^2\right) - 2c^2\left(t_1 - t_2\right)t_0 - \left(x_1^2 - x_2^2\right) + 2\left(x_1 - x_2\right)x_T$$
$$- \left(y_1^2 - y_2^2\right) + 2\left(y_1 - y_2\right)y_T - \left(z_1^2 - z_2^2\right) + 2\left(z_1 - z_2\right)z_T = 0$$
$$c^2\left(t_2^2 - t_3^2\right) - 2c^2\left(t_2 - t_3\right)t_0 - \left(x_2^2 - x_3^2\right) + 2\left(x_2 - x_3\right)x_T \qquad (12.37)$$
$$- \left(y_2^2 - y_3^2\right) + 2\left(y_2 - y_3\right)y_T - \left(z_2^2 - z_3^2\right) + 2\left(z_2 - z_3\right)z_T = 0$$
$$c^2\left(t_3^2 - t_4^2\right) - 2c^2\left(t_3 - t_4\right)t_0 - \left(x_3^2 - x_4^2\right) + 2\left(x_3 - x_4\right)x_T$$
$$- \left(y_3^2 - y_4^2\right) + 2\left(y_3 - y_4\right)y_T - \left(z_3^2 - z_4^2\right) + 2\left(z_3 - z_4\right)z_T = 0$$

Rearranging terms

$$(x_1 - x_2)x_T + (y_1 - y_2)y_T + (z_1 - z_2)z_T =$$
$$c^2(t_1 - t_2)t_0 - \frac{1}{2}c^2\left(t_1^2 - t_2^2\right) + \frac{1}{2}\left(x_1^2 - x_2^2 + y_1^2 - y_2^2 + z_1^2 - z_2^2\right)$$
$$(x_2 - x_3)x_T + (y_2 - y_3)y_T + (z_2 - z_3)z_T = \qquad (12.38)$$
$$c^2(t_2 - t_3)t_0 - \frac{1}{2}c^2\left(t_2^2 - t_3^2\right) + \frac{1}{2}\left(x_2^2 - x_3^2 + y_2^2 - y_3^2 + z_2^2 - z_3^2\right)$$
$$(x_3 - x_4)x_T + (y_3 - y_4)y_T + (z_3 - z_4)z_T =$$
$$c^2(t_3 - t_4)t_0 - \frac{1}{2}c^2\left(t_3^2 - t_4^2\right) + \frac{1}{2}\left(x_3^2 - x_4^2 + y_3^2 - y_4^2 + z_3^2 - z_4^2\right)$$

Recognizing that

$$\|\mathbf{r}_1\|^2 = x_1^2 + y_1^2 + z_1^2$$
$$\|\mathbf{r}_2\|^2 = x_2^2 + y_2^2 + z_2^2$$
$$\|\mathbf{r}_3\|^2 = x_3^2 + y_3^2 + z_3^2 \tag{12.39}$$
$$\|\mathbf{r}_4\|^2 = x_4^2 + y_4^2 + z_4^2$$

these equations can be written in matrix form as

$$\mathbf{A}\mathbf{r}_T = c^2 \mathbf{u} t_0 + \mathbf{s} \tag{12.40}$$

where

$$\mathbf{A} = \begin{bmatrix} x_1 - x_2 & y_1 - y_2 & z_1 - z_2 \\ x_2 - x_3 & y_2 - y_3 & z_2 - z_3 \\ x_3 - x_4 & y_3 - y_4 & z_3 - z_4 \end{bmatrix}$$

$$\mathbf{u} = \begin{bmatrix} t_1 - t_2 \\ t_2 - t_3 \\ t_3 - t_4 \end{bmatrix} \tag{12.41}$$

$$\mathbf{s} = \frac{1}{2}\begin{bmatrix} \|\mathbf{r}_1\|^2 - \|\mathbf{r}_2\|^2 - c^2(t_1^2 - t_2^2) \\ \|\mathbf{r}_2\|^2 - \|\mathbf{r}_3\|^2 - c^2(t_2^2 - t_3^2) \\ \|\mathbf{r}_3\|^2 - \|\mathbf{r}_4\|^2 - c^2(t_3^2 - t_4^2) \end{bmatrix}$$

The solution to this equation is given by

$$\mathbf{r}_T = c^2 \mathbf{A}^{-1} \mathbf{u} t_0 + \mathbf{A}^{-1} \mathbf{s} \tag{12.42}$$

assuming that $\mathbf{A}^{-1}$ exists, which is true here. Let

$$\mathbf{v} = \mathbf{A}^{-1}\mathbf{u} \tag{12.43}$$
$$\mathbf{w} = \mathbf{A}^{-1}\mathbf{s}$$

Substituting this solution into one of the equations at (12.34) yields

$$d_1^2 = \|\mathbf{r}_T - \mathbf{r}_1\|^2 = c^2(t_1 - t_0)^2$$
$$\|\mathbf{v}t + \mathbf{w} - \mathbf{r}_1\|^2 = c^2 t_1^2 - 2c^2 t_1 t_0 + c^2 t_0^2 \qquad (12.44)$$
$$\|\mathbf{v}t + (\mathbf{w} - \mathbf{r}_1)\|^2 - c^2 t_0^2 + 2c^2 t_1 t_0 - c^2 t_1^2 = 0$$

Let

$$\mathbf{v} = \begin{bmatrix} v_1 \\ v_2 \\ v_3 \end{bmatrix} \qquad (12.45)$$

$$\mathbf{w} - \mathbf{r}_1 = \begin{bmatrix} q_1 \\ q_2 \\ q_3 \end{bmatrix}$$

then

$$\begin{aligned}
\|\mathbf{v}t + (\mathbf{w} - \mathbf{r}_1)\|^2 &= (v_1 t_0 - q_1)^2 + (v_2 t_0 - q_2)^2 + (v_3 t_0 - q_3)^2 \\
&= v_1^2 t_0^2 - 2 v_1 q_1 t_0 + q_1^2 + v_2^2 t_0^2 - 2 v_2 q_2 t_0 + q_2^2 \\
&\quad + v_3^2 t_0^2 - 2 v_3 q_3 t_0 + q_3^2 \\
&= (v_1^2 + v_2^2 + v_3^2) t_0^2 - 2(v_1 q_1 + v_2 q_2 + v_3 q_3) t_0 + (q_1^2 + q_2^2 + q_3^2) \\
&= \|\mathbf{v}\|^2 t_0^2 - 2 \mathbf{v}^T (\mathbf{w} - \mathbf{r}_1) t_0 + \|\mathbf{w} - \mathbf{r}_1\|^2
\end{aligned} \qquad (12.46)$$

where T denotes transpose.
Therefore,

$$\|\mathbf{v}\|^2 t_0^2 - 2 \mathbf{v}^T (\mathbf{w} - \mathbf{r}_1) t_0 + \|\mathbf{w} - \mathbf{r}_1\|^2 - c^2 t_0^2 + 2c^2 t_1 t_0 - c^2 t_1^2 = 0$$
$$(\|\mathbf{v}\|^2 - c^2) t_0^2 + 2[c^2 t_1 - \mathbf{v}^T (\mathbf{w} - \mathbf{r}_1)] t_0 + [\|\mathbf{w} - \mathbf{r}_1\|^2 - c^2 t_1^2] = 0 \qquad (12.47)$$

Equation (12.47) can be solved for t_0, the time that the signal was transmitted. Two answers ensue, corresponding to the two roots of the polynomial. Other information is necessary, then, to ascertain which of the two answers is correct. Once t_0 is determined, it can then be substituted into (12.42) to determine r_T. This, or course, works for any of the original equations, not just $i = 1$.

12.6 CONCLUDING REMARKS

Hyperbolic position-fixing techniques based on measuring the TOA, TDOA, or differential Doppler at two or more widely separated platforms are presented in this chapter. These techniques have been shown in practice to yield better accuracy than the direction finding techniques presented in Chapter 11. The possible accuracies from such systems are illustrated in Figure 12.12. In general, however, longer integration times are required than for the DF approaches.

In addition to processing time, typically considerably more signal processing is required in CAF processing to compute the ambiguity surfaces. High-speed signal processors are required as the surfaces typically consist of considerable data points.

Lastly, as would be expected, accurate knowledge of the platform parameters is required to exploit the accurate PF capabilities of the hyperbolic technologies. These parameters are the three-dimensional positions, velocities, and accelerations.

References

[1] Chan, Y. T., and K. C. Ho, "An Efficient Closed-Form Localization Solution from Time Difference of Arrival Measurements," *Proceedings of IEEE MILCOM*, 1994, pp. II-393–II-396.

[2] Ho, K. C., and Y. T. Chan, "Solution and Performance Analysis of Geolocation by TDOA," *IEEE Transactions on Aerospace and Electronic Systems*, Vol. 29, No. 4, October 1993, pp. 1311–1322.

[3] Stein, S., "Algorithms for Ambiguity Function Processing," *IEEE Transactions on Acoustics, Speech, and Signal Processing*, Vol. ASSP-29, No. 3, June 1981, pp. 588–599.

[4] Ho, K. C., and Y. T. Chan, "Geolocation of a Known Altitude Object from TDOA and FDOA Measurements," *IEEE Transactions on Aerospace and Electronic Systems*, Vol. 33, No. 3, July 1997, pp. 770–783.

[5] Chan, Y. T., and K. C. Ho, "A Simple and Efficient Estimator for Hyperbolic Location," *IEEE Transactions on Signal Processing*, Vol. 42, No. 8, August 1994, pp. 1905–1915.

[6] Bard, J. D., F. M. Ham, and W. L. Jones, "An Algebraic Solution to the Time Difference of Arrival Equations," *Proceedings IEEE MILCOM*, 1996, pp. 313–319.

[7] Schau, H. C., and A. Z. Robinson, "Passive Source Localization Employing Intersecting Spherical Surfaces from Time-of-Arrival Differences," *IEEE Transactions on Acoustics, Speech, and Signal Processing*, Vol. ASSP-35, No. 8, August 1987, pp. 1223–1225.

[8] Quazi, A. H., "An Overview on the Time Delay Estimate in Active and Passive Systems for Target Localization," *IEEE Transactions on Acoustics, Speech, and Signal Processing*, Vol. ASSP-29, No. 3, June 1981, pp. 527–533.

[9] Campbell, L. L., "Asymptotics of Performance of Estimators of Arrival Time," *ISIT,* 1998.
[10] Gardner, W. A., and C. K. Chen, "Interference-Tolerant Time-Difference-of-Arrival Estimation for Modulated Signals," *IEEE Transactions on Acoustics, Speech, and Signal Processing,* Vol. 36, No. 9, September 1988, pp. 1385–1395.
[11] Rusu, P., "A TOA-FOA Approach to Find the Position and Velocity of RF Emitters," Applied Research Laboratories, The University of Texas at Austin, Austin, TX.
[12] Yost, G. P., and S. Panchapakesan, "Automatic Location Identification Using a Hybrid Technique," *VTC,* 1998, pp. 264–267.
[13] Yuan, Y. X., G. C. Carter, and J. E. Salt, "Correlation Among Time Difference of Arrival Estimators and Its Effect on Localization in a Multipath Environment," *Proceedings of IEEE MILCOM,* 1995, pp. 3163–3166.
[14] Chestnut, P. C., "Emitter Location Accuracy Using TDOA and Differential Doppler," *IEEE Transactions on Aerospace and Electronic Systems,* Vol. AES-18, No. 2, March 1982, pp. 214–218.
[15] Belanger, S. P., "An EM Algorithm for Multisensor TDOA/DD Estimation in a Multipath Propagation Environment," *Proceedings of MILCOM,* 1996, pp. 3117–3120.
[16] Schmidt, R., "Least Squares Range Difference Location," *IEEE Transactions on Aerospace and Electronic Systems,* Vol. 32, No. 1, January 1996, pp. 234–242.
[17] Shin, D. C., and C. L. Nikias, "Complex Ambiguity Function Based on Fourth-Order Stastics for Joint Estimation of Frequency-Delay and Time-Delay of Arrival," *Proceedings of MILCOM,* 1993, pp. 461–465.
[18] Smith, J. O., and J. S. Abel, "Closed-Form Least-Squares Source Location Estimation from Range-Difference Measurements," *IEEE Transactions on Acoustics, Speech, and Signal Processing,* Vol. AssP-35, No. 2, December 1987, pp. 1661–1669.
[19] Fang, B. T., "Simple Solutions for Hyperbolic and Related Position Fixes," *IEEE Transactions on Aerospace and Electronic Systems,* Vol. 26, No. 5, September 1990, pp. 748–753.
[20] Wiley, R. G., *Electronic Intelligence: The Interception of Radar Signals,* Dedham, MA: Artech House, 1985, pp. 116–121.
[21] Bard, J. D., and F. M. Ham, "Time Difference of Arrival Dilution of Precision and Applications," *IEEE Transactions on Signal Processing,* Vol. 47, No. 2, February 1999, pp. 521–523.
[22] Whalen, A. D., *Detection of Signals in Noise,* New York: Academic Press, 1971, p. 339.
[23] Rusu, P., "The Equivalence of TOA and TDOA RF Transmitter Location," Applied Research Laboratories, The University of Texas at Austin, Austin, TX.

Chapter 13

Exciters and Power Amplifiers

13.1 INTRODUCTION

Exciters and power amplifiers are included in an EW system when active EA is one of the functions the system is to perform. The signals generated by this subsystem are emitted by the system with the intent of interfering with the opponent's communications. As mentioned in Section 1.4, such interference is, of course, on a noncooperative basis. Therefore, most of the time it is necessary to inject more energy into the target receiver than the intended transmitter does. Typically, but not always, the cooperating transmitter and receiver are closer together than the EW system and the receiver are. These factors, then, dictate that the EW system must emit considerably more raw power than the cooperating transmitter.

Exciters and power amplifiers are discussed together in this chapter because they are inextricably linked in function. The exciter generates the signal that is then boosted in power by the power amplifier. In fact, there is frequently an amplifier stage integrated with the exciter, forming the first link in the amplifier chain.

13.2 EXCITERS

The exciter is the source of the signal used for EA in communication EW systems. It consists of a frequency synthesizer and modulator. The frequency synthesizer generates RF signals while the modulator imposes the desired modulation type onto the RF carrier according to the attack strategy of the communication EW system.

The exciter output is typically a fairly low-level signal that must be amplified considerably in order to be useful for EA. It is usually desirable to have the

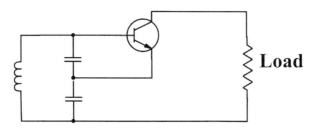

Figure 13.1 A Colpitts oscillator architecture.

exciter signal fairly clean and as free of harmonics as possible in order to minimize the detrimental effects to friendly communication signals, or other signals that are not the target of the attack that EA can cause.

13.2.1 Oscillators

This section briefly describes a few of the more popular types of oscillators used in exciters.

13.2.1.1 Colpitts Oscillator

One common type of oscillator is the Colpitts oscillator, named after its designer. A transistor version of this oscillator is shown in Figure 13.1. A feedback path is formed with the LC-circuit at the left of the diagram. The inductor and capacitors determine the frequency of operation.

13.2.1.2 Crystal Oscillator

Piezoelectric crystals can be used as the resonant device in oscillators to provide excellent frequency stability. Such a circuit is shown in Figure 13.2. Crystals are known for their stable operating characteristics over temperature ranges normally

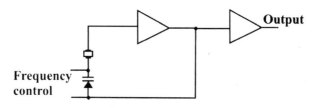

Figure 13.2 A crystal oscillator architecture.

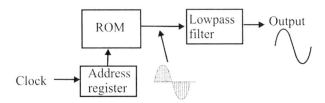

Figure 13.3 Direct digital synthesis. An address register is clocked in sequence to address a ROM where samples of a sine wave are stored.

experienced by electronic circuits. The piezoelectric effect causes the crystal to resonate at a particular frequency with a relatively high Q. That is, the width of the frequency band over which the crystal resonates is quite narrow. The resulting oscillating waveform is thus frequently quite pure, with low harmonics present.

13.2.1.3 Direct Digital Synthesis

The advances in the speed of digital electronics have facilitated another form of oscillator called the *direct digital synthesis* oscillator shown in Figure 13.3. In this case an address register is clocked so that the addresses of a *read-only memory* (ROM) are increased by one on each clock cycle. Stored within the ROM are the digital samples of a sine wave. The output of the ROM represents a digitized sine wave and, once converted to an analog signal and filtered, it is a very accurate analog waveform. The frequency of the output signal can be determined by the rate at which the address register is clocked.

As with other forms of sampled signal processing, the maximum frequency that can be generated this way is one-half the maximum clock rate. With gigahertz semiconductor technology it is possible to generate direct digital synthesizers that operate in the gigahertz range. One of the significant advantages of direct digital synthesis is that the frequency can be changed quickly.

13.2.1.4 Integrated Oscillators

The configurations shown in Figure 13.4 are popular for integrated circuit oscillators [1]. Such a configuration might be applicable to small EW systems where much of the signal processing is on a small number of integrated circuits. Normally additional signal conditioning, such as filtering, is necessary with this configuration. Thus, this circuit takes advantage of the negative resistance characteristic of the cross-coupled differential pair of transistors. The inductors are normally off-chip.

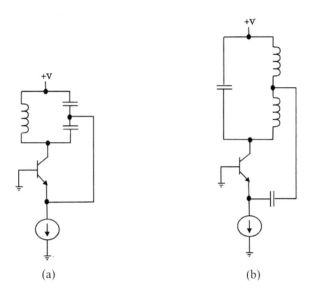

Figure 13.4 Configuration for oscillators built on an integrated circuit: (a) Colpitts and (b) Hartley. (*Source:* [1], © 1998, Prentice Hall. Reprinted with permission.)

13.2.2 Synthesizer

One form of synthesizer is shown in Figure 13.5, which is a PLL. The phase of the crystal oscillator is compared to the phase of the downconverted signal from the VCO. If there is a difference, an error signal is generated by the phase comparator. This error signal is lowpass filtered by the loop filter, and is used to adjust the VCO frequency up or down. The divide by N function is used to adjust the frequency output of the synthesizer since the output of the divide by N is what is used to compare to the crystal oscillator. The VCO then is adjusted to oscillate at N times the crystal oscillator frequency. N is adjustable over whatever range is desired.

Synthesizers in communication EW systems must be able to change frequency quickly in some applications. Such applications include time-sharing of the power amplifier and EA against frequency-hopping targets, for example. Time-sharing is when the frequency from the exciter is changed quickly so that the power from the power amplifier is quickly shifted to another frequency. Normally the frequency is quickly shifted back and forth between the two (or more) frequencies sharing the power from the amplifier. As discussed previously herein, against older analog modulations up to three targets can be attacked essentially simultaneously this way. Against modern digital communications, theoretically on the order of

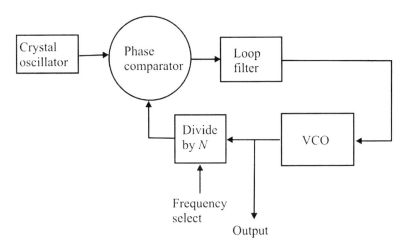

Figure 13.5 PLL as an oscillator/synthesizer.

100 targets can be simultaneously engaged, assuming the exciter can be tuned fast enough.

Frequency-hopping targets can be classified as slow hopping or fast hopping. In the former, several data bits are sent on each change of frequency of the carrier (hop). In the latter, one or fewer data bits are sent on each hop (that is, a single data bit is to be sent on one or more than one carrier frequencies). For voice digitized so that the data rate is 16 Kbps, approximately 160 data bits are sent for each hop of a 100-hps transmitter. To transmit this data rate over a fast frequency-hopping link, the transmitter would have to hop at 16,000 hps or faster.

13.2.3 Modulator

A modulator is used in an exciter just like it is used in any transmitter. Some types of target signals are better interfered with than others, with one type of EA modulation compared with others. The type of jamming modulation used then depends on the type of signal to interrupt. The modulator is used to impose the selected signal onto the carrier.

Since a frequency modulated signal is represented as

$$s(t) = A\sin(2\pi ft + k\int m(t)dt + \varphi) \tag{13.1}$$

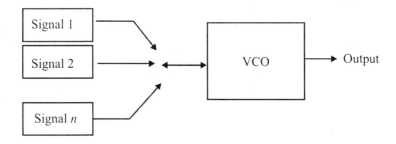

Figure 13.6 One form of a modulator that can frequency modulate several types of signals onto a carrier.

where $m(t)$ is the modulating signal and φ is a random phase angle, then the circuit shown in Figure 13.6 can be used as an FM modulator. A VCO such as used here will provide for FM of the signal by the various signals shown. These signal sources might be tones or random noise, for example.

One of the most common and effective types of modulation for EA against FM and PM signals is pseudorandom noise frequency modulated onto a carrier. Since many constant modulus communication systems, which include FM and PM, clip the amplitudes of the received waveform prior to demodulation, attempting to jam such signals with AM jamming waveforms produces poor results.

An AM signal has the mathematical form

$$s(t) = m(t)\sin(2\pi ft + \varphi) \qquad (13.2)$$

so a simple multiplier as shown in Figure 13.7 can be used as the modulator.

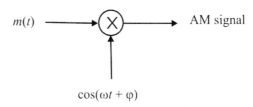

Figure 13.7 An example of an AM modulator.

13.3 POWER AMPLIFIERS

The power amplifier in an EW system is used to amplify signals generated in the exciter to significantly higher power levels for radiation via the antenna, eventually to find their way into the receivers of the adversary. Such signals disrupt the operation of the adversary's communication system.

Typically, one of the key characteristics of power amplifiers in communication EW systems is that they must operate over a wide frequency range. Communication EW systems must be designed to address a wide range of targets—the target array is typically not completely known ahead of time. A ground-based communication EW system, for example, might be designed to cover the 20–150-MHz frequency range (higher than this limit becomes impractical for ground-based systems since signals do not propagate very well close to the surface of the Earth). Airborne EW systems, on the other hand, can have a substantially broader frequency range than this since they are not necessarily limited by the signal propagation characteristics close to the Earth. This broad frequency range makes the design of such amplifiers somewhat different from those found in the commercial marketplace.

13.3.1 Amplifier Operating Characteristics

The general operating characteristic given previously for receivers in Figure 9.7 applies to amplifiers as well. That figure is repeated here, as Figure 13.8, for convenience [2].

Amplifiers used in power EW applications are frequently untuned amplifiers. That is, there are typically no tuning circuits within them. The exciter provides to the amplifier a narrowband, fairly clean, modulated signal to amplify. Any spurious noise generated in the amplifier is filtered in a separate tuned filter following the HPA. Being untuned, however, creates other problems that must be dealt with, such as optimally matching the output to an antenna.

The output stage of power amplifiers is where the most power amplification occurs. The schematic of a simple common emitter transistor amplifier is shown in Figure 13.9. Small signal analysis, where the transistor is biased in its linear region and always operates there, of the operating characteristics of this simple transistor amplifier are typical, and one is shown in Figure 13.10. Details of the bias design are precluded from this discussion. Suffice it to say that dc bias is applied to the base of such circuits to cause them to operate in the desired region of their operating characteristics.

This description assumes that a single sine wave is applied to the input of the circuit. The quiescent point, denoted by Q, is the point at which the circuit is biased to operate when there is no input signal. Thus, the circuit is at rest. Also note, that for small signal analysis, the supply voltage, V_{cc}, does not change, is thus

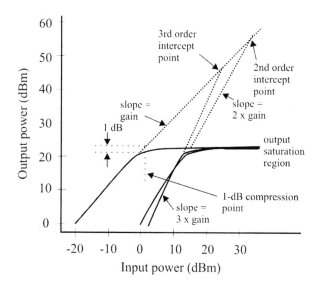

Figure 13.8 Transfer characteristics of an amplifier.

at a constant value, and from an oscillating signal perspective, is at the same potential as ground. Therefore, the signal output voltage is as shown in Figure 13.10. As the input voltage increases, the base current i_b increases, causing the collector current, i_c, to increase above its quiescent value. The collector current, i_c,

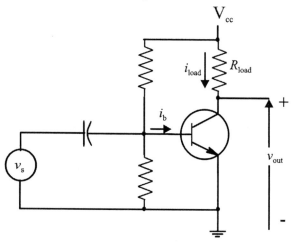

Figure 13.9 Simplified circuit diagram for an amplifier.

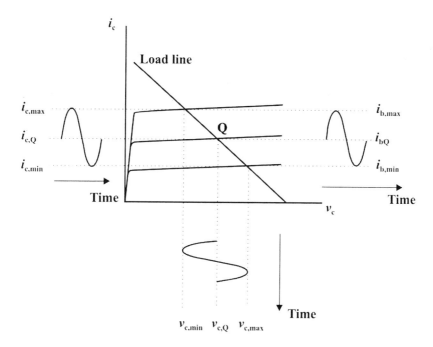

Figure 13.10 Characteristics of an amplifier biased to operate class A. Even if there is no signal to amplify, current flows through the output dissipating heat, thereby decreasing efficiency.

is also the load current, i_{load}, in this circuit. The load line characteristic is the way the voltage and current must behave in the load resistor, and is shown superimposed on the transistor transfer curves in Figure 13.10. Thus, because the transistor causes its collector current to increase, the voltage drop across the load resistor must increase, causing v_c, the voltage at the collector terminal of the transistor and also the output voltage, to decrease relative to V_{cc}, the power supply voltage for the circuit. As the input voltage increases, the output voltage continues to decrease, until the maximum input voltage is reached, at which time the process reverses, and the output voltage starts to increase. It increases until the input voltage reaches its minimum at which time the output voltage has achieved its maximum value. At that point the output voltage decreases until the quiescent point is again reached.

13.3.2 Efficiency

The *efficiency* of a power amplifier in an RF EW system reflects its ability to convert prime power, such as from a generator, into power out of the amplifier that

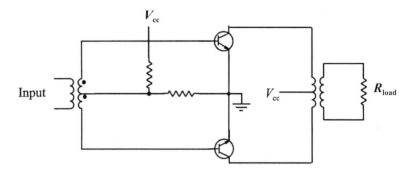

Figure 13.11 Push-pull amplifier configuration.

can be radiated into space. Typically such efficiencies are not very high. Achieving an efficiency of 50% would be quite remarkable; values of 20% to 30% or less are more common.

13.3.2.1 Heat Removal

The power represented by the inefficiency of an amplifier is manifest in heat generated. If an amplifier were perfectly efficient, then all of the prime power would be converted into signals to be output. Since there is no such amplifier as yet, some mechanism must be provided to remove the heat generated in the process. If there is a 0.5-V drop across the junctions in a transistor, which is passing a current of 100A, then the transistor must dissipate 50W of power. This power is converted into heat and is removed by properly configured heat sinks. Diamond has been found to be very good at heat removal.

Sometimes it is necessary to remove this heat with a cooling solution. Water, suitably conditioned for cold weather, has been used for this function. Refrigerants such as freon have also been used.

13.3.3 Push-Pull Architecture

A type of arrangement for power amplifiers known as a *push-pull* configuration is shown in Figure 13.11. As the input signal oscillates between positive and negative voltages, the output transistors conduct alternately. It has the advantage that the configuration of the circuit facilitates total suppression of the even harmonics in the output (assuming ideal components). Thus as the transistors are pushed to their linear limits and beyond, the only significant harmonic in the output is the third, which can frequently be removed by simple lowpass filters.

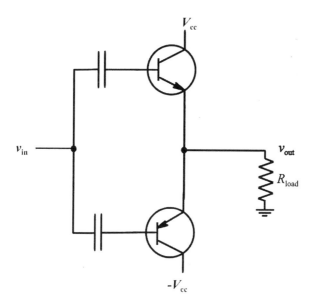

Figure 13.12 Push-pull amplifier configuration without an output transformer.

The suppression of even order harmonics is due to the way the output of the two channels combine in the output transformer.

It is also possible to construct push-pull amplifiers without using transformers. The advantage of such a configuration is that transformers frequently have limited frequency ranges. Figure 13.12 shows one such configuration. The disadvantage here is that two power supplies are needed. The bias circuitry is not shown in Figure 13.12 for clarity. Another disadvantage is this circuit requires both npn and pnp transistors. Because of the limited mobility (speed) of charges in p-doped material, these amplifiers are limited to the lower frequency ranges and cannot be used, for example, in the UHF and above ranges.

13.3.4 Classes of Amplifiers

Transistor amplifiers can be biased to operate is several regions. Which bias arrangement depends on the application at hand. One of the principal constraints is the degree of nonlinearity that can be tolerated—nonlinear amplifiers generate spurious, unwanted output signals. The region in which an amplifier is biased determines the amplifier's *class*.

13.3.4.1 Class A Amplifiers

A *class A* amplifier is biased approximately in the middle of its linear region, as shown in Figure 13.10. Input signals can vary in amplitude from their maximum to minimum and still remain in the linear region, so that the output signals have the minimum distortion possible. Because of the low distortion, the output signal has low spurious products. As such, filtering the output to minimize interference in other channels may not be necessary.

Bias current flows through the output stage continuously, whether there's a signal to amplify or not. Thus, since class A amplifiers are biased to be on all the time, they have relatively low efficiency. The power that is dissipated when there is no signal to amplify is wasted.

13.3.4.2 Class B Amplifiers

Class B amplifiers when used like class A amplifiers exhibit considerable distortion of the output signal because they are biased so that the active device is off during approximately half of the input cycle. In cases where this distortion can be tolerated, such amplifier configurations are applicable. That is not normally the case in EW systems, however, and filtering of the output signal is often required to reduce spurious emissions.

These amplifiers are biased at one end of their operating characteristic as shown in Figure 13.13. When there is no signal to amplify, there is little current flowing in the output stage. Class B amplifiers are more efficient than class A because of this no-signal low-quiescent current. The quiescent point for a class B biased amplifier is shown at Q in Figure 13.13.

When class B amplifiers are used in the push-pull arrangement discussed above, there is a small amount of time when both transistors are turned off. This produces what is called "crossover distortion." One transistor is on during half of the input cycle and the other is on for the other half. The point where one is almost off and the other is almost on is called the crossover point. Crossover distortion is illustrated in Figure 13.14.

Because of this distortion, class B amplifiers are of limited use as power amplifiers in EW systems. That is not to say that they have limited use in general, however. An example of where they are very well-suited was provided previously. In satellite communication systems where PMs versus AMs predominate, the amplitude of the output of the amplifier is of secondary importance, as the information in the signal is carried in the phase variations. In that case there can be considerable spurious outputs and the system will work quite well. Even in this case, however, the output is usually filtered to minimize the spurious signals from interfering with other communication systems.

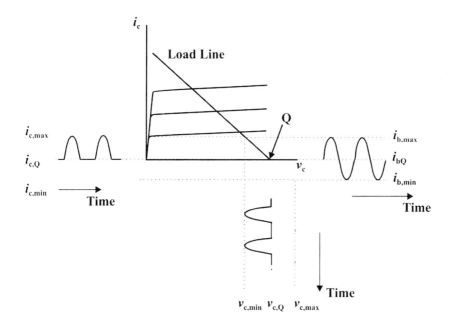

Figure 13.13 Operating characteristics of an amplifier biased as class B. During half of the input signal cycle, the amplifier is turned off. If there is no signal to amplify, the amplifier is off, saving power. Significant spurious signals are generated in such a configuration due to its nonlinear operation.

13.3.4.3 Class AB Amplifiers

Class AB amplifiers are biased between classes A and B. The characteristics of this type of amplifier are shown in Figure 13.15. This arrangement is more efficient than class A but less efficient than class B. Class AB amplifiers can be biased so that both of the transistors in the push-pull configuration overlap in their

Figure 13.14 Crossover distortion in class B amplifiers.

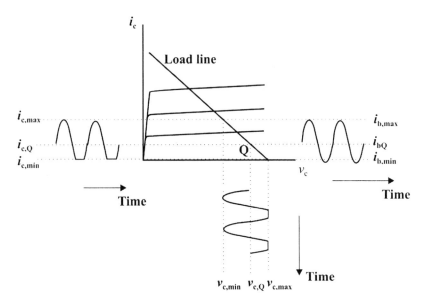

Figure 13.15 Amplifier characteristics biased as class AB. This configuration is between classes A and B. Efficiency is higher than A and less than B, and spurious signal generation is less than B and more than A.

on cycle. In this way, the crossover distortion can be controlled. Thus, the push-pull amplifier biased in class AB can be, and is, frequently used as high-power amplifiers in EW systems.

13.3.4.4 Class C Amplifiers

Class C amplifiers are biased so that more than half the time there is no output current. Class C amplifiers are biased to be not on at all during portions of the signal cycle, and thus are the most efficient of these amplifier classes. Like class B amplifiers, however, there are considerable spurious signals generated in such amplifiers, so output filtering is normally required for communication EW system applications.

13.3.5 Switching Architectures

There is a class of amplifier architectures that switch rapidly to perform the amplification. Discussed here are classes D, E, F, and S architectures, although there are others. The most significant advantage of these architectures is their

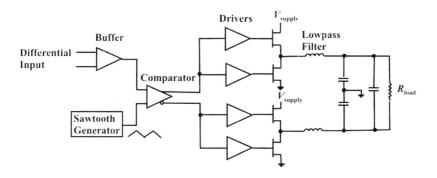

Figure 13.16 Block diagram of a class D amplifier. Switching amplifiers sequentially saturate and turn off power transistors in the output stage using pulse width modulation.

efficiency. Generally, they are much more efficient than the types discussed above.

13.3.5.1 Class D Amplifiers

A *class D* amplifier is substantially different from those of the other classes mentioned previously. An amplifier operating in class D rapidly turns the power transistors on and off in the output stage, producing a digital output of 1's and 0's. With *pulse width modulation* (PWM), the longer the transistor is on, the greater the average output power.

A block diagram of a class D amplifier is shown in Figure 13.16. The differential input of the signal to be amplified is input to one side of a comparator. A sawtooth waveform is applied to the other input as shown. When the amplitude of the input signal exceeds that of the sawtooth, the output of the comparator changes state to a logical one in this case. As seen in Figure 13.17, which displays the time waveforms involved, the lower the amplitude of the input signal, the more of the sawtooth is included in that which is less than the input signal, causing the output to remain at a high state longer. In this way the circuit adjusts the width of the output of the comparator as a function of the amplitude of the input signal. This PWM waveform then drives the high-power output stage which switches on and off according to this signal. The wider the pulse, the longer the power stage stays on and the more of the prime power is fed to the output lowpass filter. The output, being a square wave, would generate spurious signals if transmitted in that form. This output is lowpass filtered prior to transmission to extract only the fundamental frequency.

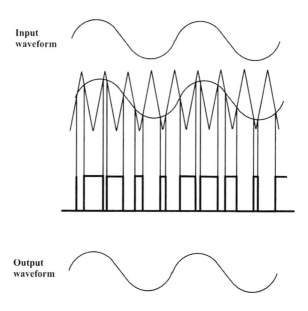

Figure 13.17 Waveforms within a class D amplifier. Every time the sawtooth waveform goes above the input waveform, the output of the comparator goes to a high state and vice versa. The comparator output, through drivers, turn the output transistors on and off.

The primary advantage of class D amplifiers is their efficiency compared to the other classes. Efficiencies greater than 80% are possible. However, because the transistors must be switched much faster (say, by a factor of 10) than the highest frequency in the signal, the use of class D amplifiers is limited to low frequencies with modern transistors. The most common usage of class D amplifiers as of this writing is in audio amplifiers for home stereos.

13.3.5.2 Class E Amplifiers

Class E amplifiers are perhaps the most efficient type of amplifier available. A typical schematic of a class E amplifier is shown in Figure 13.18 [3]. The transistor here is not really amplifying at all, but is switching on and off. Shunt capacitor C_1 and the tuned circuit consisting of C_2, L, and R_{load}, are chosen so that the phasing of the voltage across C_1 and the current through the transistor are as close to π radians out of phase as possible. If they are exactly π radians out of phase, then the transistor dissipates no power. All of the power is delivered to the load except that lost in the other nonideal components. (The other inductor is

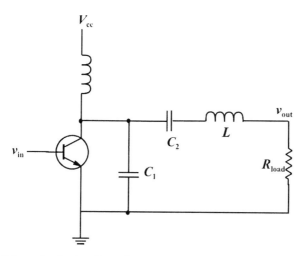

Figure 13.18 Schematic of a class E amplifier. The power transistor is switched with a square input waveform, while the output is tuned or lowpass, thus eliminating the spurious response. (*Source:* [3], © 2001, A. Nigam. All rights reserved.)

simply there for bias purposes and keeps the signal out of the power supply circuit.)

The simplified waveforms of this circuit are shown in Figure 13.19. For simplicity it is assumed that the input signal to the amplifier is a 50% duty cycle square wave, although other waveforms are possible depending on the desired

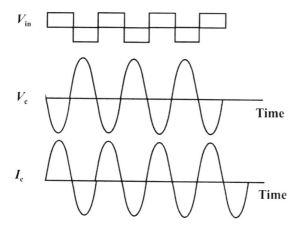

Figure 13.19 Time waveforms for the class E amplifier.

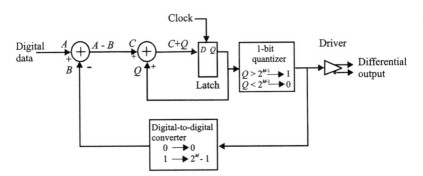

Figure 13.20 Block diagram of a sigma-delta modulator. (*Source:* [4], © 1995, Kluwer Academic Publications. All rights reserved.)

effect. As the transistor is switched on and off by the input signal, the current and voltage vary as shown. The RLC circuit on the output of the transistor prevents these output waveforms from instantaneously changing the way the input signal is changing. They must change more slowly, thus forming a continuous waveform.

Class E amplifier configurations have limited utility as the power amplifier in EW systems because they are obviously tuned amplifiers. EW systems typically require amplifiers that can quickly (instantaneously) cover a broad frequency band. A class E amplifier would need its output circuit changed quickly to function as a broadband amplifier.

13.3.5.3 Class S Amplifier

Class S amplifiers operate similarly to class D described above in that the signal to be amplified is initially in digital form and is converted to analog. In its digital state it is fed to a *delta-sigma modulator*.

Delta-sigma modulators were discovered in the field of A/D and D/A converter design. One of the difficulties with D/A converters is the precision required in some of the components to convert a digital signal to an analog form. The goal was to remove all analog components from the D/A design and the delta-sigma modulator was discovered. A simplified block diagram of perhaps the simplest of these devices is shown in Figure 13.20 [4]. The current state of the modulator, B, is subtracted from the level of the digital input signal forming $A - B$. B is a function of the output of the 1-bit quantizer. If the quantizer output is 0, then $B = 0$. If the quantizer output is 1, then $B = 2^M - 1$, where M is the number of bits in the modulator. $A - B$ is called the error signal. The second adder adds this error to the accumulated error as stored in the latch. On the next clock cycle, this updated accumulated error is clocked into the latch, forming the next state of the

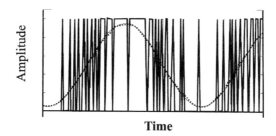

Figure 13.21 Input/output of the sigma-delta modulator driving a class S output stage.

modulator. The 1-bit quantizer outputs a zero if the accumulated error is less than one-half the maximum value allowed ($= 2^M / 2 = 2^{M-1}$), and a 1 if it is greater than half the maximum.

The net effect of the modulator is shown in Figure 13.21. When the input is largely positive, the output of the quantizer is mostly 1, whereas when the input amplitude is small, this output is mostly 0. Another way of saying this is the output remains in a 1-state, on average, longer when the input signal level is greater. The output of the quantizer is periodic.

Although overall operation produces results that appear to be PWM, as with class D amplifiers, in fact, a delta-sigma modulator produces *pulse density modulation* (PDM). The output width of the pulses is not changed per se according to the amplitude of the input signal, but the number of pulses in the periodic cycle is. The differential outputs are then used to drive the output circuit as shown above for the class D amplifier. This is shown in Figure 13.22. In this case there is no need for the sawtooth generator or the comparator, and a bandpass filter has replaced the lowpass filter.

It is desirable to keep the differences between one sample and the next relatively small. In order to do this, the sample clock frequency must be much higher than the highest input frequency. This is called *oversampling*. Nevertheless, much higher frequencies of operation are possible than class D, achieving cutoff frequencies in the gigahertz range. In addition, there are other forms of delta-sigma modulators. In particular, there are higher-order modulators. The *order* of a delta-sigma modulator is the number of delay stages it has. They tend to calculate the error better and therefore replicate the input signal better. The first-order modulator shown in Figure 13.20 is one of the simplest to illustrate their basic operation.

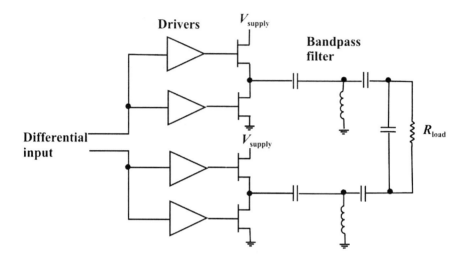

Figure 13.22 Output stage of a class S amplifier. The input takes the output of the sigma-delta modulator and turns on and off the output power transistors using PWM. The output filters can be bandpass, as shown here, or lowpass for broadband operation.

13.3.6 Amplifier Linearization

From the above discussion it should be clear that linear operation of an amplifier is an important notion in many circumstances. As discussed previously, when there are substantial amplitude variations in a signal to be amplified, such as in AM and many forms of PM, linear amplifiers are required to avoid unacceptable distortions in the amplified signal. For constant-amplitude signals, amplifiers are typically operated in saturation where they are the most efficient. Distortion is not an issue in these circumstances.

One way to efficiently amplify a signal that contains amplitude variations is called *envelope elimination and restoration* [5]. Shown in Figure 13.23 [6] is a simplified block diagram of how the process works. A limiter removes the amplitude variations in the signal and the resultant constant modulus signal is amplified through its own amplification chain. Since the signal has a constant amplitude, a saturated amplifier can be used in this path. A separate channel is provided which is preceded by an amplitude detector. This slowly varying signal is amplified with a nonsaturated, and thus less efficient, amplifier. These two components are then combined at the output of the amplifier.

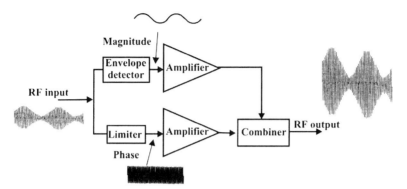

Figure 13.23 Amplifier linearization by envelope elimination and restoral. (*Source:* [6], © 1998, Hewlett-Packard, Inc. All rights reserved.)

13.3.7 Basic Power Modules

High-gain amplifier stages tend to become unstable and oscillate with age and temperature variations. These oscillations produce unwanted spurious output signals. Therefore, it is sometimes necessary to "gang" these circuits in series to achieve the necessary power output. The individual circuits that form one element of this "gang" are called *basic power modules*. Forming the high-power signal is done in stages with each stage contributing only a moderate amount of gain. Moderate gain amplifiers such as these are easier to design for stability and reliability than high gain stages. Such a configuration is shown in Figure 13.24. A parallel configuration for a high-power amplifier using relatively low-gain basic power modules and combiners for RF EW applications as shown in Figure 13.25. The input signal is routed through a splitter that could just be wires, but not

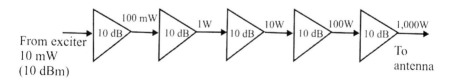

Figure 13.24 Amplification chain using basic power modules. Each stage provides modest gain, while boosting the exciter signal to a level adequate for EA operations.

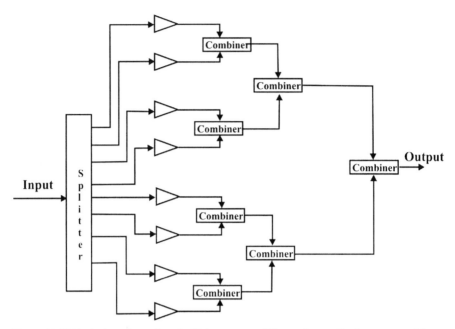

Figure 13.25 Typical configuration of a high-power amplifier made up of basic power modules.

necessarily. The splitter is often included for impedance matching. Once the signal is routed to the amplifiers, the outputs are combined to generate the high-power signal.

Sometimes switches are added to this tree of amplifiers for more flexibility. If two or more signals at different frequencies are to be amplified, the amplifier tree can be split in half or some other combination, in order to accomplish this.

It is not at all uncommon to power share the power amplifier in an EA system. *Power sharing* is splitting the available power from the amplifier between two or more signals at the same time. This allows a single amplifier to jam more than one signal simultaneously. Such sharing must be carefully managed, because the output signal power per frequency decreases approximately as the square of the number of signals sharing the available power, which is directly from the following definition of power. Suppose n signals are to share the fixed available power P from the amplifier. Therefore, the signal to be amplified is $ns(t)$. Then

$$P = \frac{1}{T}\int_0^T [ns(t)]^2 dt$$
$$= \frac{1}{T}\int_0^T n^2 s^2(t) dt \qquad (13.3)$$

$$\frac{1}{T}\int_0^T s^2(t) dt = \frac{1}{n^2} P$$

The expression on the left is recognized as the output power due to a single signal $s(t)$. Clearly the output power due to that signal is decreased by the square of the number of input signals. It is also clear that this is only an approximation to the power in each individual signal as the input signals in this case are identical. In power sharing that is not what normally occurs.

If $s_1(t)$ and $s_2(t)$ are two signals to be amplified, then the output power would be

$$P = \frac{1}{T}\int_0^T [s_1(t) + s_2(t)]^2 dt \qquad (13.4)$$
$$= \frac{1}{T}\int_0^T s_1^2(t) dt + \frac{2}{T}\int_0^T s_1(t) s_2(t) dt + \frac{1}{T}\int_0^T s_2^2(t) dt$$

The first and last terms on the right represents the output power of the two input signals. The available power, however, is shared with the middle term that represents a cross-signal product. If $s_1(t) = s_2(t)$, then the middle term represents the same power as in the other two terms and it represents an amount of power that is not available at the frequencies intended.

Power sharing places significant linearity constraints on the amplifiers in the chain. Nonlinear amplifiers generate frequency cross products that can occur at frequencies that interfere with friendly communications. Time-sharing of the power amplifier is also possible. (See Section 13.2.2.) Time-sharing implies that the frequency of the signal source can be changed very rapidly. A specific example of power sharing when amplifying four separate signals is shown in Figure 13.26. In this case each input signal is at 1W while the result of amplification and combining the output level of each is 6.25W. By the preceding analysis, for two inputs to a combiner there is approximately a 6-dB loss (10 log 0.25 ≈ –6) in the combiner.

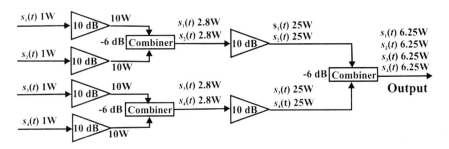

Figure 13.26 Combining several signals together in power sharing. The combiners contribute considerably to the loss in the process of putting the signals together.

13.3.8 Combiners

Combining refers to adding the outputs of two or more amplifiers together. Such ganging is fraught with design complexity, however. If the circuit components are not matched closely enough, then unbalanced amplitude and phase conditions can cause significant losses to occur. Furthermore, if one of the power amplifiers fails, in some designs the output power does not simply decrease by 50%, but can be substantially more.

Combiners, in addition to combining the signal outputs from two or more amplifiers, should provide some degree of isolation from one amplifier to another. Ideally, the characteristics of one amplifier should not influence that of another, irrespective of the circumstances. The load impedance seen by an amplifier should be constant, so the output of the amplifier can be conjugate matched to that load for maximum power transfer. This is rarely, if ever, the case, whether the output of two or more amplifiers are combined or not. Combining the outputs complicates this situation.

The configuration shown in Figure 13.27 is perhaps the simplest way to combine the output of two amplifiers. Here, the currents from the two amplifiers are summed while the voltage output of the two amplifiers is forced to be the same. This configuration provides little isolation of the outputs, and can lead to inefficiencies in operation if the two signal paths are not identical in amplitude and phase characteristics. In addition, if one of the outputs fails as a short circuit, it will route all of the current from the other amplifier to ground and there will be no output. The current flowing through the load is $2I_{out}$, assuming that the currents from the amplifiers are equal. Since the power dissipated in the load is given by $P_{load} = I^2_{load} R_{load}$, then the power output of this configuration is four times the power of a single amplifier configuration.

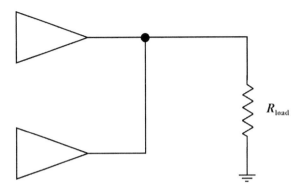

Figure 13.27 A simple way of combining the output of two amplifiers.

Another combining technique is shown in Figure 13.28. In this case the voltages from the two amplifiers are added at the secondary of the transformers. Some amount of isolation is provided in this configuration.

13.3.9 Output Filters

The purpose of filters between the power amplifier and antenna are twofold: (1) filtering the harmonics of the signal from the power amplifier thereby reducing unwanted emissions, and (2) conjugate matching the amplifier impedance to the antenna impedance as best as possible.

When operating in classes B, C, or D, the power amplifier generates significant harmonics, or signals at integer multiples of the intended signal. In

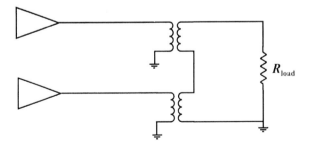

Figure 13.28 Another way of combining the output of two amplifiers. In this case, failure of one of the amplifiers does not cause catastrophic failure of the total system.

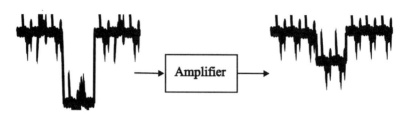

Figure 13.29 Noise-power ratio measurements of an amplifier is a thorough test of all the nonlinearities at the same time.

order to reduce interference to friendly communications and to maximize the radiation of signals at the intended frequency, filters are needed. To filter the first harmonic and higher, these filters need to pass a bandwidth that is no more than an octave; otherwise the first harmonic will pass through. Typically these filters are suboctave—less than one-half octave in bandwidth.

For RF EW applications these filters must be tunable. For power applications such as filtering the output of power amplifiers, this tuning is normally accomplished by switching in and out the proper suboctave filter.

For matching purposes, the filter tries to conjugate match the antenna load to that of the amplifier. In such a condition maximum power is transferred to the antenna and maximum radiation thus ensues. Perfect matching is difficult to achieve in these applications since the systems are typically wideband. Obtaining a VSWR of 2:1 or less is normally the design goal and is often achieved.

13.3.10 Noise-Power Ratio

The *noise-power ratio* (NPR) is a way to measure the overall distortion performance of an amplifier, or any other electronic device. It is a test procedure that measures the amount of intermodulation noise introduced by the nonlinearity of an amplifier. Conditions that simulate the total input to the amplifier are presented to the input and the output is measured. A wideband noise source is input to the amplifier, with a portion of the spectrum notched out. The inherent amplifier nonlinearities will fill the notch, due to the intermodulated signal components caused by the nonlinearities. The total output noise power is measured, to include the noise components in the notch. The ratio of the measured output noise to the input power (with noise notched out) is the NPRNPR. The NPR technique is one of the most thorough tests of an amplifier in that all of the nonlinearities are considered at the same time. The NPR measurement notion is shown in Figure 13.29.

13.4 CONCLUDING REMARKS

The basic operating principles of signal generators (exciters) and power amplifiers as they are used in EW systems are presented in this chapter. The signal used to jam a target receiver originates in the exciter, is amplified to a considerably higher level for transmission by the power amplifier, and is sometimes filtered at the output of the power amplifier to suppress energy at unwanted frequencies.

Care must be exercised in the design of high-power amplifiers used in such systems. Reliability is always a concern and unless designed properly, failure of a single component can preclude such an amplifier from operating at all. Active components used in the amplifier stage can only handle limited amounts of power.

These components must exhibit wide bandwidths, that is, they must provide reasonable gain characteristics over substantial frequency spans. This bandwidth requirement usually means that the gain at each stage is limited as the gain-bandwidth product is frequently more or less a constant. Therefore many stages of gain are required, resulting in the requirement for basic power modules, each one of which performs only a limited amount of the amplification required. Gains of 10 dB at each stage are not uncommon.

The output filters are often used to conjugate match the output of the final stage of the amplifier to the antenna. Perfect conjugate matching ensures maximum power transfer to the load (antenna in this case).

References

[1]   Razavi, B., *RF Microelectronics*, Upper Saddle River, NJ: Prentice Hall, 1998, p. 211.
[2]   Millman, J., and C. C. Halkias, *Electronic Devices and Circuits,* New York: McGraw-Hill, 1967, p. 559.
[3]   Nigam, A., "Class E Amplifier," accessed July 2001, http://ece.iisc.in/~nitin/anurag/classe.html.
[4]   Roberts, G. W., and A. K. Lu, *Analog Signal Generation for Built-In-Self-Test of Mixed Signal Integrated Circuits,* Appendix A, Boston, MA: Kluwer Academic Publishers, 1995.
[5]   Kahn, L., "Single-Sided Transmission by Envelope Elimination and Restoration," *Proceedings IRE,* July 1952, pp. 803–806.
[6]   Su, D., and W. McFarland, "An IC for Linearizing RF Power Amplifiers Using Envelope Elimination and Restoration," accessed July 2001, http://fog.hpl.external.hp.com/techreports/98/HPL-98-186.html.

Chapter 14

Early-Entry Organic Electronic Support

14.1 INTRODUCTION

In any kind of hostile activity, the U.S. military has the requirement to collect battlefield combat information about the adversary's disposition and intent. One way to assist in addressing this requirement is to intercept the tactical communications associated with the adversary using ground-based and airborne collection assets. Early-entry forces are constrained in terms of organic support available due primarily to airlift availability. Intelligence support to these forces may be furnished from higher echelons and, in particular, national assets. In some cases, however, it may be desirable due to access and other reasons to provide ES with organic assets. The collection capabilities that could be provided this way would naturally be very limited. Lightweight, manpack type of equipment would be all that could reasonably be provided [1–8].

This chapter presents how well a ground-based communications intercept system would perform against a threat communications environment that might be encountered by early-entry forces. It is based on a simulation that compares, in an operational and engineering sense, the values of various system tradeoffs that can affect performance. For example, does the requirement to prepare preformatted reporting messages, based on an intercept, significantly degrade the system output?

There were two independent models used in the simulation: a target environment model and a collection system model. The model of the signal environment did not represent any particular part of the world or any particular hostile threat. The target frequencies were randomly selected from the low-VHF military frequency range and were randomly placed geographically within an area, which was 30 km square. The model of the collection system did not represent any particular existing system but did closely and with a fair amount of detail model both the operational and systemic steps necessary to perform the communications intercept mission.

14.2 TARGET MODEL

The target environment consisted of a selectable number of simplex, clear text, VHF, PTT, ground, stationary communication nets. The tactical communications taking place within the first hour after the intercept systems were put into operation were simulated. No a priori knowledge of the target environment (such as frequencies) was assumed (which is a pessimistic assumption). The number of threat communication nets was a simulation variable: 100, 200, and 300 nets. Each net consisted of five nodes. This would approximate a brigade-sized adversary force.

In any live case, communications transpiring over tactical communication nets could take several forms. For the purposes of modeling, a particular communication structure was simulated. In each case of a communication exchange, net member one, called the NCS, transmitted first. The length of this transmission was (almost) a Gaussian random amount of time with an average of 5 seconds, a standard deviation of 1.67 seconds, and a minimum time of 1 second. This transmission was followed by a period of silence, the duration of which was randomly selected with an average value of 3 seconds, a standard deviation of 1.33 seconds, and a minimum of 1 second. Net member two then transmitted for a random time with the same statistics as the NCS. This transmission was followed by a period of silence with the same statistics as above, followed by the NCS transmitting to station three and station three responding. After that, the NCS transmitted to station four and station four responded. Finally, the NCS transmitted to station five and station five responded. Each net communicated in this fashion repetitively, the timing of which was (almost) a Gaussian random variable called the *net time*. The mean of this variable was varied between 150 seconds and 450 seconds, while the standard deviation was varied between 50 and 150 seconds. This process is shown schematically in Figure 14.1.

The target nets were placed randomly within a 30 km × 30 km area. The node separations on any given net were tactically reasonable. Thus, the emitters were not concentrated in any particular area. This may not be as accurate as it could have been since normally one would expect a greater concentration of emitters closer to the *forward line of troops* (FLOT), if there is one, than at the rear. The parameter for the Longley-Rice propagation model that was used presumed that the transmitters were randomly sited. (This has to do with how carefully the transmitters were sited for communication purposes.) Throughout the duration of the simulation, the targets were assumed to be stationary.

The targets were assumed to all have an effective radiated power of 20W with vertical polarization and the transmitter antenna heights were 5m. Signal losses with distance were computed using the Longley-Rice signal propagation model [9] using values of terrain variables that are considered average for typical ground conditions found in many parts of the world, with gently rolling terrain and soil that is not dry, but not too wet. Specifically, it was an equatorial temperate climate with a Δh value of

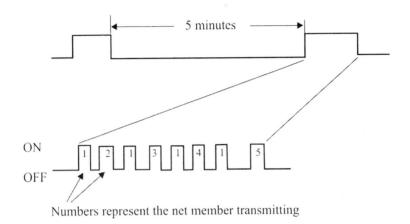

Numbers represent the net member transmitting

Figure 14.1 Target communication link timing.

100m. This model has been used extensively to approximate signal losses that occur close to the Earth, such as the case here.

The principal quality that constrains signal intercepts is the SNR of the transmission at the receive site. Therefore, so that an SNR could be calculated, a noise model was included in the simulation. The noise model used is based on the CCIR report described in Chapter 3, which includes a typical rural noise environment. The amount of noise present at the frequency of transmission was added to the signal received at the receiving system and if the SNR did not exceed a specified value, which depended on whether intercept or DF was being performed, the signal was considered as either detectable or not detectable.

14.3 INTERCEPT SYSTEM MODEL

The intercept systems were comprised of three receiving systems with DF capability. Each system had one operator, typical of manpack configurations. These receivers had the capability to step between frequencies (from a preprogrammed set) and scan between two frequencies set by the operators. If energy were detected at a channel (frequency), then the receiver dwelled there until the collection operation was concluded (defined later) at which time the next frequency was selected. The search mode used in the simulation consisted of starting with a frequency band assigned to each operator, the particular band being selected by dividing the 30–90-MHz tactical

VHF, PTT frequency band into equal increments. Each segment was then assigned to a different operator [10–12].

When the simulation started, each VHF receiver began scanning from its lowest limit and stepped in one-half *IF bandwidth* (IFBW) steps up to its maximum limit, the IFBW being a parameter in the simulation. The system dwelled 25 ms at any channel unless energy were detected there. When energy was detected at a particular frequency, that frequency was entered into a directed search list, and the operator prosecuted (copied) that particular frequency. Figure 14.2 shows an overview of the operation of the intercept systems.

Because communications are usually between (at least) two people, it is a good idea to continue to dwell for some time on a channel after a transmission has ceased, in order to hear the follow-on communication (if there is one). That capability was included in the simulation; the length of the dwell continuation is called the *bridge time*. This timer was set at 15 seconds. The bridge timer started when the signal dropped below the intercept SNR, which was a simulation parameter. For all of the simulations, a 6-dB SNR was required for the receiver to initially stop scanning and, once dwelling on a target, a 3-dB SNR was necessary to remain there. If a 3-dB SNR were not present, then the bridge timer continued running. Once the bridge timer expired, then a period of postprocessing time occurred. The length of this postprocessing time was a variable set at 15, 100, and 200 seconds. After the postprocessing, the frequency scanning continued.

Once a channel was found that contained a target signal, that channel became a priority frequency, which should be revisited more often than the other channels. In the general search mode described above it was indicated that these frequencies were saved on a list called the *directed search* list. The length of time spent in the general search mode before scanning was interrupted and the directed search frequencies were examined for activity is called the *revisit time*. The revisit time for this simulation was 7 seconds.

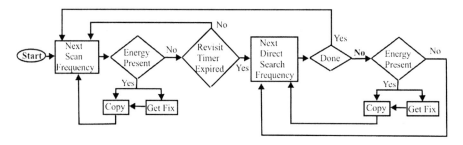

Figure 14.2 Flow diagram of the overall operation of the collection systems.

The intercept systems were netted together so that target fixes could be obtained. The DF integration time (the amount of time necessary to obtain an LOB) was set at 2 seconds and the net cycle time (the total time required to obtain a fix) was 8 seconds. If the DF net was busy when the operator requested a fix, then the request went into a queue of length one. If the queue was already full, then the request was denied. Any operator in any system could request a fix but the queue length was one for the entire net. In order for a system to obtain an LOB that may be used in a subsequent fix calculation, an SNR of –6 dB or more must have been present. The simulation model included Gaussian random errors added to the LOBs yielding an effective LOB accuracy of $0°$ average and $3°$ rms error. The fix algorithm used was that due to Brown described in Section 11.3.3.

No friendly signals interfered with the intercept process because the capability to automatically recognize the 150-Hz squelch tone present on NATO radios was assumed present. Therefore, modeling of the friendly radio frequency environment was not necessary.

Each operator was assumed to have a computer on which he or she could compose preformatted messages describing intercept operations and report them out of the intercept system to an analysis center. Modeling of the analysis center was not part of the simulation. The time it took to prepare these messages was a simulation variable called the *postprocessing time*. When it is set to 15 seconds (its lowest value), this approximates the situation in which the operator terminal operates autonomously in that target locations and frequencies are automatically sent to the analysis center without the necessity of the operator getting involved. This was the mode where the most target searching could be performed. The postprocessing times were 15 seconds, 100 seconds, and 200 seconds to illustrate the effects of having the operator prepare brief, but finite reports. Such reports would typically be, in addition to the target locations and frequencies, a brief synopsis of the transmissions intercepted on preformatted message formats. Such information, for example, might include which of the net nodes was the NCS. After the postprocessing time, the operator and equipment were then free to continue intercept operations.

The intercept systems were placed on the friendly side of the assumed FLOT at coordinates (–5, 7.5), (–8, 15), and (–5, 22.5) [The FLOT is assumed to be a vertical line running from coordinates (0, 0) to (0, 30).] All distances are in kilometers. The antenna height for the intercept systems was 7m.

In addition to the external noise modeled as indicated above, the RF receiving components in the collection system added thermal noise. This phenomenon is manifest in the equivalent system *noise factor* (NF) discussed in Chapter 9. The NF was a parameter in the simulation and was set at 10, 15, and 20 dB. The amount of external noise present and the NF together were used to set the noise values used in the SNR calculations. Although not an operational parameter that can be controlled, the NF is one of the most important design tradeoff parameters that, among other variables, set system sensitivity.

The IF bandwidth (and for these discussions also called the search bandwidth) was also a model parameter that was varied between 10 kHz and 250 kHz. The wider the search bandwidth, the faster the scanning process proceeds; however, the receive noise varies directly with the bandwidth as discussed in Chapter 3. Therefore, a tradeoff is typically necessary between search time and search bandwidth.

After a signal goes away, the receiver continued to dwell at the tuned frequency for a duration known as the *dwell time*. During the period immediately after the termination of a signal was the time when the system looked for another transmission from other members on the same net.

The antenna height of receive systems, in addition to the system sensitivity, largely determines how far into the target area the collection systems can hear. For this simulation this height was 7m. As mentioned, the system sensitivity depends on the IF bandwidth of the receivers, which was set at 10 kHz, 50 kHz, and 250 kHz.

The wider IFBW then searching the spectrum occurred at a more rapid pace, thereby increasing the probability of detecting a signal earlier in a transmission. However, the wider the bandwidth, the more noise was allowed into the passband and therefore the SNR was reduced, thus making distant targets harder to detect. Wider bandwidths also overlap several VHF frequency channels and thereby increase the possibility of cochannel interference (more than one signal in the passband, making listening to either one harder). The system sensitivities are shown in Table 14.1.

The antennas used in the simulation were "typical" antennas that might be used with ground-based communications intercept systems. They were omnidirectional in the horizontal plane and had gains that were about –20 dB at the low end of the band rising to about –3 dB at the upper end of the frequencies used in the simulation. Of course, the height of the receiving (and transmitting, for that matter) antennas, to a large extent, determines the amount of signal power that can be received. On the

Table 14.1 Approximate System Sensitivity as a Function of IFBW and Noise Factor

	Noise Factor (dB)		
IFBW	10	15	20
10 kHz	–124 dBm	–119 dBm	–114 dBm
50 kHz	–117 dBm	–112 dBm	–107 dBm
250 kHz	–110 dBm	–105 dBm	–100 dBm

other hand, the amount of noise present at the receiving site does not depend on the receive antenna height.

14.4 SIMULATION RESULTS

The parameters that were varied in this analysis were: (1) the number of target nets, (2) the search bandwidth of the receivers, (3) the system noise factor, (4) post-processing time, and (5) the duration of the simulation. A *target* herein refers to a particular communication node or radio. A *target net* or just *net* is a set of targets that are operating on the same frequency and occasionally communicate with one another. Each net had a *net control station* that communicated more than the others; normally it would be associated with a unit commander. A net was considered *detected* if one or more of the transmissions associated with the net were collected. A *target* was considered *detected* if one or more transmissions were collected from that target. Since it is routine practice to transmit one's own call sign at the beginning of a communiqué, a target was considered *identified* if one or more transmissions were collected from that target and the collection started within the first 2 seconds of the transmission. The a posteriori probabilities, probability of detection of a target transmitter (at least once), denoted P_{det}, probability of identification of a target transmitter (at least once), denoted P_{id}, probability of location of a target transmitter (at least once), denoted P_{loc}, probability of identification of a net's control station, denoted $P_{id,NCS}$, and probability of location of a net's control station, denoted $P_{loc,NCS}$, are the statistics that were collected and analyzed.

There were three reasons why a fix may not have been obtained on a target transmitter: (1) there was too little time left when the data link was acquired (the signal went away before the DF acquisition started), (2) there were too few LOBs available (< 2) because of too weak a signal at too many collection sites, and (3) the DF queue was full when the operator requested use of the DF assets. All else being equal, that is, all of the targets could be "heard" by all of the stations, if the DF assets were busy all the time, then 450 fixes per hour could be obtained.

It was possible to obtain a fix on the wrong target because the fix process took a certain period of time. When the operator requested a fix and the DF assets were busy, the request may have gone into the queue of length one. When the DF assets became available, the DF request was honored but by that time a different net member may have become active, in which case the fix was on the wrong target. Since a target was identified by its frequency of operation, even though the specific target was in error, the fix would have been on another member of the same net, however. This could cause confusion as to whether the target located was the NCS or not, but the results in tactical scenarios is not useless data in that case. The total number of wrong

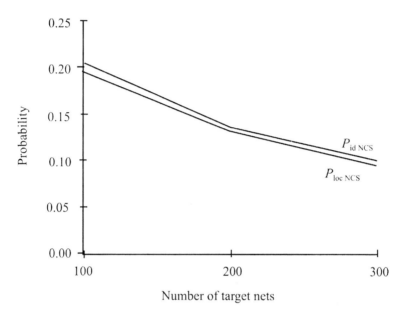

Figure 14.3 NCS collection results versus the number of target net.

fixes indicates the number of occasions of this happening.

Ten passes through the simulation were performed using different random seeds for each condition of the simulation variables. The mean values of the results were calculated and used in the analysis to follow.

14.4.1 Performance Versus the Number of Target Nets

The results as the number of target nets was varied are shown in Figures 14.3 and 14.4. The ability to find net control stations is shown in Figure 14.3. As expected, the number found decreases as the number of nets is increased. This is because (1) the operators of the intercept systems on average were busier because there were more targets with which to contend and (2) more of the target nets were probably out of range, which decreases the percentage results. Since, on average, a transmission would be detected halfway through, and considering that the dwell time was 15 seconds, with little postprocessing time an operator would spend approximately 46 seconds per intercept. That means that one operator could process about 78 targets per hour if that operator were employed full time. Note that some of these targets need not be unique; some could be duplicates. Therefore, under these conditions, a

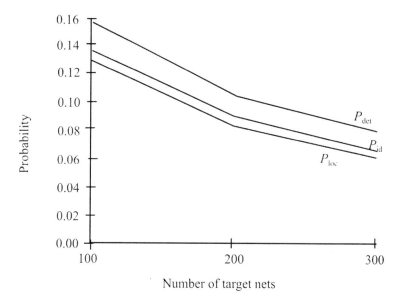

Figure 14.4 General collection performance versus the number of target nets.

maximum of 234 targets could be prosecuted. Ignoring hearability considerations, then, a maximum of 234 nets could be detected. Due to the nature of the communication on the target nets shown in Figure 14.1, $P_{id,NCS}$ would be this same level of probability since on the average a net cycle would be encountered in the middle. Also due to this, the latter net members would stand a better chance of being detected than the earlier ones, yielding an expected P_{id} about one half that of $P_{id,ncs}$.

The above analysis ignored any limiting effects of hearability. The receive antenna heights were only 7m. While this is practical from a deployment point of view, it does not facilitate much range from the intercept sites. Of course, if available, mountaintops are useful to extend this range. Such mountaintops are rarely the lone province of EW systems, however, and must be shared with other facilities that like such high ground. Nevertheless, there was not much range from such short antennas. Certainly not enough to extend to the 30-km depth of the target space. Hence, as more targets were added, since they were added uniformly throughout the battle space, more were added out of range than within range. Therefore, the collection statistics decreased as more targets were added. There were additional timing constraints for location and identification, which caused these probabilities to be less than the detection results.

The limiting effects of these timing constraints are illustrated in Figures 14.3 and

14.4. The probability of locating and identifying an NCS, shown in Figure 14.3, were about the same and varied from 0.2 at 100 target nets to about 0.1 at 300 target nets. P_{loc}, P_{det}, and P_{id} versus the number of target nets are shown in Figure 14.4. These values are somewhat lower than those for the NCS, consistent with the discussion above. P_{det} is not fully a factor of 1/2 that for the NCS. They vary from about 0.15 at 100 nets to about 0.07 or so at 300 nets. P_{det} is a little higher than P_{loc} and P_{id}, which are about the same. P_{loc} would be lower than P_{det} because of the data link cycle time of 8 seconds. Contention for DF assets caused some fixes to be lost even though the target was detected. P_{id} would be lower because not all of the targets that were detected were detected within the first 2 seconds. In fact, as discussed above, on average the earlier net members were rarely identified, while the latter net members, if the net was detected, were identified. The fact that P_{loc} and P_{id} turned out to be more or less the same is a coincidence. There is nothing in the basic parameters that would suggest this should happen.

14.4.2 Search Bandwidth

The wider the search bandwidth, the faster the search band was covered. The amount of external noise entering the system was obtained from the CCIR model described in Chapter 3. In particular, in this case, a rural high noise level was assumed. These values are in decibels above kTB, where B is the system bandwidth, as described above. Therefore, the wider the search bandwidth, the more external noise entered the system. Thus, the tradeoff is speed versus signal level, both of which can limit the overall system performance at finding target signals.

In dense target environments a search bandwidth wider than the channel allocations allows for increased cochannel interference. In fact, cochannel interference is possible even if the target environment is not dense, but this cochannel, theoretically at least, is caused by friendly radios interfering with targets. With wide bandwidths, such interference can be caused by two friendly (or two or more) targets. There are 2,400 25-kHz channels between 30 and 90 MHz. With only a maximum of 300 target nets, the possibility of two adjacent channels being occupied is small (about 12%). However, when the bandwidth is 250 kHz, which is 10 channels wide, the possibility of cochannel interference in this case is a certainty (subject to the two nets operating at the same time). In any case, while cochannel interference is an operationally important consideration, the effects were not included in this simulation.

The systems started with no a priori information known about the threat environment. Therefore, at the beginning, all the searching was from a general search list that is of the "from f_1 to f_2 in ½ IFBW steps" variety. As time progressed and more targets were found, a more directed search was executed. In reality such a "cold start" would be highly unlikely because as much information about an adversary force as possible would be provided by higher echelon sources. This information would

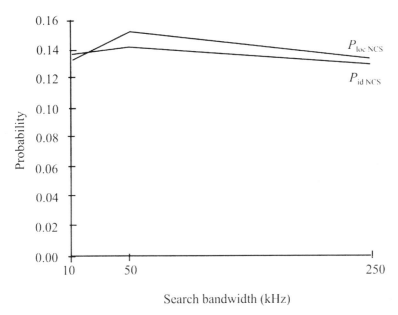

Figure 14.5 NCS collection performance versus the search bandwidth.

almost always include some data on net usage.

Shown in Figure 14.5 are the results for $P_{id,NCS}$ and $P_{loc,NCS}$ versus the search bandwidth. The variation as the bandwidth was increased is insignificant. This indicates that faster general search times do not increase the ability to find NCSs and it implies that the effects of increased noise were insignificant for this function as well. The same general characteristics for collection of target nodes in general are exhibited in Figure 14.6. The two opposing effects balanced each other out. There is a slight rise as the bandwidth increases from 10 to 50 kHz, implying that more targets were being found with a faster search rate, although the effects were slight. Yet the additional external noise is not that great. From 10 kHz to 50 kHz was an additional about 7 dB of noise. On the other hand, increasing from 50 kHz to 250 kHz, the collection performance dropped off, but again the differences are slight. This was an increase again of about 7 dB, but an increase of 14 dB from 10 to 250 kHz. The fact that these were offsetting effects can be seen from the results below on increasing the noise factor. In that case the search rate as well as the external noise were held constant because the bandwidth was constant. Increasing the noise factor did indeed decrease the collection performance. Hence, while the collection performance increased by increasing the search rate by widening the search bandwidth, it was

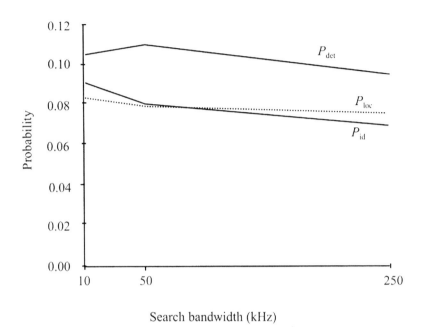

Figure 14.6 General collection performance versus the search bandwidth.

more or less offset by the decrease in system sensitivity by admitting more noise.

P_{det}, P_{loc}, and P_{id} are plotted in Figure 14.6 versus the search bandwidth. The effects were the same as those indicated above for the statistics on collection of NCSs and for the same reasons. For all practical purposes these results show that it made no difference what the search bandwidth was. This is not really the case, however, for reasons that were not modeled here. The fact is that the range from the intercept systems was so small that using the narrowest search bandwidth is indicative of the limitation caused by an inadequate SNR. The wider the search bandwidth used, the greater the amount of noise present as discussed in Section 14.3. On the other hand, using a wider search bandwidth decreased the amount of time it took to search the RF spectrum for signals, thus increasing the probability of detecting a signal sooner. Clearly, it is better to use a narrow bandwidth and search more slowly, at least in a limited target environment as modeled here.

14.4.3 Noise Factor

The system noise factor establishes the noise floor. This, in conjunction with the externally received noise, determines whether a received signal is detectable or not. If

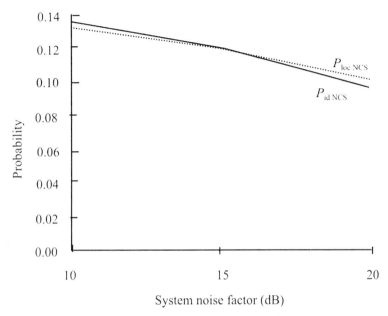

Figure 14.7 NCS collection performance versus the system noise factor.

the received signal is below the minimum detectable signal level, then there is not enough energy present. Although low noise factors are typical for higher-frequency ranges, in the range of interest here, a 15-dB noise factor is considered typical, 10 dB would be considered excellent. The noise factor is a design parameter that can be traded off with other design parameters. Shown in Figure 14.7 are the probabilities of locating and identifying an NCS versus the system noise factor. The effects of varying the system noise factor on collection in general are shown in Figure 14.8. These results show that the lower the noise factor, generally the better collection results ensued. A 10-dB noise factor generated about a 20% improvement in collection results over 15 dB.

The results shown in Figures 14.7 and 14.8 are somewhat surprising in one respect. In the frequency range considered, it is sometimes believed that external noise limits collection performance. Figures 14.7 and 14.8 show that this is not always the case, for the noise factors considered, because lowering the noise factor without changing the external noise had a significant effect on collection performance.

As expected, as the noise factor was increased the probabilities decreased due to

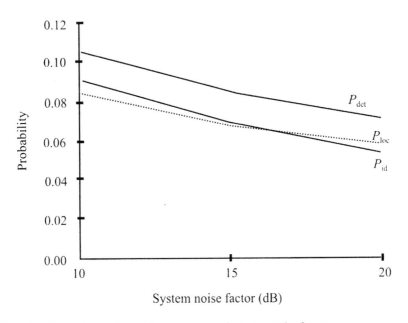

Figure 14.8 General collection performance versus the system noise factor.

the reduced detection range caused by the additional noise. However, these results indicate that the noise factor needs to be considered in system design.

14.4.4 Postprocessing Time

The effects of varying the postprocessing time on NCS collection is shown in Figure 14.9 where $P_{id,NCS}$ and $P_{loc,NCS}$ are plotted. These two variables were about the same and varied between about 0.12 and 0.08. The gradual decrease in performance as the postprocessing time was increased was caused by the operators not being available to continue general search. In fact, while postprocessing was occurring, there was no intercept going on—the signal had gone away. In early-entry operations, while in the GS mode during the first part of deployment, probably a message needs to be recorded on an intercept only if confirming the detection, location, or identification of an NCS. Otherwise, when very little is known about the target environment, reporting on every intercept is probably not necessary.

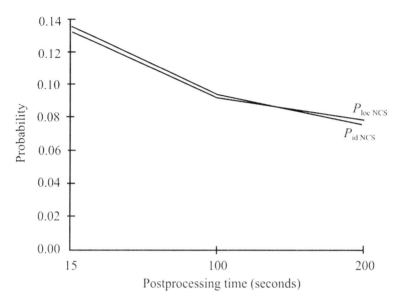

Figure 14.9 NCS collection performance versus the postprocessing time.

Collection performance against targets in general is plotted versus postprocessing times in Figure 14.10. The same general effects as far as collection of NCS targets are evident. About half the number of targets were collected at a 200-second postprocessing time versus 15 seconds. This is also an indication that it was a target-rich environment for the systems; there were still targets to be collected within range of the EW systems.

If, when a net frequency was found, the collection for that net was assigned to an available operator somewhere in the collection ensemble, and, after that time, that operator only copied that frequency, it would be equivalent to using approximately three operators / 200 nets = 0.015 probability of identification and location of a net's control station in the 200 net case (ignoring for the moment hearability concerns). At a 15-second postprocessing time, the effects of searching and sharing collection assets increased the number of NCSs collected by a factor of about eight. At a 200-second postprocessing time this was reduced to a factor of about five. These factors indicate that the processing flow described earlier did indeed produce work enhancements.

14.4.5 Mission Duration

All else equal (all targets are within range), as the duration of the mission is increased,

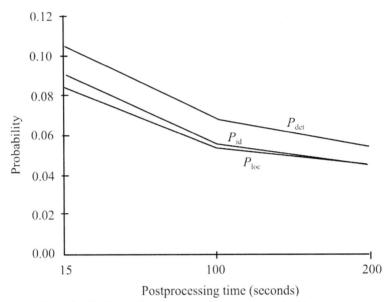

Figure 14.10 General collection performance versus the postprocessing time.

more of the targets should be collected and that was indeed the case as shown in Figure 14.11. Approximately 75% more net control stations were collected at 3 hours versus 1 hour. Similar results were obtained for all of the targets as shown in Figure 14.12. These results indicate that within the first hour, neither all of the target net control stations nor all net members were detected, even though they were within hearability distance. In other words, as indicated above, it was a target-rich environment. There were still targets to collect and identify after the initial hour or two.

Recall that every so often, as determined by the revisit timer, the channels (frequencies) where targets had already been found were revisited. This tends to decrease the number of new target nets found, especially during the latter part of the collection period, and therefore tends to decrease such probabilities computed here. In an operational scenario, as a net is determined to be of no interest it should be removed from the directed search list so it does not slow down the search process. Such frequencies go onto a "lock-out" list of frequencies. The revisit process, as determined by the revisit timer, is normally intended to provide intercept on frequencies known to contain activity. If it is not of interest it should be precluded.

This process is complicated by target frequency changes, however. It is common practice for communication networks to change their operating frequency on a regular basis to preclude or minimize targeted collection efforts such as those described here.

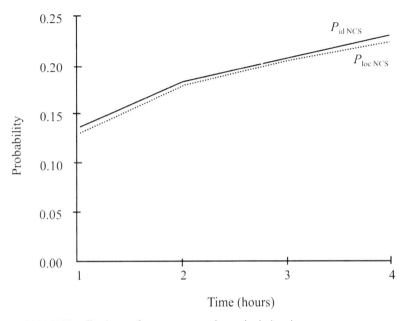

Figure 14.11 NCS collection performance versus the total mission time.

It is necessary to recover the new frequencies once they have changed.

14.5 CONCLUDING REMARKS

One of the conclusions based on the results presented herein is that ground-based communication intercept systems should use a postprocessing time in the range of 20–100 seconds and no longer against limited target environments. The processing time after a signal has disappeared is comprised of two major components: (1) the requirement (or not) of preparing messages to be reported out of the collection system, and (2) whether the operator has to listen to a recording of the intercept in order to prepare the report.

The first component can be made smaller by not requiring extensive information to be reported from the collection system. This can be facilitated by having an off-line (to the collection mission), separate system with appropriate operators perform the reporting mission. This would necessitate the exchange of collected signals between the collection system and the reporting system, which could be

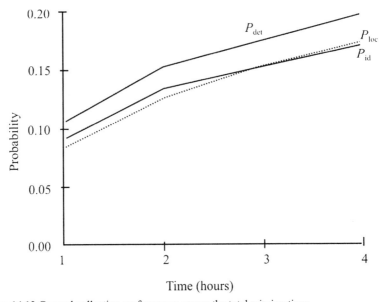

Figure 14.12 General collection performance versus the total mission time.

accomplished in a variety of efficient ways. The second component can be made smaller by supplying the collection operator with more automated means to retrieve collected signals than analog tape recorders. Sophisticated digital storage means are available to allow instantaneous recall with random access to any speech segment. Providing the stored signal to a separate system from these digital recorders could easily be accomplished.

An alternative is not to require any reporting on signal content at all from such collection systems. In this case, the operator may simply edit fix data and nothing more, or raw fixes may be reported as is, with appropriate editing being performed elsewhere, if at all. In this case, postprocessing times would come very close to the optimum 20–100-second range indicated in the charts herein. Although this may be attractive because of its simplicity, it is highly questionable if such fix data alone would be of much value to anyone. In order to establish an adversary's intent and disposition, it is necessary to do more than just locate the targets on the surface of the Earth. Identification of a target, in general, requires that the contents of the transmission be understood. Other forms of automation support to collection operations, however, are also useful which collect external information about signals. Such techniques could include automatic translation and signal type identification.

Without a priori information about the target frequencies, it is necessary to search the frequency spectrum to find the active targets. No such knowledge was assumed for the simulation. Each operator was assigned an equal portion of the 30–90-MHz range and scanned that frequency band until targets were found.

Using a search bandwidth that is narrow is key to providing efficient collection. Herein, the effects on collection performance were essentially independent of the search bandwidth. Cochannel considerations as well as range considerations thus dominated the selection of search bandwidth, and these mutually required narrower bandwidths. In this case, searching more slowly is more important for noise elimination than searching rapidly by using a wide search bandwidth and allowing more noise.

The results of the simulation showed that external noise was not the limiting noise source in the frequency range considered (30–90 MHz). The noise figure of the receiving system should be as small as possible, and certainly should not be higher than 10 dB.

References

[1]   TRADOC Pamphlet 525-5, "Force XXI Operations," Chapter 3, August 1994.
[2]   FM 100-12, Passive Defense, Chapter 7.
[3]   United States Army Posture Statement 2001, Chapter 3, The Army Vision and Force Modernization, accessed July 2001, http://www.army.mil.
[4]   U.S. Department of Defense Joint Publication 3-0, Ch. IV, Joint Operations in War, June 1996.
[5]   U.S. Army TRADOC Pamphlet 525-69, Concept for Information Operations, HQ, USA TRADOC, Ft. Monroe, VA, August 1995.
[6]   U.S. Army Vision 2010, Dominant Maneuver.
[7]   U.S. Army Field Manual 5-100, Chapter 12, Contingency Operations.
[8]   U.S. Army TRADOC Pamphlet 525-75, Intelligence XXI, November 1996.
[9]   *A Guide to the Use of the ITS Irregular Terrain Model in the Area Prediction Mode*, U.S. Department of Commerce, National Telecommunications and Information Administration, NTIA Report 82-100, April 1982.
[10] Adamy, D., *EW 101: A First Course in Electronic Warfare*, Norwood, MA: Artech House, 2001.
[11] Neri, F., *Introduction to Electronic Defense Systems*, Norwood, MA: Artech House, 1991.
[12] Torrieri, D. J., *Principles of Secure Communication Systems*, Norwood, MA: Artech House, 1992, Chapter 4.

Chapter 15

Detection and Geolocation of Frequency-Hopping Communication Emitters

15.1 INTRODUCTION

Low probability of intercept communication techniques have been developed to try to thwart attempts at executing EW. One of those techniques is known as frequency hopping. In this technique, the frequency of the transmitter is changed often so that ES systems do not know where the frequency of transmission is at any point in time [1–3].

Described in this chapter is a technique for detecting and geolocating such frequency-hopping targets. The technique is first described analytically and then the results of a simulation of the technique for four different configurations of ES systems are presented.

15.2 ANALYSIS

The technique consists of stepping a digitally controlled receiver from one channel to the next, dwelling on any one channel just long enough to measure the energy there. If the energy level exceeds an SNR threshold, then signal detection is declared. Once this occurs, other processing would be invoked, including geolocation of the target. This subsequent processing is discussed elsewhere herein.

The fundamental tradeoff in this technique is the scan rate of the ES receiver versus receiver bandwidth. As previously discussed, the wider the bandwidth, the lower the receive SNR for a constant signal level. Therefore, the wider the bandwidth, the fewer the number of targets that could be detected, and the lines of position are less accurate because of the lower SNR. However, more channels are covered in a given period of time, thus searching faster. On the other hand, the narrower the bandwidth, the slower the scan rate and therefore the fewer the number of channels covered during a given time period, taken as 1 ms and 10 ms for now. In this case, the SNRs are higher, however.

One of the more important questions surrounding a receiving system of the type described above is whether a scanning receiver will ever detect a signal and, if so, with what probability. By the nature of the processing, a single detection of a signal is adequate for locating the transmitter associated with it. Therefore, the probability of locating a target is equivalent to the probability of detecting it. This probability is calculated in this section.

A scanning superheterodyne receiver essentially forms what is called in communication theory a *partial-band filter-bank combiner* (PB-FBC) [4]. This is because at any given instant in time the receiver is dwelling at a single frequency with a given bandwidth. In the 10 ms that a 100-hps target dwells on a channel, the receiver covers one or more of these channels depending on the scan rate, as previously discussed. Therefore, the partial band corresponds to that portion of the 30–90 MHz frequency band over which the receiver is scanned in this time increment. The probability of a hit is given by

$$P_{hit} = \frac{N_s}{N_T} \tag{15.1}$$

where N_T is the total number of channels (25-kHz-wide channels in our case) in the total bandwidth and N_s represents the number of channels scanned in one time interval. Since the targets are assumed to be randomly hopping over the entire 60-MHz bandwidth and they do not use "hop sets," then $N_T = 2,400$. The values of N_s and the corresponding values for P_{hit} are given in Table 15.1.

The probability of getting n hits per transmitted message (for specificity a message corresponds to a 5-second transmission) is given by the binomial probability distribution, namely,

$$P_B(n) = \frac{N_h!}{n!(N_h - n)!} P_{hit}^n (1 - P_{hit})^{N_h - n} \tag{15.2}$$

which gives the probability of detecting exactly n target frequencies from a given transmitter. N_h is the number of hops in the message (in the case of interest here, N_h is equal to 500 since the messages are 5 seconds long and the targets are hopping at a 100-hps rate). This can be used to compute the probability of detecting at least one hop in a 5-second transmission for the various configurations because

Table 15.1 Partial-Band Bandwidths and Corresponding Values of P_{hit}

Dwell Time (ms)	Bandwidth (kHz)	N_s	P_{hit}
1	25	10	0.0042
1	100	40	0.0167
1	200	80	0.0333
10	25	1	0.00042
10	100	4	0.00167
10	200	8	0.00333

$$P(n \leq N-1) = \sum_{n=0}^{N-1} P_B(n) \tag{15.3}$$

and

$$P(n > N) = 1 - P(n \leq N-1) \tag{15.4}$$

These results are illustrated in Figures 15.1 and 15.2. The interpretation of these charts is as follows. The value on the curve at a value of the abscissa, say, n_1, represents the probability that n_1 (or more) hops will be detected in a 5-second transmission. The detection rules used herein only require the detection of one

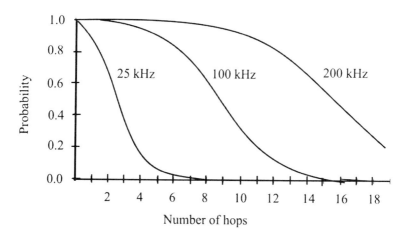

Figure 15.1 Probability of detection (location) of a target for a 1-ms dwell time.

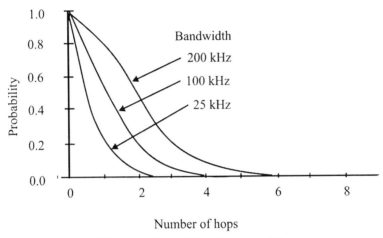

Figure 15.2 Probability of detection (location) for a 10-ms dwell time.

transmission to locate the targets; hence, the probability values beyond $n_1 = 1$ are of academic interest only. If a way could be found to correlate one fix with another, then multiple fixes on the same target could be used to improve the location accuracy, however. Clearly, when the dwell time is 1 ms, we are almost always assured of detecting at least 1-hop out of a 5-second transmission independent of the bandwidth used. For 10 ms, the probability of detection, and therefore location, is not as good for a single 5-second transmission but for the wider bandwidths is still respectable. It can be concluded that there is a good chance of detecting and therefore locating targets with a scanning receiving system that scans at a rate of between 1 ms and 10 ms per channel operating against 100-hps targets after only 5 minutes of collection. Furthermore, this can be accomplished using reasonable instantaneous scanning bandwidths to reduce problems caused by cochannel interference and sensitivity reductions due to noise.

15.3 SIMULATION

In order to ascertain the significance of the above analysis, a simulation was developed. The simulation assumed four different configurations of collection/geolocation systems. These four configurations are illustrated in Figure 15.3. For configuration 1, a low-flying airborne system operating with three ground standoff systems was simulated. The airborne system was at a standoff range of 20-km at an altitude of 5,000 feet. The ground systems were deployed in a lazy V configuration, equally separated vertically, and at standoff ranges of 5 km

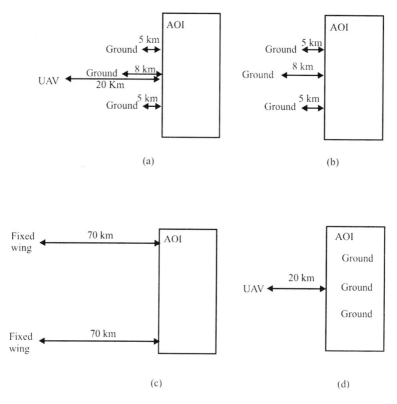

Figure 15.3 System configurations considered in the simulation: (a) configuration 1, three ground systems with one low-flying airborne sensor; (b) configuration 2, three standoff ground sensors; (c) configuration 3, two high-flying standoff sensors; and (d) configuration 4, three ground stand-in sensors.

and 8 km as shown. In configuration 2, the airborne system was not deployed. The ground systems were deployed as in configuration 1. For the third configuration, two higher-flying airborne systems were simulated at a standoff range of 70 km and at an altitude of 20,000 feet. The last configuration is similar to configuration 1 except the ground systems were deployed in the midst of the *area of interest* (AOI), thus simulating a nonlinear battlespace.

The target networks in the simulation were based on the threat disposition shown in Figure 15.4, which was an area 50 km wide and 20 km deep. The nodes

478 Introduction to Communication Electronic Warfare Systems

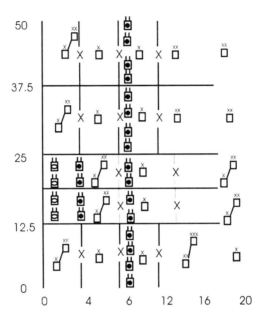

Figure 15.4 Target scenario for the simulation.

shown were configured into specific networks which comprise 19 specific networks of radios. The networks are graphically portrayed in Appendix B. The target network communications are described in Section 14.2.

15.3.1 ES System Operation

The receiving systems consisted of stepping narrowband receivers, energy detectors connected to the output of these receivers, thresholding devices that measured the output of the energy detectors, and azimuth/LOP measuring devices. The processing time, that is, the time it took to do the stepping from one RF channel to the next, to perform the energy measurement and thresholding function and to measure the LOP was varied according to the configuration described below, between 1 ms and 10 ms. The basic operation is the same as described in Section 14.3, except the only function performed is direction finding.

The operational modes of configurations 1 and 4 are different from those in 2 and 3. The result of the propagation differences is that the ground stations do not receive much signal energy. The stations scanned the frequency spectrum in a synchronous way, that is they stepped to each channel at precisely the same time.

n those cases where ground stations were included with an airborne element configurations 1 and 4), each ground station, at each channel, computed an azimuth whether they detected the signal or not. This is feasible because, for correlation interferometry as well as some other types of DF techniques, usable LOPs can be obtained on signals that are weaker than can be automatically detected (the distinction here is signal detection versus computing an LOP) [5].

In configuration 2, even though an LOP was computed at the ground sensors regardless of the SNR present there, an adequate SNR was required at the DF net control station (the middle one) for signal detection, otherwise the systems would not know if there was a signal present or not. The randomly computed fixes that would result would be worthless information.

The accuracy of a computed LOP normally will vary proportionally to the inverse of the square root of the SNR. Depending on the particular DF approach used, a DF system may compute an LOP on signals that are too weak to be automatically detected. Some approaches even provide a capability at levels below 0-dB SNR. This is achieved by making the processing bandwidth very small, smaller than a VHF channel, thereby substantially decreasing the processing noise. For the purpose of the subsequent analysis, it was assumed that the Cramer-Rao accuracy bound was achieved in the LOP systems, an optimistic assumption. Note that wider receive bandwidths produce better accuracy; this, then, is another system tradeoff that must be considered, because the wider the system bandwidth the more noise enters, and therefore the lower the SNR.

The collection equipment on the low-flying system in configurations 1 and 4, which was flying at a considerably higher altitude than the altitude of the ground stations, operated in a similar fashion except that energy detection was first performed. Every station transmitted its time-frequency-LOP information to the computer at a ground site, with the airborne sensor serving as a communication relay for the forward-deployed ground stations. The ground site only computed and displayed fixes at those frequencies at which the airborne sensor indicated there was energy present. For configuration 3, the operation was identical to that indicated above for the low-flying system with energy detection being performed at each site before an LOP was computed. In all of these configurations, each receive site computed LOPs and sent this information to a central location where a fix was computed.

The digital receivers were stepped with instantaneous bandwidths of 25, 100, and 200 kHz. The analysis provided above calculated the probability of detecting a hopping target with such a technique. The scan rate was different for configurations 1 and 4 from 2 and 3. This is because large quantities of data must be transferred when azimuths are reported at every scanned channel and the data link would not support the same scan rates. For configurations 1 and 4, a 10 ms dwell time per channel was used, whereas in configurations 2 and 3, the systems dwelled for 1 ms per channel. To put these numbers in perspective a receiver with a 10 ms dwell time could scan one channel in 10 ms (the time a frequency hopper stayed at a single frequency) while with a 1-ms dwell time, 10 channels could be measured. In the latter case, when the bandwidth was 25 kHz, a total of 250 kHz could be covered, when the bandwidth was 100 kHz, 1 MHz could be covered, and when the bandwidth was 200 kHz, 2 MHz could be covered.

A target location was obtained on each cell according to the rules explained above, as long as adequate data link capacity were present. The rate of the data link that interconnected the systems together limited the number of LOPs that could be transferred among systems. Since all the systems were time synchronized and covered the same instantaneous frequency band at the same time, a target detected at one system at a frequency would also be detected at the same time at all collection systems, subject to signal level constraints (the signals must have been of an adequate level as well as an adequate SNR to be detected). If too many signals were detected at a time for the data link to transfer the LOPs to the analysis station, the remaining signals were considered not detected for the purpose of computing locations.

The data link data rate used herein was 200 Kbps for configurations 2 and 3 and was 400 kbps for configurations 1 and 4. This is not the sustained rate, because all stations shared this capacity and in some cases relaying was necessary. The maximum number of LOPs that could be transferred over this data link was data rate/100 × 1/hop rate assuming that 100 bits per LOP record would be required. Since the hop rate was always 100 hps, then the maximum number of fixes that could be computed was 20 during any individual dwell of the receiver for a 200-Kbps rate and 40 for a 400-Kbps rate. The data link was assumed to be using a frequency of 1.7 GHz, and its propagation characteristics were included in the model.

The collection system and target transmit antenna heights are among the variables that determine the amount of power received at the receive sites. The receive antenna heights of the airborne systems were the altitude of the aircraft indicated previously. For the ground systems, in configuration 1 the receive antenna height was 10m, while for configuration 4 the receive antenna heights were 5m.

A fix may not have been obtained on a hop for a variety of reasons, probably the most important being that the channel was not covered by the receive systems at that dwell. Other reasons why an individual hop may not have collected were: The SNR was too low [an SNR of 15 dB was necessary which is approximately that required for an ideal energy detector to obtain a probability of detection, P_d, of 0.9 with a false alarm probability, P_{fa}, of 10^{-6}, and varying amounts of background noise as well as the receive system noise figure (10 dB) were added to the signals], or the signal level was too low (the receiver dynamic range was a simulation

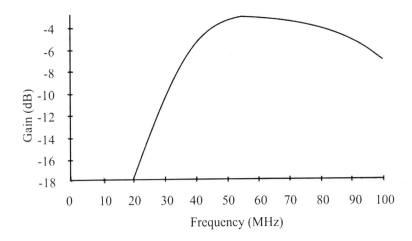

Figure 15.5 Receive antenna gain used in the simulation.

variable and close interferers combined with this dynamic range set the minimum signal levels that could be detected). In all cases a dynamic range of 72 dB for the receive systems was used. In the case of the airborne systems, the closest interferer was assumed to be a distance away equal to the altitude of the aircraft. For the standoff ground stations, the closest interferer was assumed to be 0.5 km away and for the cross-FLOT ground stations, the closest interferer was assumed to be 15 km away (close transmitters are not interferers but are targets, although close targets could still desensitize the receiver).

The receive antenna gain used in the simulation is shown in Figure 15.5. An antenna with this gain characteristic is typical for the frequency range of interest herein. This antenna performance could be attained with a half-wavelength dipole tuned to a frequency of about 60 MHz, for example. This would be typical of a ground configuration but would probably be difficult to achieve with a small airborne platform, such as a UAV.

15.3.2 Results and Analysis

The results of the simulation are presented in this section. Two figures of merit were calculated for each configuration: the probability of location, P_{loc}, and the average fix miss distance. Both of these parameters are dependent on system timing parameters, which are the primary variables of interest in this analysis.

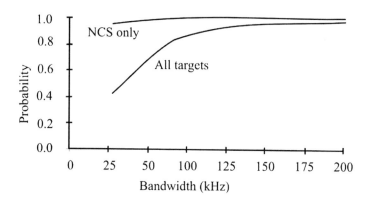

Figure 15.6 Probability of location for configuration 1, low-flying airborne with three ground standoff platforms.

15.3.2.1 Probability of Location

Figure 15.6 shows P_{loc} for configuration 1 (low-flying airborne and three ground standoff systems) plotted versus bandwidth. The upper curve shows P_{loc} for the net control stations while the lower curve corresponds to all of the targets. In the simulation the net control stations broadcast much more frequently than the rest of the net members, which accounts for the difference (see Figure 14.1). Clearly, very good location performance is obtained in all cases except perhaps for using a 25-kHz bandwidth.

Figure 15.7 shows the same data for configuration 2 (ground standoff systems only). In this case the performance was not very good. This is because of the low SNRs available at the ground stations for many (probably most) of the targets. The ground systems computed an LOP regardless of the SNR at the sensor; at the PF net control station the signal had to be of adequate SNR to be detected, thus the low probabilities of location (detection). One would expect these curves to be monotonically increasing functions of bandwidth, while Figure 15.7 shows a dip as the bandwidth is increased. The reason for this is the tradeoff between search speed; the wider the bandwidth, the faster the search and the higher the noise level thus decreasing the SNR.

For configuration 3 (two higher-flying aircraft), the results are shown in Figure 15.8. Here again, P_{loc} was very good; every NCS and almost every target was located regardless of the bandwidth. This was due to the relatively high SNRs that such a configuration would receive even though it was located 70 km to the rear of the FLOT. One would also expect that the LOP accuracy was good in this configuration. The fact that reasonable performance was achieved even at a

Detection and Geolocation of Frequency-Hopping Communication Emitters 483

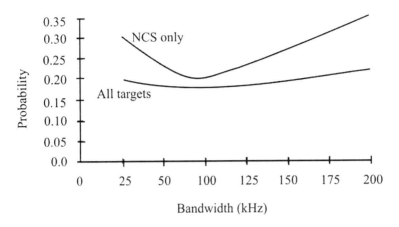

Figure 15.7 Probability of location for configuration 2, three ground standoff sites.

bandwidth of 25 kHz is important because this configuration would experience the highest probability of cochannel interference.

Figure 15.9 shows the performance for configuration 4 (low-flying aircraft with three ground systems deployed 15 km across the FLOT). As with configuration 1 in this case, the ground sites did not have to detect energy first, but computed an azimuth on every channel. The low-flying aircraft performed the energy detection function. The performance was similar to as configuration 1, which was expected because the detection functions of the systems are the same.

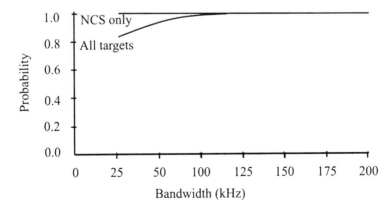

Figure 15.8 Probability of location for configuration 3, two fixed-wing aircraft.

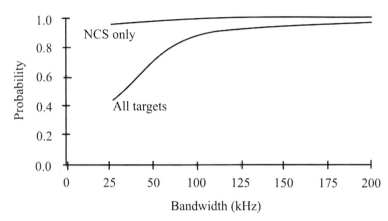

Figure 15.9 Probability of location for configuration 4, airborne sensor with three cross-FLOT sensors.

15.3.2.2 Miss Distance

One would expect that at least for configurations 1 and 4 the accuracy of computing fixes would be significantly degraded due to the fact that the ground sites could not receive an adequate SNR in many cases. A comparison of the average miss distances computed during the simulation is shown in Figure 15.10. The data in this chart represents the radius of a circle within which one-half of the computed fixes would lie and one-half would be outside. Configurations 1 and 4 produced the least accurate fixes, while configurations 2 and 3 produced the best. The decreasing LOP accuracy effects of the bandwidth (and therefore the SNR) are most evident for configuration 1 as the miss distance about doubles as the bandwidth increases from 25 kHz to 200 kHz. In the other cases apparently the SNRs were adequate most of the time to not significantly degrade the LOP performance.

As discussed in Section 15.3.2.1, one would expect these curves to be monotonically increasing with bandwidth because the LOP accuracy is inversely related to the square root of the SNR. The average miss distance for configuration 4 does not follow this pattern.

15.3.3 Discussion

One should not conclude which configuration performed the best overall based on one of these charts alone; they must be considered together. Configuration 2 produced the highest accuracy but missed many of the targets, as shown in Figure

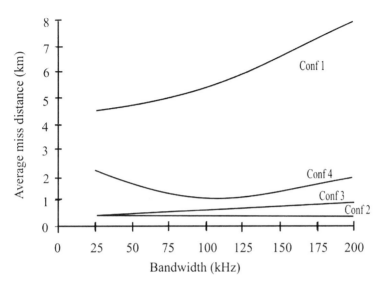

Figure 15.10 Average miss distance.

15.7. The best overall results were obtained with the higher-flying aircraft configuration while the second best with the low-flying aircraft working with cross-FLOT ground systems.

One of the advantages of a DF system of the kind described herein is that its performance is not limited by the size of the target environment, except for when cochannel interference occurs. Collection systems that try to track LPI targets exhibit a characteristic that there is a point where the processing equipment cannot keep up with the load and they become saturated. Since LOP gates are often used as a sort parameter, basic system performance such as LOP accuracy also limits the sorting and therefore tracking performance of collection systems as well. Since the equipment analyzed here is not trying to track, it simply computes an azimuth at a frequency channel and reports the result, no such limitations exist.

While a 1-km miss distance when locating targets is probably not adequate for targeting purposes, it is adequate for target and situation development. Since the total target area covered was 20 km × 50 km, this corresponds to a total area of 1,000 km^2, the area covered by a 1-km circle is approximately 3 km^2. Therefore, it is possible with a system such as that described herein to localize targets to a granularity of 0.3% of the AOI at least half the time.

As expected, the case where the ground standoff systems computed an LOP on every channel regardless of the level of energy present produced the least accurate fixes. It is also the case, as expected, that when there was only ground standoff sensor systems that produced the fewest number of fixes because the distant targets could not be heard. These results indicate that whenever possible it is always desirable to have an elevated sensor component as part of the sensor array.

15.4 CONCLUDING REMARKS

The results contained herein indicate that (1) theoretically there is a very high probability that 100-hps frequency-hopping targets can be detected by a scanning receiving system and (2) there are reasonable tactical configurations of receiving systems that can locate these targets. The accuracy of these locations varies, but one would expect that on the order of 1-km accuracy is possible.

The scan rate (bandwidth) versus SNR results can be summarized as follows. Any configuration with an aircraft involved produces $P_{\text{loc}} \rightarrow 1$ for some bandwidth, indicating that an adequate SNR was obtained for the scan rates used. For the ground-only case, SNR limits performance in most cases for any bandwidth.

References

[1] Simon, M., et al., *Spread Spectrum Communications Handbook,* New York: McGraw-Hill, 1994.

[2] Torrieri, D. J., *Principles of Secure Communication Systems,* 2nd ed., Norwood, MA: Artech House, 1992, Chapter 1.

[3] Petersen, R. L., R. D. Ziemer, and D. E. Borth, *Introduction to Spread Spectrum Communications,* Upper Saddle River, NJ: Prentice Hall, 1995.

[4] Dillard, R. A., and G. M. Dillard, *Detectability of Spread Spectrum Signals,* Norwood, MA: Artech House, 1989.

[5] Jemkins, H. H., *Small-Aperture Radio Direction-Finding,* Norwood, MA: Artech House, 1991.

Chapter 16

Signal Detection Range

16.1 INTRODUCTION

This chapter examines the distances over which communication ES systems can detect target signals. This range is dependent on several factors, including ES system sensitivity, antenna gains, power levels, and frequency. Determining the nominal detection ranges of typical systems against typical targets provides some insight to the limiting factors involved.

The detection range is also strongly dependent on how high the antennas involved are elevated. Herein, only ground-based targets will be addressed, so the only possible elevated antenna is on the receiving system. Four configurations are examined—the first is a high-flying aircraft. The second configuration is a low-flying aircraft, consistent with the flight characteristics of a UAV. The last two configurations are ground-based. For the first of these, the antenna is mounted on top of a relatively short mast—say, 10m or so in height. This configuration would be consistent with a stationary deployment. The second ground-based configuration has a short antenna mast such as that which would be used for OTM operations [1–8].

The advantages of the utilization of UAVs in many circumstances are obvious. They can overfly unfriendly areas with minimal risk of loss of life. They can be flown for long periods of time without the human-needs problems of manned aircraft. They can carry a varied mix of sensors, and indeed, the sensors can be changed relatively easily.

The advantage of ground-based systems is their all-weather capability. Their most significant shortcoming is the limited detection range they provide. Even when placed on the highest hills available in the area, their range is exceptionally limited compared to their airborne counterparts. The rare exception would place a ground-based system on the edge of an available cliff with low-lying plains over which to watch—not too likely an occurrence, and even if it did occur, in conflict situations an adversary would not permit such systems to remain in operation very long.

Up until recently, the most significant advantage of high-flying, large aircraft was their weight-carrying capability. With the advent of large UAVs, however, this advantage is disappearing. In some cases the ES system operators fly on the aircraft and in some cases they do not. In the latter case, the aircraft mission payload must be electronically linked somehow to wherever the operators are located.

Relatively high-flying fixed wing aircraft have been used for ES missions in the military services for several years. The higher above the surface of the Earth an ES system flies, the more signals can be seen, all else being equal. There are some limitations to using this configuration for ES, however. Its primary advantage can sometimes also be its biggest nemesis. Flying higher, more of the friendly transmitters can be seen, as well as targets. This creates cochannel interference. Depending on the type of signals involved and the geographical relationships among the emitters, this cochannel interference can sometimes be suppressed.

When possible, the high-flying systems are usually flown in a standoff posture—out of harm's way. In that case, the systems must look through many blue emitters before seeing the target array. To further exacerbate the problem, the signals from these blue emitters are frequently stronger than those from the red targets. In nonlinear situations there are also many blue signals mixed in with the targets. In mountainous terrain, the terrain can block communication signals. This is true for the intended receiver as well as an ES system. This can create coverage problems for standoff systems. In order to cover valleys in such terrain, over-flight is often the only way. Terrain effects can be helpful or harmful, depending on where the targets are located. Herein terrain is not taken into account; smooth Earth is assumed.

In the future, low-power PCSs based on commercial technology are expected to proliferate in the military and paramilitary forces worldwide. Such systems are characterized by very low-powered handsets. In addition, these systems intentionally reuse the frequency spectrum. Currently in the United States, every fourth cell in any straight line utilizes the exact same set of frequencies as illustrated in Figure 16.1. In Figure 16.1 the three cells marked with an X all utilize the same frequency set. The hexagonal cells, which of course in real systems only approximate hexagonal, arranged in this pattern, are maximally separated in distance from the next one that reuses the same frequencies. The seven cells surrounded in bold lines make up a "cluster" of cells. All the frequencies in a region are used in each cluster. If an ES system sees two or more of these cells that reuse frequencies, the same cochannel problem mentioned above

Signal Detection Range

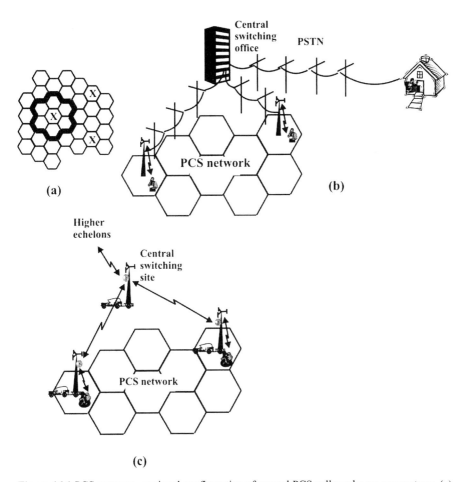

Figure 16.1 PCS systems—regional configuration of several PCS cells make up one system: (a) Those marked with X all reuse the same frequencies. The cells in groups of seven use all the frequencies in a region. (b) The physical makeup of a PCS system connects mobile users to other mobile users or mobile users to the PSTN. (c) Military and paramilitary PCS systems are similar to their commercial counterparts.

occurs, degrading both the ability to intercept as well as PF a target. Therefore, low-flying systems or ground-based systems are required in such instances [4, 9].

PCS base stations will have substantially more power available than the handsets. For commercial communication systems the base stations are fixed sites. When military they will be mobile but not normally moving when in operation. Thus, they can facilitate operation from generators, which provides considerably more power than the batteries in the handsets do. Here, the PCS base station has an ERP of 10W at 1 GHz. The antenna height for the base station will be on the order of 10m or so.

16.2 NOISE LIMITS ON DETECTION RANGE

The amount of noise present at the receiver compared to the amount of signal present determines the detection performance of ES systems. The noise level is situation-dependent. For any kind of reasonable engineering analysis of range performance, the level of noise is either measured or assumed. Herein it will be assumed as explained below. The SNR, given by

$$\text{SNR} = \frac{P_R}{N_{total}} \qquad (16.1)$$

and frequently expressed in decibels, must be above some minimum value in order to detect signals. This level depends on the modulation type and what is trying to be done with the detected signal, but for simplicity here the required SNR level will be 10 dB, which is representative of a broad range of digital modulation types, to include PCS signals. It is, however, pessimistic (too large) for effective collection of analog (AM, FM) modulations [10, 11].

N_{total} is the total effective noise at the input to the ES system as discussed in Chapter 3. Due to the altitude of the airborne collection systems considered here, it is reasonable to assume that there is little man-made noise to limit their performance. Whereas such noise sources are very important (and limiting) for ground-based military systems embedded with military forces operating equipment that is radiating, such as vehicles, the aircraft are far enough away that these sources are insignificant. Of course, the noise generated by the aircraft itself must be considered but this can usually be filtered out. At least it is known ahead of time. Therefore, the limiting noise source in most airborne cases will be galactic noise. From Figure 3.1 the following noise values are used for this analysis when considering airborne configurations.

$$N_{external} = \begin{cases} 18 \text{ dB} & f = 30 \text{ MHz} \\ 6 \text{ dB} & f = 100 \text{ MHz} \\ 0 \text{ dB} & f > 200 \text{ MHz} \end{cases} \qquad (16.2)$$

When the ES system is ground-based, it is sometimes embedded among many systems that generate man-made noise, such as vehicles and generators. In these circumstances it is more accurate to model the noise environment as suburban man-made limited in Figure 3.1 rather than galactic noise limited. Thus, $N_{external}$ is given by

$$N_{external} = \begin{cases} 35\text{ dB} & 30\text{ MHz} \\ 24\text{ dB} & 100\text{ MHz} \\ 2\text{ dB} & 1\text{ GHz} \end{cases} \quad (16.3)$$

16.3 TARGETS

Three notional target types are considered for comparison purposes here. The first is a 10-W ERP PTT transmitter typical of the low-VHF frequency range common to military ground forces but also typical of PCS base stations, which is the second target type. The third target has typical characteristics of a PCS handset with an ERP of 0.5W at a frequency of 1 GHz. In these cases the transmit antennas are assumed to be omnidirectional.

In Figures 16.2–16.11, the sensor system is placed at $x = 0$ and $y = 0$, and the distances are displayed in kilometers. Since a nonlinear battlespace is the most likely to be encountered by military forces, the detection performance in a 360° sector around the sensor is important.

Military targets in the low-VHF frequency range would typically employ vertical whip antennas, either dipoles or monopoles. The result is predominantly vertical polarization. Ground-based ES system antennas, when stopped and deployed, can readily be built for vertical polarization. Antennas for PCS targets will need to accommodate both polarizations since the handset antennas will be in about any attitude at any time. The base stations will typically be built with vertical polarization.

In addition to these targets, the detection range performance against the sidelobes of two directional antenna types is examined. In particular, two popular directional antennas, the Yagi and horn-fed parabolic dish, are considered.

16.4 DETECTION RANGE WITH THE REFLECTION PROPAGATION MODEL

The reflection propagation model described in Section 2.9 given by (2.43) will be used to calculate the signal levels. As previously indicated, at short ranges this model tends to underestimate the amount of power at the receiver. However, at the limits of the intercept range, it is a more accurate model than many others. The limits of the ranges are where this analysis is targeted.

In this chapter, isotropic antennas are assumed for the receiver; thus, $G_R = 1$. As shown previously, actual antennas, optimized for a particular frequency, could have a higher or lower gain than this. Also, reasonable tactical configurations for antennas would not exhibit this gain characteristic. It is assumed here for simplicity and can serve as a basis of departure when actual antenna gains are known.

Since terrain is not considered here, the coverage areas will be circles with the sensor at the center. The radii of the circles will be the calculated detection ranges.

16.4.1 Airborne Configurations of ES Systems

Two airborne configurations are considered. The first is a high-flying system, at an altitude of 10,000-m AGL. In the second configuration the aircraft is at 3,000m AGL.

In Figures 16.2–16.11 the SNR required for adequate interception is represented by a horizontal plane in three-dimensional plots. On the other hand, the SNRs will vary depending on where in the target area the target emitter is located. The SNR will also vary with frequency because the noise varies with frequency. Thus, the signal level will be strongest and the SNRs the highest when the targets are near to the sensor system, and weaker and lower further away. Anywhere the SNR contours rise above these planes the targets can be detected.

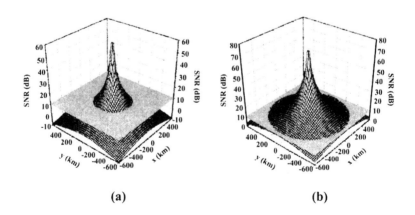

Figure 16.2 Noise-limited range performance for high-flying system: (a) 30 MHz, 10-W ERP; and (b) 100 MHz, 10-W ERP.

16.4.1.1 High-Flying/Standoff

Results are presented for a single sensor system. Usually two or three such systems are flown at the same time; however, including results in the analysis for more than one platform adds no additional information for detection range determination. The coverage area for PF, however, is the conjunction of the coverage regions for two or more sensors. Therefore, in those investigations the overlap of the coverage charts is important.

Figure 16.2(a) shows the range performance when the frequency is 30 MHz and the transmitter power is 10-W ERP. Note that the geographic scale is ±600 km. The 10-dB SNR signal detection range is approximately 300 km. When the frequency is 100 MHz and the power is 10-W ERP, the detection range performance is as shown in Figure 16.2(b). The detection range is larger than when the frequency was 30 MHz. While the signal at a higher frequency does not propagate as well as one at a lower frequency, the noise is less at the higher frequency, yielding greater range performance. Approximately, the RLOS (425 km) is the range over which detection is possible at 100 MHz.

For the PCS frequency range at around 1 GHz, the handset detection performance is shown in Figure 16.3(a) where the ERP is 0.5W. The detection range in this case is about 380 km. Detection of a PCS base station is possible at ranges out to RLOS as shown in Figure 16.3(b).

The results presented in Figure 16.3(a) assume that the handset antenna pattern is omnidirectional, thus facilitating equal radiation in all directions. PCS handset antennas have far from omnidirectional antenna patterns, and vary with orientation to the user's body, reflections off local objects such as stop signs, and a

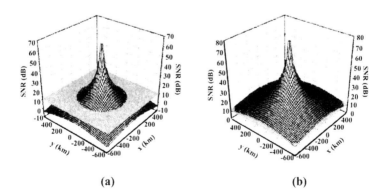

Figure 16.3 Detection range performance of high-flying system: (a) PCS handset at 1 GHz and (b) PCS base station at 1 GHz.

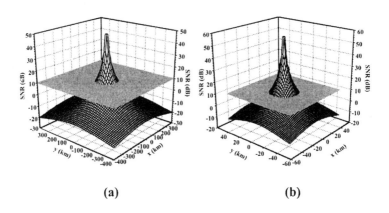

Figure 16.4 Intercept range of a high-flying system when the system is in the sidelobe of a directional antenna: (a) Yagi at 1 GHz and (b) horn-fed parabolic dish at 4 GHz.

myriad of other factors. Unless the target is stationary and in a fairly benign RF environment, the antenna pattern is probably randomly varying. These results, therefore, are optimistic and will not always represent correct detection ranges.

The results in Figure 16.3(b) also assume that the base station antennas are omnidirectional, a poor assumption since PCS base station antennas are typically not. In fact, the expected "smart" base station antennas steer nulls in directions other than toward the particular handset being tracked. Directional antennas at the base station implies that considerably less than 10-W ERP will arrive at the second sensor platform even if the first sensor platform is in the correct antenna beam. That would tend to substantially decrease the intercept range of the second platform, degrading the ability to perform real-time PF.

When the high-flying system is attempting to intercept the sidelobes of directional antennas, the range is limited. Using the power levels determined in Section 8.8 for typical Yagi and parabolic dish antennas, and with a 1.5-MHz bandwidth, meaningful at the multichannel frequency ranges considered here, the intercept ranges shown in Figure 16.4 result. In Figure 16.4(a) the target antenna is the Yagi, with an ERP in the direction of the sensor of 0.14W at 1 GHz. Note the scale is ±400 km. The intercept range is approximately 75 km. The target antenna in Figure 16.4(b) is the horn-fed parabolic dish with an ERP of 2×10^{-4}W in the direction of the sensor. The scale here is ±60 km. The intercept range is about 24 km. With a standoff high-flying sensor system, neither of these targets would be detected.

Signal Detection Range

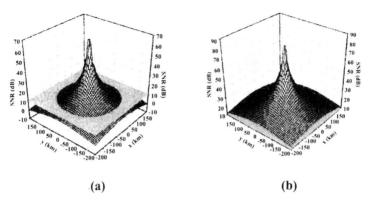

Figure 16.5 UAV detection range: (a) 30 MHz, 10-W ERP; and (b) 100 MHz, 10-W ERP.

16.4.1.2 UAV

A UAV can be used to augment the collection capability of the high-flying ES system. Such systems have the ability to overfly areas that might be prohibitive for the larger systems for a variety of reasons, some of which were mentioned previously. If the high-flying systems are piloted or otherwise manned, it may be too dangerous in some situations to overfly certain areas. If the system is carrying sensitive equipment for special missions, it may be unwise to risk the loss of such equipment. In mountainous terrain, a UAV can be used to augment the larger system, filling in gaps in the coverage in valleys, for example.

Detection range performance at 30 MHz and 10-W ERP is shown in Figure 16.5(a). Note the scale of ±200 km. The SNR is adequate for signal detection to a range of about 160 km or so. The range at 100 MHz and 10-W ERP is beyond the RLOS of 200 km as shown in Figure 16.5(b). At 1 GHz, the low power target at 0.5-W ERP can be detected at a range of 200 km or so as shown in Figure 16.6(a). Against the base station, the range increases to beyond RLOS as shown in Figure 16.6(b).

A UAV when overflying a target area has the opportunity to fly into and loiter within the main lobe of a point-to-point directional communication system. This is one of the advantages of a UAV. On the other hand, just as for the higher-flying standoff system, interception of the sidelobe signal is also possible. The intercept range for a UAV flying at 3,000m is shown in Figure 16.7. Figure 16.7(a) is a sidelobe intercept of the Yagi antenna at 1 GHz, with a signal (and noise) bandwidth of 1.5 MHz. The scale in that case is ±60 km and the intercept range is approximately 55 km. Figure 16.7(b) is for sidelobe intercept of the parabolic dish where the scale is ±20 km. The intercept range is about 10 km.

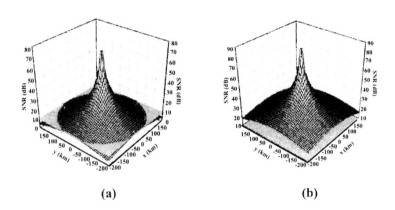

Figure 16.6 Detection range performance from a UAV: (a) PCS handset at 1 GHz and (b) PCS base station target at 1 GHz.

Another implication of these results is that if real-time *position fix* (PF) is desired, two or more aircraft must be flying within the footprints shown. This implies that to cover the gaps in the entire target area, a considerable amount of time may be required, depending on the terrain. This, in turn, implies that a UAV ES system such as this should be used to cover specific areas rather than an entire corps *area of operations* (AO). Of course, real-time PF is not always required, and if delays can be tolerated so that the UAV can be allowed to fly some reasonable distance, then a single aircraft would suffice. GDOP (see Section 11.5) causes this distance to be on the order of the same as the range to the target, 5–10 km in this case, in order to produce reasonably accurate locations. At 100 km per hour aircraft speed this may create collection problems, as the target may go away in the 3 minutes required to fly 10 km.

Extensive range against PCS cell base stations and handsets implies that the sensitivity of the ES system must be decreased in order to minimize cochannel interference. The sensitivity should be set so that no more than one set of the seven cells making up a group in Figure 16.1 are covered at a time. A typical cell radius is 10 km.

Likewise, the extensive intercept range at low-VHF could require decreasing the sensitivity there as well. Since low-VHF frequency-hopping targets utilize much of the spectrum, these intercept ranges could cause cochannel interference, especially since the spectrum is shared by red, blue, and gray forces/populations.

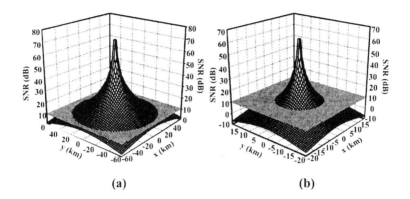

Figure 16.7 Sidelobe intercept range of a UAV: (a) Yagi at 1 GHz and (b) parabolic dish at 4 GHz.

16.4.2 Ground-Based Detection Ranges

Two ground ES system configurations are considered. The first has an elevated antenna on a short mast at $h_R = 10$m reflecting the performance of a stationary deployment. The second antenna height is $h_R = 3$m, reflecting expected performance while OTM.

As shown in Figure 16.8, at 30 MHz and 100 MHz, the detection range is about 3 km and 6 km, respectively, from the sensor for a target ERP of 10W and $h_R = 10$m. The scale in these figures is ±20 km. The limited range is because of

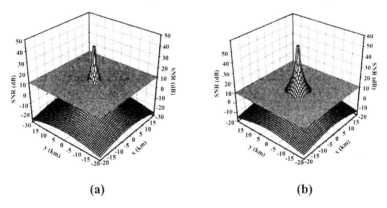

Figure 16.8 Ground-based detection range for $h_R = 10$m: (a) 30 MHz, 10-W ERP; and (b) 100 MHz, 10-W ERP.

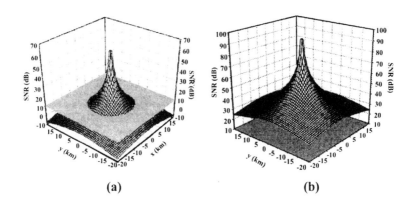

Figure 16.9 Ground detection range for h_R = 10m: (a) PCS handset at 1 GHz with a 0.5-W ERP; and (b) PCS base station at 1 GHz with an ERP of 10W.

the excessively high noise levels at these low frequencies in the suburban manmade noise environment.

This noise environment would not be encountered in all cases. It is probably appropriate for road marches or when there is other significant battlespace activity. When preparing for hostile action, but prior to it, a circumstance where ES systems are normally employed is that the noise would be substantially less, especially at night. Nevertheless, this analysis points out that the range of ground-based communication ES systems is very limited.

In the higher-frequency range around 1 GHz there is insignificant external noise impinging on the receive antenna (about 2 dB_{kTB}). Thus the noise effects are small compared to the low VHF. These effects are shown in Figure 16.9. The handset range is about 10 Km and the base station is out to RLOS (16 km). One of the principal functions of communication ES is to support operations OTM. When moving, the antenna height cannot be too high. These results assume that the receive antenna is erected just above the vehicle, at a height from the ground of 3m. The results are shown in Figure 16.10(a) for a target frequency of 30 MHz and a target ERP of 10W. The scale here is ±10 km. The detection range is about 2 km. At 100 MHz, where the man-made noise is less the detection range, from Figure 16.10(b), is about 3 km.

OTM performance against the PCS system are shown in Figure 16.11. Against the 1-GHz, 0.5-W ERP handset the detection range is about 6 km, reflecting the relatively low external noise at this frequency (2 dB_{kTB}). Against the 10-W ERP base station the detection range is beyond RLOS.

Signal Detection Range 499

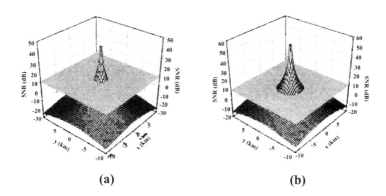

Figure 16.10 Ground based detection range while OTM when h_R = 3m: (a) 30 MHz, 10-W ERP; and (b) 100 MHz, 10-W ERP.

16.4.3 Discussion

Generally speaking, the condition that must be met for signal detection is that the signal must have a certain SNR or higher at the receiver. When external noise is not considered, an E-field level can equivalently specify that SNR. The detection ranges concluded from this analysis are summarized in Table 16.1.

Because in the overflight configuration both the high-flying and low-flying configurations cover a significant portion of the low-VHF band, near-optimum geometries are possible for PF. The best geometry for CAF TDOA and DD PF

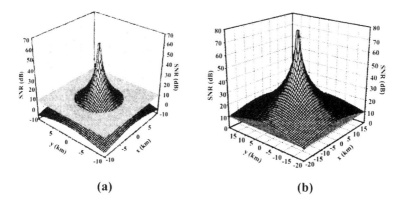

Figure 16.11 Ground detection range while OTM when h_R = 3m: (a) PCS handset at 1 GHz, 0.5-W ERP; and (b) PCS base station at 1 GHz, 10-W ERP.

Table 16.1 Detection Ranges for the System Configurations Considered (km)

	Target			
	0.5W, 1 GHz	10W, 30 MHz	10W, 100 MHz	10W, 1 GHz
High-flying	380	300	RLOS (425)	RLOS (425)
Low-flying	200	160	RLOS (200)	RLOS (200)
Ground	10	3	6	RLOS (18)

processing (see Section 12.4) has the targets more or less between the sensor systems. This can be accomplished by flying a UAV on the other side of the primary target area, placing the targets for which precision locations are required between the sensors. Not only is the geometry near optimum this way, but also the SNR is higher at the UAV, which can substantially increase the system throughput by decreasing the processing time necessary to compute the locations.

If CAF processing is too slow or not available, a PF geometry consisting of an L-shaped configuration is much better than PF with a linear baseline. The optimal geometry for DF processing is to have the LOPs cross at 90° angles. This occurs more frequently for L-shaped baselines than for linear baselines. An L-shaped baseline is available with two high-flying sensors on a linear baseline with a UAV on the other side or to the rear of the primary target area.

Intercept range performance by detection in directional link sidelobes is sometimes possible with either airborne system. In Table 16.2 the range performance is summarized. With these ranges, overflight of the target area is required.

Ground systems have limited range, limited by noise as well as RLOS. The latter of these should be applied carefully, however. On the ground, as opposed to the air, detection range is sensitive to system elevation relative to surrounding terrain. Both air and ground configurations, however, must be concerned with signal blockage. In OTM operations the RLOS is about 12 km over smooth Earth. Over hilly terrain or roads, the detection range would be limited more by blockage and temporary elevation advantages than detection range limits imposed by reflections and RLOS. In general, this is true for any ground deployment. These

Table 16.2 Sidelobe Detection Range (km)

	Yagi	Parabolic Dish
High-flying standoff	75	24
UAV	55	10

P_T = 10W, Yagi at 1 GHz and parabolic dish at 4 GHz

Table 16.3 Ground-Only Detection Ranges (km)

	0.5W, 1 GHz	10W, 30 MHz	10W, 100 MHz	10W, 1 GHz
Stationary	10	3	6	RLOS (18)
OTM	6	2	3	RLOS (12)

results are summarized in Table 16.3, which compares the stationary versus OTM detection ranges.

16.5 CONCLUDING REMARKS

This chapter presented some expected performance information on airborne and ground-based ES systems. Two airborne configurations are considered—one at high altitude and one at a lower altitude. The flight characteristics of the latter are compatible with those expected with a UAV. Two ground system configurations were also examined—one stationary and one OTM.

The analysis presented herein indicates that augmenting the high-flying ES collection system with a low-flying UAV can in many cases improve the collection performance. Indeed, in some scenarios the UAV could be the only alternative—in circumstances where it is desired to not use the high-flying system for safety or other reasons. Either system can cover the corps AO if there are no terrain blockages as shown in Figure 16.12. With terrain blockage, the UAV can overfly such terrain to obtain the targets. The UAV has enough intercept range to detect signals in these coverage zones. The UAV also has sufficient range to augment the DF/CAF processing of the high-flying system. When the targets are PCS/cell communication systems, the coverage is extensive enough from either airborne system so that the sensitivity must be decreased to avoid hearing multiple cells at the same time.

There are clear scenarios where not only is a UAV augmentation to the high-flying ES system desirable, but it is also required. In low-power and directional target situations the high-flying system simply cannot detect the targets without flying very close to them. For these targets, a standoff configuration cannot detect them.

The high-flying systems, being manned platforms, are vulnerable to enemy countermeasures. Since it is possible to perform ES collection with unmanned platforms, that should be the method of choice.

There are additional situations where the geometry provided by a UAV augmentation facilitates significantly enhanced PF of targets. Keeping a target between two airborne platforms is the optimum geometry for CAF computations.

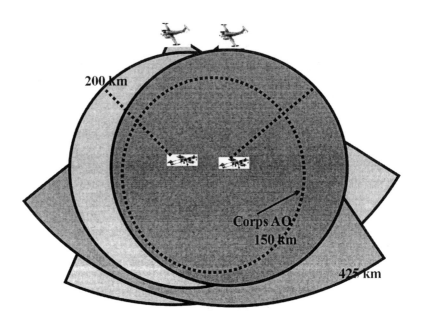

Figure 16.12 Coverage zones of a corps AO with both high-flying standoff systems augmented with overflying UAVs.

Furthermore, the SNRs available in that situation improve system throughput significantly.

A scenario where a UAV may be the only choice is when ground access is not feasible. A ground-based PCS ES system is a reasonable alternative to airborne solutions if access is available. The ground equipment must be placed somehow, and may have to be guarded. If security is not an issue, then the ground assets can be air-dropped close to base stations and remotely controlled and monitored. If there are security concerns, then manning the ground assets is probably required, even if collection is remote. A much simpler solution to this problem is the UAV ES system.

The UAV should be employed as a precision sensor system as opposed to a general search system. Their largest advantage is to overfly the target area picking up low-power and masked targets. In order to do PF, this will normally require at least two platforms. They must fly fairly close together to simultaneously detect low-power and masked targets.

One of the advantages of the UAV augmentation is that in some scenarios, in particular those where the high-flying system is in a standoff, rear deployment, the UAV sensor need not look through and sort through many friendly emitters to get to the targets. An emission in this case must first be considered friendly especially if it is stronger. The UAV can fly among the targets in this case, where every emission is likely to be a target as opposed to friendly. The coverage of high-flying standoff systems augmented with UAVs is shown in Figure 16.12 Both the standoff systems and the UAVs have adequate range to cover the entire AO.

When the targets are directional point-to-point communication links, the high-flying system in a standoff posture has difficulties with intercept range. If the communication system uses Yagi antennas, there is insufficient leakage into the sidelobes to intercept these systems at an adequate distance into the corps AO. Against such links that are using parabolic dishes, there is also inadequate leakage into the sidelobes to intercept them from a standoff posture. A UAV, on the other hand, can either fly into the main beam and loiter there if required, or intercept either of these links in a sidelobe by getting close enough.

This analysis considered whether targets could be detected versus range from the sensor system. It did not consider timing concerns, such as system throughput, or timing considerations associated with geolocating or collecting LPI targets.

References

[1]   Shigekazu, S., *A Basic Atlas of Radio-Wave Propagation,* New York: John Wiley & Sons, 1987.
[2]   Hall, M. P. M., *Effects of the Troposphere on Radio Communications,* London: Peter Peregrinus, Ltd., 1979.
[3]   Braun, G., *Planning and Engineering of Shortwave Links,* 2nd ed., New York: John Wiley & Sons, 1986.
[4]   Parsons, J. D., *The Mobile Radio Propagation Channel,* New York: John Wiley & Sons, 1992.
[5]   Volland, H. (ed.), *Handbook of Atmospherics, Vols. I and II,* Boca Raton, FL: CRC Press, 1982.
[6]   Davies, K., *Ionospheric Radio,* London: Peter Peregrinus, Ltd., 1989.
[7]   Schnker, J. Z., *Meteor Burst Communications,* Norwood, MA: Artech House, 1990.
[8]   *Reference Data for Radio Engineers,* 6th ed., Indianapolis, IN: Howard W. Sams & Co., 1975.
[9]   Calhoun, G., *Digital Cellular Radio,* Norwood, MA: Artech House, 1988.
[10] Gagliardi, R. M., *Introduction to Communications Engineering,* New York: John Wiley & Sons, 1988.
[11] *Reference Data for Radio Engineers,* 6th ed., Indianapolis, IN: Howard W. Sams & Co., 1975, Ch. 29.

Chapter 17

Electronic Attack: UAV and Ground-Based

17.1 INTRODUCTION

Denying communications is the function of EA. This can be accomplished by interjecting more RF energy into the target receivers than the intended transmitter does. This chapter presents results of static calculations of jammer-effective range from a few configurations of jamming platforms. This range is the distance from the jammer within which the jammer will remain effective.

For effective communications the SNR at the receiver must be above some minimum level in the absence of a jammer. The SNR is

$$\text{SNR} = \frac{P_R}{N_{total}} \qquad (17.1)$$

which is frequently expressed in decibels. P_R is the power from the receive antenna. Without a jammer

$$N_{total} = N_{noise} \qquad (17.2)$$

where

$$N_{noise} = N_{external} + N_{internal} \qquad (17.3)$$

as discussed in Chapter 3.

With a jammer present the total noise is expressed as

$$N_{\text{total}} = N_{\text{noise}} + J \qquad (17.4)$$

where J represents the jammer power. Under many electronic attack scenarios $J \gg N_{\text{noise}}$ and N_{noise} can be, and is, neglected in such calculations. In these cases the J/S ratio is the quantity of interest, which is the reciprocal of (17.1).

As discussed in Chapter 7, modern digital forms of communication are easier to jam than the older, analog modulations. A J/S ratio of 0–6 dB was necessary against analog radios. With digital modulations, creating a BER higher than about 10^{-2} is sometimes all that is necessary. This is making only 1 bit out of 100 in error. It has been shown that this can be accomplished at considerably less than 0-dB J/S. Here it is assumed that effective jamming occurs at zero J/S ratio.

Probably the most significant advantage of EA from a UAV platform is the ability to overfly the target area and put the jamming energy where it is intended to be, rather than causing fratricide in friendly radios. This, of course, does not strictly apply to a truly nonlinear situation where friendly communications are in the same geographical area as the targets, but even then the jamming energy can be placed closer to the target than in standoff configurations. In the latter, the jammer has to emit energy over friendly communications before ever getting to the target area, always creating the possibility for fratricide. The results presented here are in the form of jamming range—that is, the range from the jammer within which, if the target receiver is located, the jammer will be effective at denying communications. Three jammer configurations were considered—the first is when the jammer is mounted in a UAV; the second is a ground-based vehicle configuration; and the last is a ground-based expendable configuration.

There are many jammer configurations possible. The ones included here are representative and are likely in land warfare scenarios. Similar analysis can be undertaken for any given configuration and for other types of propagation conditions—over water, for example, [1–10].

17.2 SIGNAL PROPAGATION AT LONG RANGES

For the purposes here the reflection model from Section 2.9 given by (2.43) will be used to calculate the signal and jammer powers at the receiver. Using this expression for the received power at all ranges will tend to underestimate the power at close ranges but beyond distance, d, given by

$$d = \frac{4h_T h_R}{\lambda} \tag{17.5}$$

where the following factors more accurately reflect the total power received: h_T = transmit antenna height *above ground level* (AGL) and h_R = receive antenna height AGL. As in Chapter 16, since terrain is not considered, a circle defines the effective jamming region with the jammer at the center and with a radius defined by the calculated effective range.

17.3 JAMMING ERP

When there is adequate room and the expense makes sense, antennas should be tuned according to the frequency at which they are used. Untuned antennas over 2 octaves (30–60 MHz, 60–120 MHz), the low-VHF range considered here, have substantial loss at either end of the frequency range. Losses of 15 dB or more would not be uncommon. Tuning these antennas matches the power amplifier output impedance to the antenna depending on the frequency. When tuned, losses of 3 dB across the frequency band are not unreasonable to expect. With only 3-dB loss due to the antenna, a 100-W power amplifier produces an ERP of 50W, entirely reasonable for both the UAV and ground-vehicle configurations. With typical conversion efficiencies of 20% or so, that means that around 500W of prime power would be required for the system.

The expendable jammer case is different. Cost does not justify tuning the antenna. Furthermore, these jammers are powered by batteries that can deliver only so much power at a time and so much energy total. Thus, the expendable jammer must tolerate the antenna loss, and a power amplifier delivering 3W to an antenna with a 15-dB loss creates about 100-mW ERP. That is the power level used herein for the expendable jammer. It should be noted, however, that at midband the ERP would typically be much higher than this, so these results lean toward the worst case. The 3-W power amplifier, at 20% conversion efficiency, would require about 15W of prime power. Two zinc carbon D-cells (normal flashlight batteries) can deliver 1.5V at about 4.5 amp-hours each at room temperature for a total of 13W for about 1 hour, sufficient for this jammer to operate for almost that hour.

The two mounted systems could be configured so that the frequency coverage could be from the low-VHF to beyond the PCS range, albeit different antennas may be required over the entire range. For the expendable jammer, one antenna probably could not provide the performance required to cover the whole range, so each jammer would probably be optimized (as much as possible) for only a portion of this range.

The expression for P_R (2.43), has a dependency on the height AGL of the jammer antenna, denoted here by h_J. It is the same as h_T in (2.43). For the UAV, h_J = 1,000m. For the mounted ground jammer h_J = 3m, while for the ground expendable jammer h_J = 1m.

17.4 TARGETS

Three types of targets are considered here. One of these is in the low-VHF range, the frequency range used by almost all military and paramilitary organizations worldwide. This target was at a frequency of 60 MHz, the middle of this range. The other target considered is the emerging PCSs expected to be used by these same forces in the future. In this case the handset is assumed to have an ERP of 0.5W, while the base station has an ERP of 10W.

The ability to jam a communication net is a function of RF link distances—from transmitter-to-receiver relative to the jammer-to-receiver link distance. For specificity, herein it is assumed that the transmitter-to-receiver link distances are 5 km and 10 km, and that the antennas are omnidirectional in the horizontal plane. The ability to jam also depends on the transmit power relative to the jammer power. Herein it is assumed that in the VHF range, the PTT target transmit power is 10-W ERP. The jammer power is 50-W ERP, which reflects a 100-W jammer PA and an antenna with a 3-dB loss. In the higher-frequency range of the PCS, P_J = 50W as well.

When the target is the 10-W PTT network, $h_T = h_R = 2$m, reflecting ground-to-ground communications. When the target is the PCS handset, $h_T = 10$m, $h_R = 2$m and $P_T = 10$W. When the target is the PCS base station, $h_T = 2$m, $h_R = 10$m, and $P_T = 0.5$W.

17.5 RLOS

Radio waves traveling close to the Earth propagate for only relatively short distances as discussed in Chapter 2. These distances are referred to as the RLOS, and due to refractive properties of the troposphere, do not typically correspond to the visual line of sight. Such refraction typically increases the radius of the Earth by a factor of one-third or so. For the heights used here, the RLOS is given in Table 17.1. Included are the RLOSs for the communicating nodes as well, which indicates that considering link distances of 5 and 10 km is reasonable.

Of course, there are propagation modes that allow waves to travel less than or longer than the RLOS—such modes are not considered here. The results

Table 17.1 RLOS for the Antenna Heights Considered

Target	h_T (m)	h_R (m)	h_J (m)	RLOS (J→R) (km)	RLOS (T→R) (km)
PTT	2	2	Exp 1	5.5	6.4
	2	2	Gnd 3	7.1	6.4
	2	2	UAV 1,000	75	6.4
PCS Base	2	10	Exp 1	9.4	10.3
	2	10	Gnd 3	11.1	10.3
	2	10	UAV 1,000	78	10.3
PCS Handset	10	2	Exp 1	5.5	10.3
	10	2	Gnd 3	7.1	10.3
	10	2	UAV 1,000	75	10.3

presented herein then, can be considered nominal, with deviations either way possible.

17.6 UAV JAMMER

Consider first the jamming range performance of a jammer mounted in a UAV. For the low VHF the jamming effectiveness regions are shown in Figure 17.1. Here the target transmitter power is 10-W ERP. Note that the scale extends to ±100 km from the jammer. In Figure 17.1(a) the link distance is 5 km, while in

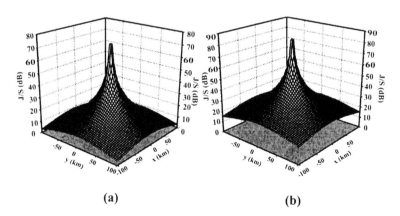

(a)   (b)

Figure 17.1 UAV jammer range when the target is a 10-W PTT network and the link distance is (a) 5 km and (b) 10 km.

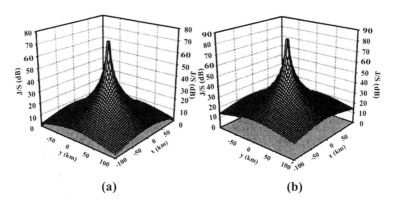

Figure 17.2 Jamming range for a UAV jammer when the target is a PCS base station and the link distance is (a) 5 km and (b) 10 km.

Figure 17.1(b) it is 10 km. The horizontal plane represented by the lighter color in Figure 17.1 represents the case when J/S = 0 dB, or the jamming power just equals the signal power at the target receiver. As discussed above J/S = 0 is conservative when considering digital communications. Therefore, any time the plane is above the signal level contour, the jammer is ineffective—that is, the intended communication is getting through. On the other hand, wherever the J/S contour is above the plane shown, the jamming is effective and communication is denied. In this case, the jammer is effective out to a radius of the RLOS (75 km) from the UAV for both link distances.

When the target is a PCS base station, the jamming range performance is as shown in Figure 17.2. Since jamming the base station implies $P_T = 0.5$W from the

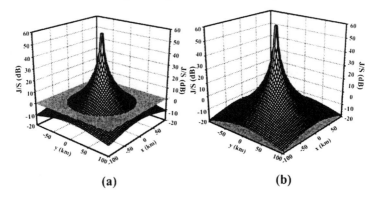

Figure 17.3 Jamming range performance of a UAV jammer when the target is a PCS handset and the link distance is (a) 5 km and (b) 10 km.

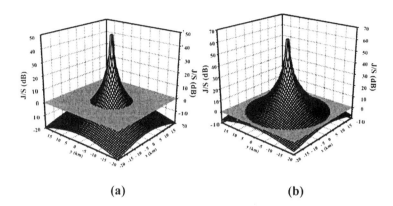

Figure 17.4 Ground jamming range when the target is a 10-W PTT network and the link distance is (a) 5 km and (b) 10 km.

handset, the base station can be jammed out to RLOS for both link distances. When jamming the handset, $P_T = 10$W and the effective jamming range is reduced somewhat, as seen in Figure 17.3; however, the range is still RLOS for both link distances.

1.7.7 GROUND-BASED JAMMER

The other mounted jammer system configuration considered is one that is ground-based. Again, a 50-W ERP jamming waveform is assumed with an antenna that is omnidirectional in the horizontal plane.

When the link distance is 5 km for low-VHF targets, the effective jamming region is as shown in Figure 17.4(a). Note the scale change to ±20 km. The jammer is effective to a range of about 8 km in this case, which implies that the range is RLOS (7.1 km). At a link distance of 10 km that range is extended to about 17 km or so, again limited to RLOS at 7.14 km, as seen in Figure 17.4(b).

When the target is the PCS base station, the transmitter power is 0.5W and the effective jamming region is a circle with a radius of about 16 km as shown in Figure 17.5(a). For a 10-km link distance, the radius of the region is about 35 km. Thus, in both cases RLOS (11.1 km) limits the range.

The mounted ground-based jammer range when the target is the PCS handset is shown in Figure 17.6. Note the scale in this figure is ±20 km. Since the base station transmits more power than the handsets, the effective jamming range is less than for jamming the base station. With a 5-km link distance, the jamming range is about 3 km or so as seen in Figure 17.6(a). With a 10-km link distance it was

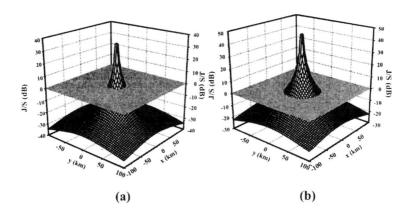

Figure 17.5 Ground jammer when the target is a PCS base station and the link distance is (a) 5 km and (b) 10 km.

about double this, to 6 km. Thus, in this case the jammer range is not limited by the RLOS, but by the jammer parameters.

It is highly desirable to be able to jam communication targets while the EA system is OTM. This implies significant restrictions on the type of antennas that can be used for this function. At 30 MHz, the wavelength is about 10m or so and a half-wave dipole or monopole would be about 5m long. While such an antenna could securely be mounted on a tactical vehicle, keeping it vertical while OTM

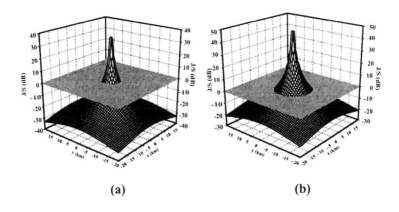

Figure 17.6 Ground jammer range when the target is a PCS handset and the link distance is (a) 5 km and (b) 10 km.

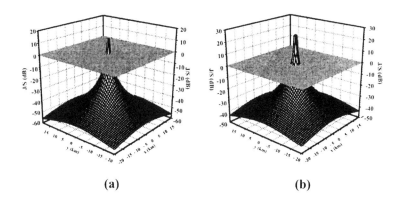

Figure 17.7 Expendable jammer range when the target is a 10-W PTT net and the link distance is (a) 5 km and (b) 10 km.

would be a significant trick. Furthermore, a 5-m antenna, sticking up vertically from a vehicle, violates many international traffic agreements, and would not clear highway overpasses in many countries, including the United States.

The jammer antenna height used here is $h_J = 3$m with a 3-dB loss. The 3-dB loss implies a certain antenna length and/or antenna tuning. Limiting h_J to 3m would allow for OTM operation, however.

17.8 EXPENDABLE JAMMER

Another type of ground-based jammer is one that is hand-emplaced, delivered by a UAV, or delivered by some other means, but is of relatively low power and is expendable. Such jammers are inexpensive enough so that they need not be retrieved once deployed. They are normally battery operated so they perform EA for only a limited amount of time.

The antennas for these jammers are normally stowed until they are put into use. If hand-emplaced, the soldier emplacing the jammer could erect the antenna much like a retractable car antenna works—simply by pulling and extending the sections. If air-delivered, the antenna would have an automatic deployment mechanism. In either case, for the low-VHF range, the antenna would be about a half-wavelength long at midrange. Since the antenna would not be tunable, it would have considerable losses at the low and high ends. For the PCS frequency range, reasonably efficient antennas could be included. Because of the necessity of powering these devices with batteries and the inefficiency of the antenna as indicated in Section 17.3, such jammers would typically have an ERP of 100 mW or so average across the low-VHF frequency band.

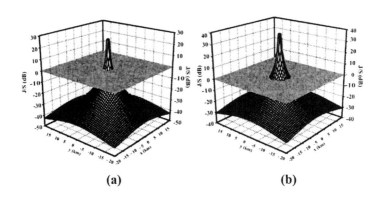

Figure 17.8 Expendable jammer range when the target is a PCS base station and the link distance is (a) 5 km and (b) 10 km.

The results of simulating one of these jammers against the low-VHF targets are shown in Figure 17.7. With a 5-km link the jamming range was less than 1 km while with a link distance of 10 km, the range increases to about 2 km.

Against the PCS base station when $P_T = 0.5W$, the jamming range was about 2 km when the link distance was 5 km as illustrated in Figure 17.8(a). When the link distance was extended to 10 km, about the typical edge of a PCS cell, the jamming range was about 5 km as seen in Figure 17.8(b).

When the target was the PCS handset, the jammer was ineffective at any range for the 5-km link as shown in Figure 17.9. For a 10-km link somewhat less than 1 km was the effective range.

17.9 CONCLUDING REMARKS

The results of the analysis of effective jamming ranges for the various system configurations are summarized in Table 17.2. In most cases the jamming range is limited only by the RLOS between the jammer and the target receiver.

The UAV jammer is quite effective, with a range of 75 km in all cases. This is the RLOS at the altitude at which the jammer was flown (1,000-m AGL). This range would be longer if the UAV were to fly higher. A range of 75 km, however, is probably too long for most scenarios since unwanted fratricide would likely ensue at even 75 km.

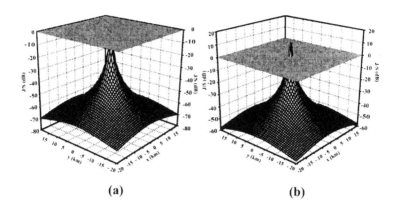

Figure 17.9 Expendable jammer range when the target is a PCS handset: (a) 5-km link distance and (b) 10-km link distance.

When the 50-W ERP jammer is in the ground mounted configuration, again the limitation was the RLOS in most cases. It is assumed here that the jammer is configured for OTM operation so the jammer antenna height was compatible with a tactical vehicle that was moving—that is, the antenna was not elevated. A jammer antenna height of 3m is assumed. This height significantly restricts the RLOS range of the jammer.

Table 17.2 Jamming Ranges for the System Configurations Considered

	Target					
	10-W PTT		PCS base station		PCS handset	
Link distance (km)	5	10	5	10	5	10
UAV	RLOS (75)	RLOS (75)	RLOS (78)	RLOS (78)	RLOS (75)	RLOS (75)
Ground mounted	RLOS (7.1)	RLOS (7.1)	RLOS (11.1)	RLOS (11.1)	3	6
Ground expendable	<1	2	2	5	0	<1

As expected, the ground expendable jammer has the shortest effective range. With an ERP of only 100 mW, the range must be limited. In some cases it is less than 1 km and, in the case of the PCS handset with a 5-km link distance, cannot jam the target at all. (Of course, if the jammer were immediately adjacent to the handset, the latter would not be able to communicate.)

These results assume a sweeping narrowband jammer simulating a barrage jammer. Timing issues were not considered. The analysis in Chapter 7 showed that the timing could be arranged so that the jammer can be effective.

Jamming the handset end of the PCS link is the more difficult problem because the base station's transmit power is substantially higher than that of the handset and the height of the base station's antenna is substantially higher than that of a ground-based jammer. Although not considered here, directional antennas are normally used with such base stations. This would extend the range, reduce the amount of cochannel interference in the PCS system, and further limit the effectiveness of the jammer by null steering, for example, or employing the smart antenna concepts in Section 8.6.2.2.

References

[1] TRADOC Pamphlet 525-5, "The Objective Force, Operational Concepts, Organizational Design Constructs, and Materiel Needs Implications," U.S. Army Training and Doctrine Command, Fort Monroe, VA.

[2] U.S. Army Field Manual FM 100-6, Information Operations, Chapter 3 Operations, August 27, 1996.

[3] U.S. Army Field Manual FM 34-40-7, Communication Jamming Handbook, October 9, 1987.

[4] Neri, F., *Introduction to Electronic Defense Systems,* Norwood, MA: Artech House, 1991, pp. 337–416.

[5] Waltz, E., *Information Warfare Principles and Operations,* Norwood, MA: Artech House, 1998, Ch. 8.

[6] Adamy, D., *EW 101: A First Course in Electronic Warfare,* Norwood, MA: Artech House, 2001, pp. 177–222.

[7] Schleher, D. C., *Electronic Warfare in the Information Age,* Norwood, MA: Artech House, 1999, pp. 31–57.

[8] Mosinski, J. D., "Electronic Countermeasures," *Proceedings of IEEE MILCOM,* 1992, pp. 191–195.

[9] "A Jamming Primer," *EW Design Engineers' Handbook,* Association of Old Crows, 1985; also contained in Hovanessian, S. A., *Radar System Design and Analysis,* Dedham, MA: Artech House, 1984.

[10] "Airborne ECM Tactics," *EW Design Engineers' Handbook,* Association of Old Crows, 1985; also contained in Van Brunt, L. B., *Applied ECM, Vol. 1,* 1978.

Appendix A

Probability and Random Variables

A.1 INTRODUCTION

Communication signals represent probabilistic processes. Only in the simplest of cases can a communication signal be considered deterministic. A probabilistic process is one that can only be described with probability terms—the exact nature of such a process is not known a priori. A deterministic process, on the other hand, can be described exactly. Therefore it is useful for understanding communication signals and EW systems to have a basic understanding of random variables and probability.

Basic information on probability and random variables is included in this appendix. It certainly is not a detailed discussion, but it should be enough to understand what is necessary for the material herein.

A.2 MEANS, EXPECTED VALUES, AND MOMENT FUNCTIONS OF RANDOM VARIABLES

Given a set of N numbers $S = \{X_1, X_2, \ldots, X_N\}$, it is sometimes necessary to add all or some of these numbers together. A shorthand notation for representing this is given by Σ. Thus,

$$\sum_{i=1}^{N} X_i = X_1 + X_2 + \cdots + X_N \tag{A.1}$$

The *average*, or *mean*, value of this set of numbers is one number that is used in a sense to characterize the set that indicates where the "middle" is of the set of numbers. Herein it is denoted as μ. It is given by

$$\mu = \frac{\sum_{i=1}^{N} X_i}{N} \qquad (A.2)$$

It is also known as the arithmetic mean of the set.

Suppose that X_1 is repeated n_1 times, X_2 is repeated n_2 times, ..., and X_N is repeated n_N times. Then the mean can be calculated as

$$\mu = \frac{\sum_{i=1}^{N} n_i X_i}{N} \qquad (A.3)$$

where, of course, $n_1 + n_2 + \ldots + n_N = N$.

A characteristic of the set of numbers, which measures the "dispersion," or "spread" of the numbers around the mean value is the *standard deviation* of the set. It is denoted herein by σ and is given by

$$\sigma = \sqrt{\frac{\sum_{i=1}^{N} (X_i - \mu)^2}{N}} \qquad (A.4)$$

σ^2 is known as the *variance*. Again, if there are repeated values in the set, then the standard deviation can be computed as

$$\sigma = \sqrt{\frac{\sum_{i=1}^{N} n_i (X_i - \mu)^2}{N}} \qquad (A.5)$$

Note that the standard deviation can be computed as

$$\sigma = \sqrt{\mu_S - \mu^2} \qquad (A.6)$$

where μ_S is the mean of the set of numbers S_S consisting of the squares of the original set, namely, $S_S = \{X_1^2, X_2^2, \cdots, X_N^2\}$.

Generalizing the above, the *r*th *moment about zero*, μ_r, of the set of numbers is given by

$$\mu_r = \frac{\sum_{i=1}^{N} X_i^r}{N} \tag{A.7}$$

Similarly, the *r*th *moment about the mean* is given by

$$\mu_r = \frac{\sum_{i=1}^{N} (X_i - \mu)^r}{N} \tag{A.8}$$

The probability of a number, denoted p_i, for number X_i of the set S is given by the fraction of N corresponding to X_i (Kosko's the whole contained in the part). Thus, in the above notation, $p_1 = n_1/N$, $p_2 = n_2/N$, and so forth. Note that by definition $\Sigma p_i = 1$.

The rms is a measurement of the offset of the average of the set from zero as well as the spread around this average value. If μ is the average value, N is the number of data points, and the data values are denoted by X_i, then

$$\text{rms} = \sqrt{\frac{1}{N} \sum_{i=1}^{N} (X_i - \mu)^2} \tag{A.9}$$

The *root sum squared* (rss) is sometimes used to specify measurement errors. It is similar to the rms calculation except that the mean value is not removed. Thus, it specifies more of the actual errors involved in measurements. The rss is given by

$$\text{rss} = \sqrt{\frac{1}{N} \sum_{i=1}^{N} X_i^2} \tag{A.10}$$

X is called a *random variable* if it can take on one of the values in a set of numbers (the set could be infinite in size) at one point (for example, at time t_n) and another value at another point, and the values it takes can only be described probabilistically. That means that ahead of time it is not known for certain what value X will take on at some future point. The expected value, or simply expectation, of a number X_i that has an associated probability of occurring in a given set is given by $E\{X_i\} = p_i X_i$. If X is a random variable that can take on the values from the set $S = \{X_1, X_2, \ldots, X_N\}$ where each X_i has associated probability of occurrence $p_1, p_2, \ldots, p_N$, then the expected value of X is given by

$$E\{X\} = \sum_{i=1}^{N} p_i X_i \tag{A.11}$$

Note that with the previous notation with the set S, $p_i = n_i/N$. Substituting this into (A.11) yields

$$E\{X\} = \sum_{i=1}^{N} \frac{n_i}{N} X_i \tag{A.12}$$

which is the mean value of the numbers as indicated previously.

A.3 PROBABILITY

Suppose a die is rolled and the results of each roll is a number $x \in \{1,2,3,4,5,6\}$ where $\in$ means "is an element of" and "{ }" denotes a set. If the die is rolled 1,000 times and some x, say, $x = 5$, turns up 200 times, then it is said that the probability of the occurrence of $x = 5$ in this experiment is $200/1,000 = 0.2$.

Suppose that a card is pulled from a fair deck of cards (fair = not marked and all cards present). Then since there are 52 cards in a deck, what is the probability of drawing the jack of clubs? It is $1/52 \approx 0.019$ since there is one such card in the deck. What is the probability of drawing an ace? It is $4/52 \approx 0.077$ since there are four aces in the deck.

A process that has a random output x is called a *stochastic process* and x is referred to as a *random variable*. Such variables are described by their statistics, because a priori, we do not know the exact value the variable will take in any given instance. There are many instantiations of random variables in communication theory. The amplitude of an AM signal corrupted by noise is a random variable. The demodulated bit in a digitally modulated signal that is corrupted by phase noise is a random variable.

A.3.1 Conditional Probability

Suppose in a card experiment the question is the probability of drawing aces from a fair deck of cards. Furthermore suppose that on the first draw an ace has been drawn. What is the probability that a second ace will be drawn on the second try? This situation is described by *conditional probabilities*. Bayes, an eighteenth-century English clergyman, was the first person known to examine conditional probabilities. The probability of an event e occurring given that event a has already occurred is denoted as $P(e|a)$. This conditional probability is given by the expression

$$P(e|a) = \frac{P(a,e)}{P(a)} \tag{A.13}$$

The numerator on the right side in this expression, $P(a, e)$, is the probability of both events a and e occurring. Rearranging this expression

$$P(a,e) = P(e|a)P(a) \tag{A.14}$$

If e and a are independent events, neither depends on the occurrence of the other. Therefore the probability of occurrence of e does not depend on whether a occurs or not. Thus $P(e|a) = P(e)$ and

$$P(a,e) = P(e)P(a) \tag{A.15}$$

In general,

$$P(e,a) = P(e)P(a) - P(e \, or \, a) \tag{A.16}$$

Since $P(e, a) = P(a, e)$ then

$$P(e|a)P(a) = P(a|e)P(e) \tag{A.17}$$

so that

$$P(e|a) = \frac{P(a|e)P(e)}{P(a)} \tag{A.18}$$

In words, this expression says that the probability of e occurring given that a has already occurred is given by the ratio of the product of the probability of a occurring, given that e has already occurred times the a priori probability of e occurring at all, to the a priori probability of a occurring at all.

Two events a and e are *mutually exclusive* if the occurrence of one means that the other did not occur. In that case $P(a, e) = 0$ and $P(e|a) = P(a|e) = 0$.

A.3.2 Random Variable Distribution and Density Functions

Suppose that you have a pair of dice. Furthermore, you do not know if the dice are "fair" in the sense that any one of the 6 × 6 = 36 combinations of numbers could occur equally. Now suppose you throw those dice 1,000 times and record the sum of the dice that occurred on each throw. Then you plot these results on a graph, with the results as shown in Figure A.1. The lines on the left are the experimental results while those on the right are the theoretical results if the dice were perfectly

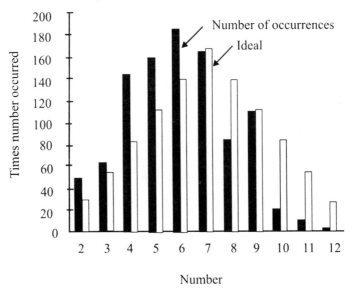

Figure A.1 Densities for the experiment.

fair. Could you then say with some degree of confidence that the dice are "fair"? Probably not (the actual results are clearly skewed to the low side indicating one or more of the die is weighted low), but that is not the point here.

Now suppose that instead of plotting the number of occurrences of each sum themselves, the numbers are "normalized" so that each "times number occurred" represents a fraction of the total times (1,000) the dice were thrown. That leads to a similar graph as shown in Figure A.2 (plots of the ideal results have been removed for clarity). A graph such as this is known as a probability density function for the discrete random variable "number." "Number" is discrete because it can take on only specific values. Note that the "area" under these bars totals 1, because $n_1/N + n_2/N + \ldots + n_N/N = (n_1 + n_2 + \ldots + n_N)/N = N/N = 1$.

If the random variable is allowed to take on continuous values, then a similar result ensues, except that the histogram is now a continuous curve. The area under this curve corresponds to probability mass and the area between two values of X, say, X_1 and X_2, corresponds to probability $(X_1 < X < X_2)$, as shown in Figure A.3.

For both discrete and continuous random variables there is a function known as the cumulative distribution function which represents the probability that the random variable x is less than some specified value, say, X. For the discrete case above, it is shown in Figure A.4. Since the continuous density function as shown above is based on the same data as the discrete density function, the distribution function for the continuous case here would look the same as Figure A.5. In mathematical terminology, the density function $p(x)$ and distribution functions $P(x)$ are related by the equations

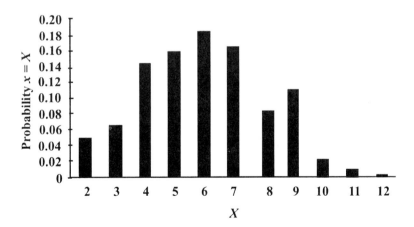

Figure A.2 Experiment results that have been normalized to the total number of trials.

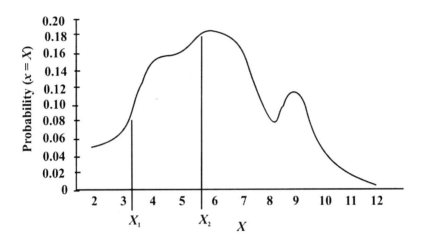

Figure A.3 Continuous density function.

$$p(x) = \frac{dP(x)}{dx} \tag{A.19}$$

and

$$P(x) = \int_{-\infty}^{x} p(\xi)d\xi \tag{A.20}$$

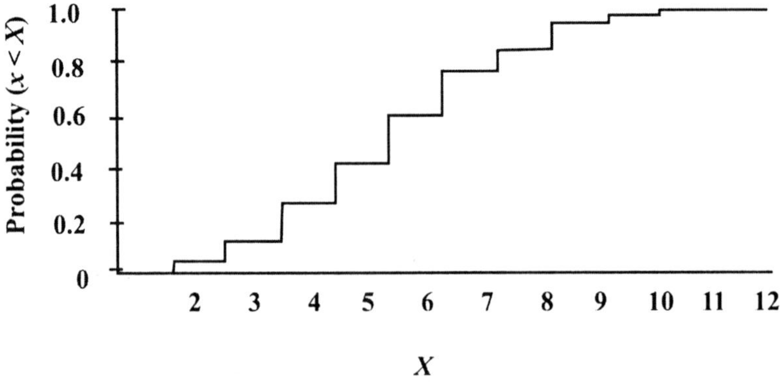

Figure A.4 Discrete probability distribution function.

Probability and Random Variables 525

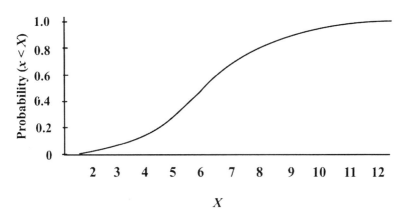

Figure A.5 Continuous probability distribution function.

That is, $P(x)$ is given by the area under the density function from $-\infty$ to x, and $p(x)$ is the slope of $P(x)$ at $x = X$.

A.4 GAUSSIAN DENSITY FUNCTION

One density function of particular interest to the analysis of communication signals and EW systems is the *Gaussian*, or *normal*, density function. Gauss was a seventeenth-century mathematician who discovered, among other things, that there is a probability density function that approximates the behavior of random variables in many circumstances. This is shown in Figure A.6. To interpret this function, like all probability density functions, the abscissa represents the values of a random variable, in this case denoted as x. Then the probability that x falls between x_1 and x_2 is given by

$$P(x_1 < x < x_2) = \int_{x_1}^{x_2} p(x)dx \quad (A.21)$$

The average, or mean, amplitude is denoted as μ and the standard deviation is given by σ. This distribution is often used to describe the amplitude of noise signals. When it is, then the power in the noise signal is given by σ^2.

A random variable described by such a density function is frequently encountered in practice. For example, it describes reasonably well the noise voltage and current associated with resistors. The density is given by the equation

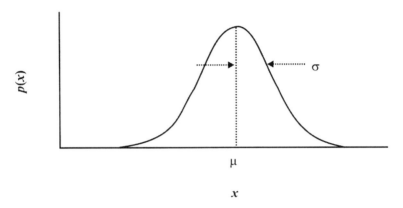

Figure A.6 Gaussian, or normal, probability density function.

$$p(x) = \frac{1}{\sqrt{2\pi}\sigma} e^{-\frac{1}{2}\frac{(x-\mu)^2}{\sigma^2}} \qquad (A.22)$$

where μ is the mean and σ is the standard deviation.

The density function is shown in Figure A.6, while the distribution function is shown in Figure A.7. In both of these figures $\mu = 0$ and $\sigma = 1$, but the results are easily generalized to other mean and standard deviation values.

For the Gaussian density, 68.27% of the probability mass lies within $\pm 1\sigma$ of the mean value, while 95.45% of the probability mass lies within $\pm 2\sigma$ of the mean and 99.73% of the probability mass lies within $\pm 3\sigma$ of the mean.

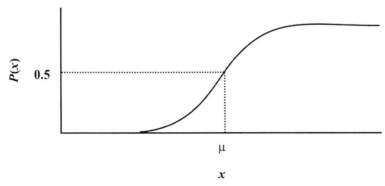

Figure A.7 Gaussian, or normal, continuous probability distribution function.

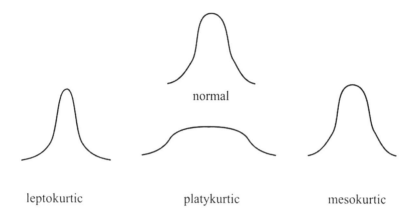

Figure A.8 Non-normal probability densities.

A.5 KURTOSIS

The *kurtosis* is a measure of the peakedness of a probability density, usually relative to the normal density. A density that is peaked considerably more than the normal density is referred to as a *leptokurtic density*, while one that is relatively flat-topped relative to normal is referred to as a *platykurtic density*. A density that is about the same as the normal density is referred to as *mesokurtic*. Examples of such densities are shown in Figure A.8.

One measure of the kurtosis of a probability density is given by

$$\kappa = \frac{\mu_4}{\sigma^4} \tag{A.23}$$

where μ_4 is the fourth moment about the mean of the random variable, as defined previously, and σ is the aforementioned standard deviation.

A.6 SKEWNESS

The *skewness* of a probability density is a measure of the asymmetry of the density. Herein it is denoted by ψ. If the tail to the right compared to the left of the mean is larger and/or longer, then the density is said to be skewed to the right or have *positive skewness*. This is similar for the left tail. The normal definition of the skewness is given by

$$\psi = \frac{E^3\{x-\mu\}}{\sigma^3} \tag{A.24}$$

A.7 USEFUL CHARACTERISTICS OF PROBABILITIES, DENSITY FUNCTIONS, AND DISTRIBUTION FUNCTIONS

Some useful properties of probabilities and their functions are presented in this section. The probability of the certain event is 1. There can be no probability larger than this. Thus, $P \leq 1$. Furthermore, the probability of the occurrence of an event cannot be less than zero; thus, $P \geq 0$.

The probability distribution function $P(x)$ for a random variable x is monotonically increasing. It is zero for some value of x and increases to 1 for some larger value of x. It cannot be larger than one or less than zero.

Since

$$P(x) = \int_{-\infty}^{x} p(\xi)\,d\xi \tag{A.25}$$

then $P(x_1) \leq P(x_2)$ whenever $x_1 \leq x_2$. Furthermore, the total area under $p(x) = 1$.

A.8 CONCLUDING REMARKS

This appendix provides the reader with an introduction to the topics in probability and random process theory necessary to comprehend the remainder of this book. Statistical processes is a large topic with a huge body of literature associated with it. It is worthy of study in its own right.

Communication systems, and the EW systems designed to counter them, process signals that are statistical in nature. This is due in part to the random noise that is invariably added to such signals as they are transmitted and processed. Some signals, however, exhibit random characteristics in and of themselves. The randomness is built into the communication signals by design. A simple example of this is encrypted signals, but there are many other examples. Therefore, an understanding of the fundamentals of statistical processes is important for a thorough understanding of the design of EW systems.

Reference

[1]   Kosko, B., *Fuzzy Thinking*, New York: Hyperion, 1993, pp. 44–64.

Appendix B

Simulated Networks

B.1 INTRODUCTION

The specific nets used in the simulations described in the text are shown in this appendix. In those cases where there were only these 19 nets, no other nets were simulated. In those cases when more than 19 nets were simulated the additional nets were added to the region shown at random, but at tactically significant ranges.

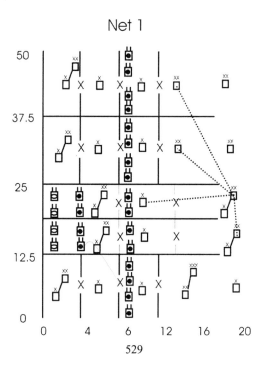

Net 1

530   Introduction to Communication Electronic Warfare Systems

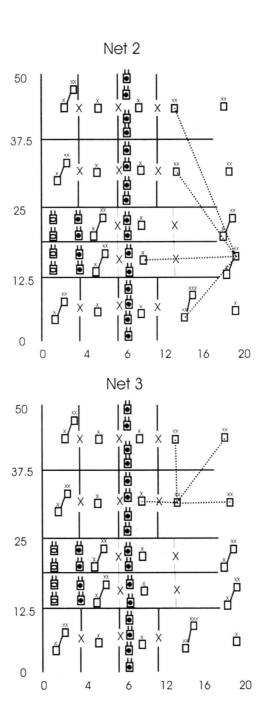

Simulated Networks 531

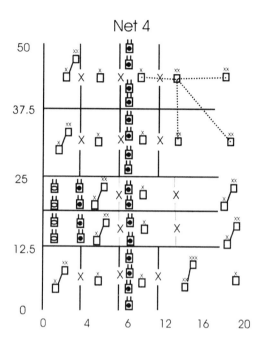

Net 4

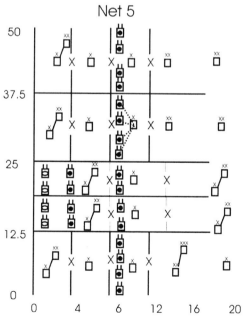

Net 5

532   Introduction to Communication Electronic Warfare Systems

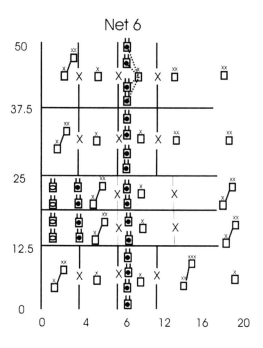

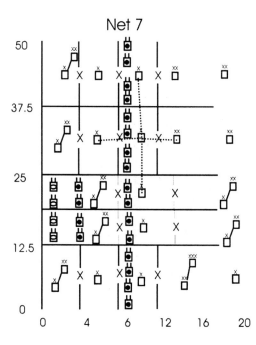

Simulated Networks 533

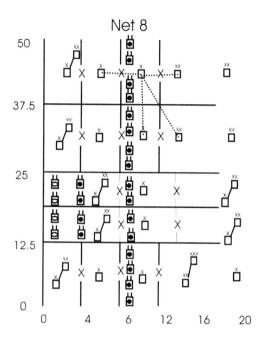

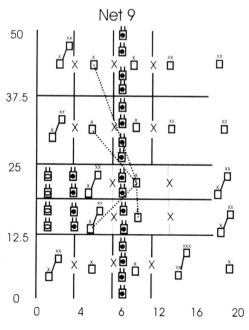

534   Introduction to Communication Electronic Warfare Systems

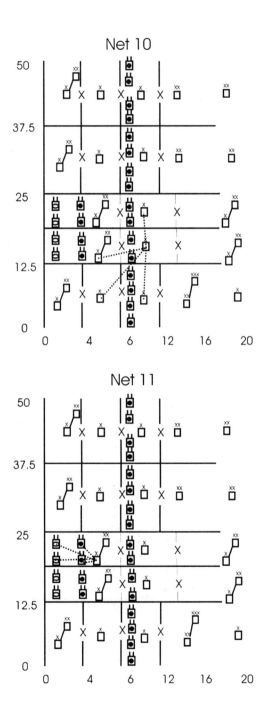

Simulated Networks 535

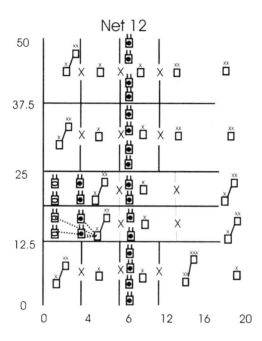

Net 12

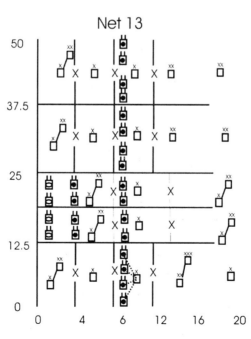

Net 13

536 Introduction to Communication Electronic Warfare Systems

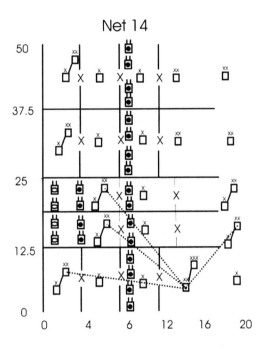

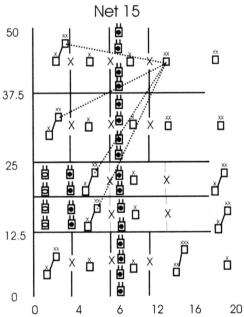

Simulated Networks 537

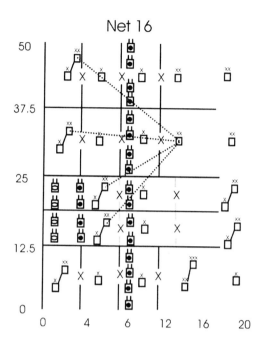

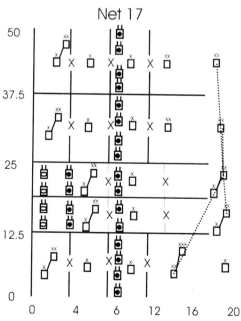

538 Introduction to Communication Electronic Warfare Systems

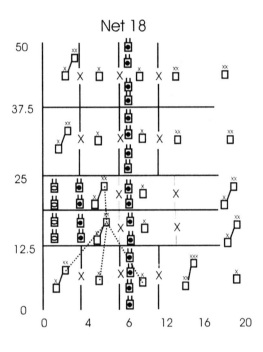

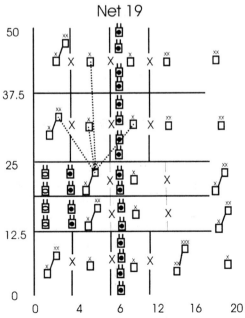

List of Acronyms

A/D	analog-to-digital
ACK	acknowledgment
ADPCM	adaptive differential pulse code modulation
A_{eff}	effective area of antenna
AGC	automatic gain control
AGL	above ground level
ALE	automatic link establishment
ALR	average likelihood ratio
AM	amplitude modulation
AMPS	American mobile phone system
AOA	angle of arrival
ARDIS	advanced radio data information service
ASK	amplitude shift key
BCH	Bose, Chaudhuri, and Hocquenghem
BER	bit error rate
BPE	best point estimate
BPM	basic power module
bps	bits per second
BPSK	binary phase shift key
BW	bandwidth
C2W	command and control warfare
CAF	cross-ambiguity function
CDMA	code division multiple access
CDMA/CD	CDMA with collision detection
CDPD	cellular digital packet data
CNR	carrier-to-noise ratio
COP	common operational picture

CRC	cyclic redundancy check
CWT	continuous wavelet transform
DAMA	demand assignment multiple access
dB	decibel
DBK	dominant battlespace knowledge
DCT	discrete cosine transform
DD	differential Doppler
DES	data encryption standard
DF	direction finding
DSB	double sideband
DMT	discrete multitone
DoD	Department of Defense
DPCM	differential pulse code modulation
DSA	digital signatures algorithm
DSSS	direct-sequence spread spectrum
DWT	discrete wavelet transform
EA	electronic attack
EES	escrow encryption standard
EHF	extra-high frequency
EM	electromagnetic
EMC	electromagnetic compatibility
EMCON	emission control
EMI	electromagnetic interference
EO	electro-optical
EOB	electronic order of battle
EP	electronic protect
ERP	effective radiated power
ES	electronic support
ESM	electronic support measures
EW	electronic warfare
FAIT	first article integration and test
FCC	Federal Communication Commission
FCS	frame check sequence
FDD	frequency division duplexing
FDDI	fiber distributed data interface
FDM	frequency division multiplexing
FDMA	frequency division multiple access
FEC	forward error correction

FET	field effect transistor
FFH	fast frequency hopping
FFT	fast Fourier transform
FH	frequency hopping
FHSS	frequency-hopping spread spectrum
FLOT	forward line of own troops
FM	frequency modulation
FSK	frequency shift key
FTR	fixed-tuned receiver
GDOP	geometric dilution of precision
GHz	gigahertz (10^9 hertz)
GMSK	Gaussian minimum shift key
GSM	global system mobile
HF	high frequency
HFE	human-factors engineering
HOS	higher order spectra
HPA	high-power amplifier
Hz	hertz (cycles per second)
HUMINT	human intelligence
IEEE	Institute of Electrical and Electronic Engineers
IF	intermediate frequency
IFBW	intermediate-frequency bandwidth
IM	intermodulation distortion
INTELSAT	international telecommunication satellites
IR	infrared
ISI	intersymbol interference
ISDN	integrated services digital network
ITU	International Telecommunication Union
IW	information warfare
J/S	jammer power-to-signal power ratio
JPEG	Joint Pictures Expert Group
Kbps	kilobits per second (10^3 bps)
kHz	kilohertz (10^3 hertz)
KLT	Karhunen-Loeve transform

LEAF	law enforcement access field
LF	low frequency
LFSR	linear feedback shift register
LO	local oscillator
LOB	line of bearing
LOP	line of position
LPD	low probability of detection
LPE	low probability of exploitation
LPI	low probability of intercept
LSB	lower sideband
LUF	lowest usable frequency
MAP	maximum a posteriori probability
Mbps	megabits per second (10^6 bps)
MDS	minimum detectable signal
MF	medium frequency
MFSK	multiple frequency shift key
MHz	megahertz (10^6 hertz)
Mil Std	military standard
MIT	Massachusetts Institute of Technology
MLR	maximum likelihood ratio
MOS	military occupational specialty
MPEG	Motion Pictures Expert Group
MPSK	multiple phase shift key
MRC	major regional contingency
ms	millisecond (10^{-3} seconds)
MSK	minimum shift key
MTBF	mean time between failures
MTI	moving target indicator
MTTR	mean time to repair
MUF	maximum usable frequency
MUSIC	multiple-signal classification
NACK	negative acknowledgment
NCS	net control station
NIST	National Institute of Standards and Technology
NPR	noise-power ratio
NSA	National Security Agency
NVIS	near-vertical incidence sky wave

List of Acronyms

O&S	operations and support
OFDM	offset frequency division multiplexing
OOTW	operations other than war
OPSEC	operational security
OQPSK	offset quadrature phase shift key
OTM	on the move
PA	power amplifier
PAM	pulse amplitude modulation
PC	personal computer
PCM	pulse code modulation
PCS	personal communication system
PF	position fix
PGP	pretty good privacy
PLL	phase-locked loop
PM	phase modulation
PN	pseudonoise
PSD	power spectral density
PSK	phase shift key
PSP	per-survivor processing
PSTN	public switched telephone network
PSYOPS	psychological operations
PTT	push-to-talk
QAM	quadrature amplitude modulation
QPSK	quadrature phase shift key
RF	radio frequency
RLOS	radio line of sight
rms	root mean square
ROC	receiver operating characteristic
rss	root sum square
SAR	synthetic aperture radar
SC	suppressed carrier
SEP	system engineering process
SFH	slow frequency hopping
SHA	secure hash algorithm
SHF	superhigh frequency
SIGINT	signals intelligence

SNR	signal-to-noise ratio
SSB	single sideband
STFT	short-term Fourier transform
SVD	singular-value decomposition
TCM	trellis-coded modulation
TDD	time division Duplexing
TDMA	time division multiple access
TDOA	time difference of arrival
TOA	time of arrival
TOC	tactical operations center
TPC	turbo-coded channel
TTIP	transparent tone in band
TV	television
UAV	unmanned aerial vehicle
UHF	ultrahigh frequency
USB	upper sideband
USSR	United Soviet Socialist Republic
VCO	voltage-controlled oscillator
VHF	very high frequency
VLF	very low frequency
WGN	wideband Gaussian noise

About the Author

Dr. Richard Poisel was born and raised in the Midwest. He received a B.S. in electrical engineering from the Milwaukee School of Engineering in 1969 and an M.S. in the same discipline from Purdue University in 1971. He spent 3 years in military service from 1971 to 1973. After his service he attended the University of Wisconsin, where he received a Ph.D. in electrical and computer engineering in 1977. From 1977 to the present he has been with the same government organization, which has had several different names and is currently known as the U.S. Army Communication Electronics Command, Research, Development and Engineering Center, Intelligence and Information Warfare Laboratory. During the 1993–1994 academic year, Dr. Poisel attended the MIT Sloan School of Management as a Sloan Fellow, receiving an M.B.A. Initially a research engineer, Dr. Poisel eventually rose to the role of director of the laboratory on an acting basis from 1997 to 1999. He was appointed as chief scientist in 1999 and was relocated to the Intelligence Center at Ft. Huachuca, Arizona, where he is currently serving as a technical advisor to the command group.

Index

1-dB compression point, 239
2^n PSK, 89

Access methods, 73
  ALOHA, 75
    Slotted, 76
  Carrier-sensed multiple access (CSMA), 76
  Code division multiple access (CDMA), 77
  Demand assigned multiple access (DAMA), 77
  Frequency division multiple access (FDMA), 74
  Time division multiple access (TDMA), 74
Acknowledgment (ACK), 42
Active antennas, 219
Advanced radio data information system (ARDIS), 85
American mobile phone system (AMPS), 71
Amplifier
  classes, 435
    A, 436
    AB, 437
    B, 436
    C, 438
    D, 439
    E, 440
    S, 442
  Efficiency, 435
  Linearization, 444
  Operating characteristics, 431
  Power, 430
  Push-pull, 434
Advanced radio data information service (ARDIS), 85
ALOHA, 75
  Slotted, 76
Amplitude direction finding, 353
Amplitude modulation (AM), 63, 242
Amplitude shift key (ASK), 82
Angle modulation, 66
Angle of arrival (AOA), 331
Antennas, 12, 207
  Aperture, 228, 332
  Directivity, 208, 209
  Effective area, 208
  Efficiency, 209
  Gain, 208
  Isotropic, 18, 207
  Near-field, 208
  Wire, 210
  Yagi, 215
Antenna arrays, 223
  Axial mode, 214
  Parabolic dish, 222, 494
  Phased array, 223
  Smart antennas, 224
Aperture antennas, 220

Array-processing bearing estimation, 363
  Beamforming, 374
  MUSIC, 364
Articulation index, 197
Assaleh-Farrell-Mammone, 310
Asymmetric encryption, 142
Atmospheric noise, 57
Authentication, 142
Automatic gain control (AGC), 241
Automatic link establishment (ALE), 52
Automatic level control (ALC), 246
Automatic repeat request (ARQ), 118, 121
Availability, 160
Avalanche criteria, 145
Average, 519

Bandwidth efficiency, 81
Balance property, 102
Basic power modules, 447
Basis functions, 264
BCH codes, 127
Beamforming, 374
Best point estimate (BPE), 380
Biconical antenna, 214
Binary phase shift key (BPSK), 86
Binary polynomials, 120
Binary symmetric channel, 116
Bispectrum, 290
Block codes, 123
Block encryption, 147
Boresight, 207
Bragg cell receiver, 253
Built-in test equipment (BITE), 161
Burst error protection, 122
Butler matrix, 345

Capture effect, 59, 197

Carrier-sensed multiple access (CSMA), 80
Carrier signal, 63
Carrier-to-noise ratio (CNR), 242
Carson rule, 70
Cepstrum, 320
Cellular digital packet data (CDPD), 85
Characteristic impedance, 22
Checksum, 120
Chip sequence, 100
Chrominance subsampling, 114
Circular error probable (CEP), 182, 388
Clipper, 245
Clipper chip, 149
Cochannel interference, 59
Compressive receiver, 254–261
Code division multiple access (CDMA), 77
Coding, 111
  Channel, 115
  Source, 111
  Speech, 111
Combiners, 448
Combining error contours, 388
Command and control warfare (C2W), 2
Common operational picture (COP), 185
Communication electronics operating instructions (CEOI), 202
Communication environments, 5
Communication jamming, 191
Communication security, 143
Compression, 114
Compressive receivers, 254
Concatenated codes, 132
Confidentiality, 142

Index 549

Cramer Rao Lower Bound (CRLB) 335, 409 (handwritten annotation)

Confusion, 147
Constant modulus, 83
Constellation, 82
Constituent code, 132
Constraint length, 126
Convolutional codes, 126
  Parallel-concatenated, 132
Critical failure, 163
Cross-ambiguity function (CAF), 413
Crossover distortion, 436
Cumulants, 290
Cyclic codes, 125
Cyclic redundancy check (CRC), 118
Cyclostationary signal processing, 286

Data encryption, 141
Data encryption standard (DES), 144
Deception, 2
Delay spread, 45
Delta-sigma modulator, 439
Demand assignment multiple access (DAMA), 77
Demodulation, 240
Detected SNR, 98
Depression angle, 187
Detection, 242
  AM, 242
  FM, 243
  PM, 243
Differential Doppler, 410
Diffraction, wave, 37
Diffusion, 147
Digital receiver, 261
Digital signaling, 79
Digital signatures, 144
Dipole antenna, 210

Direct-sequence spread spectrum (DSSS), 7, 98–106
Direct wave, 27
Direction-finding signal processing, 14
Discrete cosine transform (DCT), 112, 273
Discrete multitone (DMT), 94
Discrete wavelet transform, 277
Dominant battlespace knowledge (DBK), 4
Doppler direction-finder, 361
Ducting, 41
Duplexing, 77
  Time division (TDD), 77
  Frequency division (FDD), 78
Dynamic range, 237

Effective radiated power (ERP), 192
Efficiency
  Bandwidth, 80
  Power, 80, 433
  Spectral, 92
Eigenvalue, 365
Eigenvector manifold, 369
Electromagnetic compatibility (EMC), 57, 151
Electromagnetic interference (EMI), 57, 151
Electronic attack (EA), 4, 10, 189
Electronic mapping, 185
Electronic order of battle (EOB), 178
Electronic protect (EP), 4, 55
Electronic support (ES), 3, 5, 177
Elliptical error probable (EEP), 182, 388
Emission control (EMCON), 4
Emitter identification, 325
  Specific, 326

Emitter identification (continued)
  Type, 325
Encryption
  Asymmetric, 143
  Block, 145
  Data encryption standard
    (DES), 144
  Public key, 142
  Symmetric, 142
Entropy, 113
Envelope elimination and restoral, 444
Equalization, 96
Error correction, 122
Error detection, 116
Escrow encryption standard, 148
Escrow encryption system, 147
Excess delay, 45
Exciter, 14, 425
Expected value, 521
Externals, 178

Facsimile, 140
Fading, HF, 50
Failure modes, 163
Fast frequency-hopping, 110
Fast transforms, 283
Field strength, 22
Filters, 15
  Output, 449
First article integration and test (FAIT), 152
Fix
  Accuracy, 393
  Coverage, 397
Fixed-tuned receiver (FTR), 248
Fortezza encryption system, 146
Forward error correction (FEC), 120, 121
Fourier transform, 268
  Short term, 268

Frame, 118
Frame check sequence, 118
Free-space path loss, 26
Free-space propagation, 24
Frequency division duplexing (FDD), 77
Frequency division multiple access (FDMA), 73
Frequency-hopped spread spectrum (FHSS), 7, 109
Frequency modulation (FM), 243
Frequency shift key (FSK), 82
Fresnel zone, 35
Friis' expression, 26

Galactic noise, 57
Gaussian MSK, 86
Gaussian density function, 526
Geometric dilution of precision (GDOP), 393, 417
Genetically designed antennas, 225
Geolocation, 179
Gibb's phenomenon, 270
Global system mobile (GSM), 84
Gold codes, 104
Ground reflection propagation model, 39
Ground wave, 41

H.261, 115
Haar transform, 275
Haar wavelet, 283
Hamming,
  Codes, 123
  Distance, 123
  Weight, 123
  Window, 260
Hard decision decoding, 132
Hartley transform, 272

Helix antenna (continued)
  Axial-mode, 219
  Normal-mode, 218
HF fading, 49
High-power amplifier (HPA), 88
Higher-order statistics, 288
Homodyne receiver. *See* zero IF receiver
Human factors, 170
Huygens principle, 34
Hypothesis testing, 292

Identification, 318
  Closed, 320
  Open, 320
In-phase component, 81
Information warfare (IW), 1
Integrated services digital network (ISDN), 113
Intelligibility, 193
Intercept, 177
Intercept points, 238
Interference, 59
Interferometry, 333
  Four-element, 342
  Triple-channel, 336
Interframe coding, 113
Intermediate filtering, 241
Intermediate frequency (IF), 233, 238–240
Intermodulation distortion, 88, 237
Internals, 178
Intersymbol interference (ISI), 46, 94
Intraframe coding, 113
Ionospheric layers, 47
  D-layer, 47
  E-layer, 47
  E_s-layer, 48
  F-layers, 48
Ionospheric propagation, 46

Jammer
  Barrage, 190, 198
  Deployment, 195
  Expendable, 513
  Follower, 202
  Low probability of intercept (LPI), 201
  Narrowband, 192, 199
  Partial band, 199
  Stand-in, 190
  Standoff, 190
  UAV, 511
Joint Pictures Expert Group (JPEG), 113

Karman-Loeve transform (KLT), 112
Kasami sequences, 104
Kim-Polydoros, 315
Knife-edge diffraction, 34
Kurtosis, 527

Language identification, 324
Law enforcement access field (LEAF), 148
Likelihood function, 132
Likelihood ratio, 294
  Average, 300
  Maximum, 302
Line of bearing (LOB) 180, 331, 376
Line of position (LOP) 180, 331
Linear feedback shift registers (LFSRs), 102, 122
Linear modulations, 92
Local oscillator, 233, 240
Log-periodic antenna, 218
Look-through, 194
Loop antenna, 213
Lossless coding, 114
Lossy coding, 113

Low probability of detection (LPD), 7
Low probability of exploitation (LPE), 7
Low probability of intercept (LPI), 7
Lowest usable frequency (LUF), 51
LZ-77, 114
LZ-78, 114

m-sequences, 103
Man-made noise, 57
Maximum a posteriori (MAP) processing, 134, 293
Maximum usable frequency (MUF), 51
Mean, 519
Mean time between failure, (MTBF), 159
Mean time to repair (MTTR), 159
Meteor burst, 42
Minimum detectable signal (MDS), 235
Minimum shift key (MSK), 83
   Gaussian (GMSK), 84, 85
Mixer, 233
Mixing, 239
MOBITEX, 85
Modems, 134
Modulation, 63
   Amplitude (AM), 63, 65
   Angle, 63, 66
   Double sideband (DSB), 66
   Frequency (FM), 63
   Lower sideband (LSB), 66
   Phase (PM), 63
   Pulse amplitude (PAM), 139
   Quadrature, 71
   Single sideband (SSB), 66
   Suppressed carrier (SC), 66
   Upper sideband (USB), 66
Modulation index, 71
Modulating signal, 63
Modulator, 429
Moment functions, 521
Monopole antenna, 212
Monopulse direction finder, 347
Morse code, 323
Mother wavelet, 279
Motion Pictures Expert Group (MPEG), 113
MPEG 1, 113
MPEG 2, 113
MPEG 4, 113
Multipath interference, 54
Mutually exclusive, 524

Nandi-Azzouz, 306
Narrowband receiver, 247
Near-far problem, 195
Near vertical incidence sky wave (NVIS), 49
Negative acknowledgement (NACK), 42
Net control station (NCS), 62, 450
Noise, 57,
   Atmospheric, 57
   Effects on digital modulations, 94   WHITE 335
   External, 57
   Internal, 56
   Thermal, 56
Noise factor, 57, 234–236, 465
Noise figure, 57, 234–236
Noise power ratio (NPR), 236, 450
Nonrepudiation, 145

Omnidirectional, 211
On the move (OTM), 498

Operational security (OPSEC), 2
Order 0 model, 111
Order 1 model, 111
Orthogonal functions, 266
Orthogonal signaling, 71
Orthogonal frequency division
　multiplexing (OFDM), 94
Orthogonal signaling, 71
Oscillators, 426
　Colpitts, 426
　Crystal, 426
　Direct digital synthesis, 427
　Integrated, 427
Over-the-air rekeying, 148

Parity, 117
Parabolic dish antenna, 223
Parallel-concatenated convolution
　codes, 134
Parity, 119
Parseval's relationship, 268
Partial-band filter bank combiner
　(PB-FBC), 474
Pattern recognition, 305
Phase difference averaging, 377
Phase-locked loop 245, 427
Phase modulation 67, 246
Phase shift key (PSK), 86
　2^n, 89
　Binary (BPSK), 86
　Quadrature (QPSK), 87
　　Offset (OQPSK), 87
Physical destruction, 2
Personal communication system
　(PCS), 479
Per-survivor processing, 131
Polarization, 22
Position fixing 179, 331, 380
Postprocessing, 462
Power amplifier 14, 431
Power efficiency, 79

Power density, 22
Power sharing, 446
Postdetection SNR, 246
Poynting theorem, 22
Preselection, 234
Preselection filtering, 233
Pretty good privacy (PGP), 146
Principal (encryption), 143
Probability, 519
　Conditional, 523
　Density functions, 524
　　of detection, 296
　　of error, 295
　　of false alarm, 295
　　of missed detection, 295
Propagation loss, 180
Pseudoinverse, 371, 406
PSYOPS, 2
Pulse amplitude modulation, 141
Pulse code modulation (PCM), 78
　Adaptive differential
　　(ADPCM), 79
　Differential (DM), 79
Pulse density modulation, 443
Push-pull architecture, 434
Push-to-talk (PTT), 7

Quadrature amplitude modulation
　(QAM), 89
Quadrature component, 81

Radio frequency band
　designations, 18
Radiation safety limits, 177
Radio line of sight (RLOS), 32,
　506–507
Radiometer, 297
　Channelized, 298
Random variables, 519
Range to horizon, 31
Receivers, 229, 230

Receiver operating
 characteristic (ROC), 234
Recognition, 317
Reed-Solomon codes, 134
Reed-Solomon/Viterbi (RSV), 133
Reflected wave, 34
Refraction, 48
Reliability, 160, 161
Root mean square (rms), 521
Root sum squared (rss), 519
R^n model, 29
Run property, 102

Scattering, 43
 Tropospheric, 43
Search bandwidth, 459
Search receiver, 13
Secure hash algorithm, 144
Selectivity, 239
Sensitivity 23, 236
Set-on receiver, 13
Shadowing, 44
Signal classification, 301
Signal detection, 291
 Range, 487
Signal distribution, 13
Signal processing, 12, 13, 266
Signal propagation, 17
Sills, 315
Singular value, 369
Singular value decomposition, 368
Single site location, 391
Slow fade, 45
Slow frequency hopping, 110
Skewness, 527
Soft decision decoding, 133
Soft decoding, 132
Sounder
 Ionospheric, 392
 Oblique, 392
 Vertical, 392

Source recognition, 320
Speaker recognition, 321
Spread spectrum, 97
Direct sequence (DSSS), 98
 Frequency hopping (FHSS), 106
  Fast, 107
  Slow, 107
Spiral antennas, 218
Spreading codes, 101
Standard deviation, 518
Standing wave ratio, 208
Stansfield fix algorithm, 382
Stastical modeling, 319
Steganography, 114
Stochastic process, 520
Stream cipher, 147
Subrefraction, 33
Subspace
 Noise, 371
 Signal, 372
Superhetrodyne receiver, 233, 247
 Scanning, 248
Superrefraction, 33
Superresolution direction finding, 361
Surface wave, 41
Symbol, 81
Symmetric encryption, 142
Syndrome, 125
Synthesizer, 428
System control, 10
System engineering process, 154

Thermal noise, 56
Time difference of arrival (TDOA), 403
 Averaging, 376
Time division multiple access (TDMA), 74
Time of arrival (TOA), 418
Time sharing, 198

Transforms, 268
Transparent tone in band (TTIB), 94
Trellis-coded modulation, 139
Trellis diagram, 127
Triangulation, 180, 330
Trigonometric transforms, 268
Tri-spectrum, 290
Turbo codes, 132
Two-tone dynamic range, 239

Unattended aerial vehicle (UAV), 2, 184, 490, 498
User nonrepudiation, 143

Variance, 520
Verifier (encryption), 143
Viterbi decoding, 128
Voltage-controlled oscillator (VCO), 71, 84, 246, 426
Voltage standing wave ratio (VSWR), 208

Walsh functions, 105
Walsh-Hadamard codes, 104
Watson-Watt direction finder, 354
Wave diffraction, 33
Wavelet transform, 276
  Continuous, 279
  Discrete, 280
Whelchel-McNeill-Hughes-Loos, 311
Wideband receivers, 251
  Channelized, 252
Wigner-Ville distribution, 271
Windows,
  Bartlett, 260
  Blackman, 260
  Hamming, 260
  Hanning, 260
  Rectangular, 260

Yagi antenna, 216

Zero IF receiver, 248

The Artech House Information Warfare Library

Electronic Intelligence: The Analysis of Radar Signals, Second Edition, Richard G. Wiley

Electronic Warfare for the Digitized Battlefield, Michael R. Frater and Michael Ryan

Electronic Warfare in the Information Age, D. Curtis Schleher

EW 101: A First Course in Electronic Warfare, David Adamy

Information Warfare Principles and Operations, Edward Waltz

Introduction to Communication Electronic Warfare Systems, Richard Poisel

Principles of Data Fusion Automation, Richard T. Antony

For further information on these and other Artech House titles, including previously considered out-of-print books now available through our In-Print-Forever® (IPF®) program, contact:

Artech House
685 Canton Street
Norwood, MA 02062
Phone: 781-769-9750
Fax: 781-769-6334
e-mail: artech@artechhouse.com

Artech House
46 Gillingham Street
London SW1V 1AH UK
Phone: +44 (0)20-7596-8750
Fax: +44 (0)20-7630-0166
e-mail: artech-uk@artechhouse.com

Find us on the World Wide Web at:
www.artechhouse.com